Modelling, Simulation and Control of Urban Wastewater Systems

Springer
London
Berlin
Heidelberg
New York
Barcelona
Hong Kong
Milan
Paris
Singapore
Tokyo

Manfred Schütze, David Butler and M. Bruce Beck

Modelling, Simulation and Control of Urban Wastewater Systems

With 81 Figures

Springer

Manfred R. Schütze, Dr Dipl-Math, PhD, DIC
ifak e. V. Magdeburg, Institute for Automation and Communication,
D-39179 Barleben, Germany

David Butler, PhD, DIC
Department of Civil and Environmental Engineering,
Imperial College of Science, Technology and Medicine, London, SW7 2BU, UK

M. Bruce Beck, PhD
Warnell School of Forest Resources, The University of Georgia,
Athens, Georgia, 30602-2152, USA

British Library Cataloguing in Publication Data
Schutze, Manfred R
  Modelling, simulation and control of urban wastewater
  systems
  1.Urban runoff - Mathematical models 2. Sewage disposal -
  Mathematical models
  I.Title II.Butler,David, 1959- III.Beck, M. B.
  363.7'284
  ISBN 185233553X

Library of Congress Cataloging-in-Publication Data
Schütze, Manfred R., 1966-
  Modelling, simulation and control of urban wastewater systems / Manfred R. Schütze,
  David Butler, and M. Bruce Beck.
    p. cm.
    Includes bibliographical references and index.
    ISBN 1-85233-553-X (alk. paper)
    1. Sewage--Purification--Mathematical models. 2. Sewage disposal
  plants--Mathematical models. 3. Sewerage--Mathematical models. I. Butler, David
  1959- II. Beck, M. Bruce, 1948- III. Title.
  TD745 .S37 2002
  628.3'09173'2--dc21   2001055108

Apart from any fair dealing for the purposes of research or private study, or criticism or review, as permitted under the Copyright, Designs and Patents Act 1988, this publication may only be reproduced, stored or transmitted, in any form or by any means, with the prior permission in writing of the publishers, or in the case of reprographic reproduction in accordance with the terms of licences issued by the Copyright Licensing Agency. Enquiries concerning reproduction outside those terms should be sent to the publishers.

ISBN 1-85233-553-X Springer-Verlag London Berlin Heidelberg
a member of BertelsmannSpringer Science+Business Media GmbH
http://www.springer.co.uk

© Springer-Verlag London Limited 2002
Printed in Great Britain

The use of registered names, trademarks etc. in this publication does not imply, even in the absence of a specific statement, that such names are exempt from the relevant laws and regulations and therefore free for general use.

The publisher makes no representation, express or implied, with regard to the accuracy of the information contained in this book and cannot accept any legal responsibility or liability for any errors or omissions that may be made.

Typesetting: Electronic text files prepared by editors
Printed and bound by the Athenæum Press Ltd., Gateshead, Tyne & Wear
69/3830-543210 Printed on acid-free paper SPIN 10830782

Πάντα ῥεῖ

Heraclitus

(c. 544 - 483 BC)

Foreword
by Professor Poul Harremoës

Environmental engineering has been a discipline dominated by empirical approaches to engineering. Historically speaking, the development of urban drainage structures was very successful on the basis of pure empiricism. Just think of the impressive structures built by the Romans long before the discipline of hydraulics came into being. The fact is that the Romans did not know much about the theories of hydraulics, which were discovered as late as the mid-1800s. However, with the Renaissance came a new era. Astronomy (Galileos) and basic physics (Newton) started the scientific revolution and in the mid-1800s Navier and Stokes developed the application of Newtons laws to hydrodynamics, and later, St. Venant the first basic physics description of the motion of water in open channels. The combination of basic physical understanding of the phenomena involved in the flow of water in pipes and the experience gained by "trial and error", the engineering approach to urban drainage improved the design and performance of the engineering drainage infrastructure. However, due to the mathematical complications of the basic equations, solutions were available only to quite simple cases of practical significance until the introduction of new principles of calculation made possible by computers and their ability to crunch numbers. Now even intricate hydraulic phenomena can be simulated with a reasonable degree of confidence that the simulations are in agreement with performance in practice, if the models are adequately calibrated with sample performance data. This development started in the 1970s and has the potential to significantly improve the design of urban drainage systems and structures.

On top of this development of the combination of theory and practice came the potential of optimising the performance in operation based on on-line data on the status of the system at any particular point of time. For this development to become effective the on-line provision of adequate performance data at selected points in the sewer system and tools for optimising performance on the basis of on-line data.

Since the mid-1800s European cities have been provided with infrastructures for water supply and sewerage, consistently separating the clean and dirty water. This is the basis for the fact that modern cities of the developed world are virtually without waterborne diseases. This infrastructure is a huge investment that is hidden under ground, more often than not taken for granted and ignored by the public and politicians. That investment has to be upgraded to modern demands for better environmental performance, on top of the ongoing demand for flood control. These indispensable infrastructures have to be maintained (refurbished when ignored for too long) and operated to get the most out of the investment. That problem will be with us for as far as we can look ahead. The challenge is to make the system and the investments sustainable. Advanced tolls for mathematical simulation of the systems, adequate data on performance and on-line measurements, in combination with advanced procedures for optimisation will be indispensable in the approach to drainage of urban cities with an established water infrastructure.

As an after-thought, it is important to realise that most cities, in fact almost all of the megatropolises of the world, are not in a similar position. The route for sustainable development may take a completely different direction. Copying the European approach does not provide the basis for development, due to lack of funds required for the huge investments. While appreciating, maintaining and optimising the drainage of cities in the developed world, the provision of alternatives for the growing megacities of the developing world has to be kept in sight.

Lyngby, July 2001
Poul Harremoës
Professor at Technical University of Denmark, Lyngby, Denmark

Foreword
by Professor Peter Vanrolleghem

This book, which originated from the comprehensive PhD work of Dr Manfred Schütze combined with the many years of related investigations of the co-authors Dr David Butler and Dr Bruce Beck, deals with the modelling and control of the entire urban wastewater system. I very much welcome its publication since it is an important step forward towards making the quantitative analysis and optimisation of the urban wastewater system an objective process in which design/optimisation options can be evaluated in a step-wise fashion. This does not mean, however, that the creativity, skill and challenging fun of designing/optimisation is to be lost. It does mean that engineers making choices can be inspired by the examples and procedures provided in this book and can be more confident that their solutions are thoroughly evaluated with respect to the needs and constraints which represent the ambitions of their potential customers and, indeed, any one that will come into contact with that (implemented) solution.

This book is all about integration. Foremost, it deals with the integration of the urban wastewater system. No longer can the sewer system be considered separate from the treatment plant, let alone the river system. The same holds for the treatment plant that needs to be regarded as an integral part of the urban wastewater system. The work reported here can be seen as an anticipation of what, for instance, the recently adopted European Water Framework Directive is imposing. From now on it is not the separate emissions from the subsystems that need to be looked after, but rather the integrated input of the different pollution sources must be looked at in terms of the effect on river quality (in all its aspects). The urban system is clearly one of the main pollutant sources into the river and should be tackled as a whole. The important contribution of the work reported in this book is that it is shown to be possible to make comprehensive models of the whole system that can be used within optimisation studies that focus on river water quality objectives. A number of hurdles that had to be taken were jumped, such as the problem of linking existing mathematical descriptions that use different state variables, the problem of model complexity and the necessity of having bi-directional information flow between the subsystems for real-time control. This has led to what is probably the first simulator that can simultaneously solve a truly integrated urban wastewater system model.

The next aspect of integration pertinent to this book is the integration of disciplines. Going through the book the reader will be confronted with such widely varying disciplines as civil engineering, software engineering, (micro)biology, numerical analysis, control engineering, mathematical modelling, statistics, environmental and chemical engineering, and probably some more. It is remarkable that the authors have been able to master all of these disciplines and have brought them to synergy. This is quite an accomplishment and should set an example for any future work (not only in environmental issues!) that will increasingly rely on the multi-disciplinarity of individuals and teams.

Finally, an integration of backgrounds (schools of thought) was established in this work. In this era of globalisation, it is good to realise that this book is the result of a collaboration of a German PhD student with an English and an American supervisor. However, the effective supervision of this book is probably much wider. Dr Schütze started his research within MATECH, the European Centre for MAthematics and TECHnology of Urban Water Pollution, financed by the EC Human Capital and Mobility project. It links different research groups in Europe active in the field of Urban Water Pollution, concentrating on the modelling aspects. It goes without saying that tapping into such a network allows the efficient absorption of the necessary multi-disciplinary expertise to move forward quickly. In addition, it provides the necessary peer review options to (re)direct research. Consequently, through this comprehensive network, this book can also be seen as a result of an integrative effort within the profession.

At the time when I started my own research in the field of modelling and simulation, my then supervisor Dr Jan Spriet, introduced me to the methodological thinking of Dr Beck. Much of the fundamentals of my modelling work can still be traced back to his key publications. I feel fortunate to have the luck to discuss many of the methodological aspects with him, and to be able together, within the IWA Specialist Group on Systems Analysis and Computing in Water Quality Management, to highlight the methodological angle of quantitative thinking related to water quality problems. In view of my historical bias towards biological wastewater treatment, I feel very complementary to Dr Butler's background, particularly his work on sewer system processes and the characterisation of the dynamics of inhabitant equivalent inputs into the sewer system. These studies have had a major impact on my thinking of the sewer system as a part of the whole that can be manipulated to improve wastewater treatment.

I have been privileged to know Dr Schütze from the beginning of his PhD studies and have had many in-depth discussions with him on the different aspects of his work, in particular – to name a few – optimisation algorithms, the simultaneous integrated modelling approach, the real-time control options he proposed and the way they were evaluated. I have admired his multi-disciplinary knowledge base, his eagerness to learn and his sense of organisation. Also, having Manfred present in conferences or in EU COST (European Cooperation in the field of Scientific and Technical Research) Working Group meetings was always a pleasure. He was always driven to make progress in the convergence of thinking in the discipline. I feel that this book – to which he is the main contributor – is a reflection of this focus on common thinking and the efforts in this book to structure knowledge, work with clear terminology and methodology are illustrative of this. I am convinced the book will allow this terminology and methodology to further spread in the profession and will eliminate one of the main limitations to scientific progress, *i.e.,* the absence of common language and working procedures.

With this book the three authors share their significant contribution to the field of urban wastewater system modelling and control. They have much to tell us, and there is still more to come as they continue their research and apply it across the world.

Gent, July 2001

Peter A. Vanrolleghem

Professor at BIOMATH, Ghent University, Belgium

Acknowledgements

This book essentially presents the results of a PhD project, which I conducted at Imperial College of Science, Technology and Medicine (University of London) under the supervision of my coauthors, Professor Bruce Beck and Professor David Butler. Although the thesis work was carried out in the years 1994 to 1998, this book also includes some recent developments since then.

My special thanks are expressed to my coauthors for supervising this work. Without continuous interest and encouragement, numerous fruitful discussions and support this work would not have been possible.

Furthermore, the stimulating discussions, useful comments on this work and support of various kinds also of many colleagues and friends are acknowledged with gratitude and appreciation. Particular mention is made of Hans Aalderink, Maria do Céu Almeida, Jenny Blight, Andreas Cassar, Eran Friedler, Chantal Fronteau, Jian Hua Lei, Jonathan Parkinson, Petr Prax, Wolfgang Rauch, Mohammed Saidam, John Tyson, Peter Vanrolleghem, Hans-Reinhard Verworn, and Kala Vairaavamoorthy. Important contributions to this project were also made by a number of MSc and project students, to whom thanks are expressed: Sibel Gülen, Okan Güven, Christophe Leclerc, Eva Vazquez-Sanchez, and Wai Meng Wan.

Overall, the discussions I enjoyed at my various working places over the years, including Hannover, Dübendorf, London, Dundee, Brno, Gent and Magdeburg, proved (and continue to prove) to be fruitful sources of inspiration for this and for further work. Many cups of coffee were emptied whilst the tray of ideas was filled.

Acknowledgements are also made to the University of Hannover and to Wageningen Agricultural University for the rights to use their software and to the Environment Agency of the German state of Niedersachsen for the permission to use the Fuhrberg rainfall data set within this project.

The financial support of the European Union's Human Capital and Mobility programme is greatly appreciated, which funded the thesis work under the MATECH network.

Furthermore, I am deeply indebted to my family and friends for their unreserved and continuous support throughout the project. Particular mention should be made here to Jonathan and Luiza, who always reminded me on the existence of life besides simulation.

Finally, I would like to express my special thanks to Simone Hildebrandt and to Sabine, my wife, for their assistance in preparation of the final manuscript. I am particularly grateful to Sabine for her patience on those long evenings and weekends I spent on the final manuscript rather than with her.

Manfred Schütze
Magdeburg, July 2001

Table of Contents

1. Introduction 1

    1.1 Motivation of this Book..1
        1.1.1 Administrative Responsibilities 4
        1.1.2 Standards 4
        1.1.3 Computer Software 5
        1.1.4 Design and Operation 5
    1.2 Outline of Chapters...7

2. The State of the Art 11

    2.1 Components of the Urban Wastewater System: Basic Processes and
        Modelling Concepts..11
        2.1.1 Urban Catchment Runoff and Sewer System................................11
            2.1.1.1 Surface Runoff in Urban Areas 12
            2.1.1.2 Flow in the Sewer System...13
            2.1.1.3 Pollutant Transport in the Sewer System 17
            2.1.1.4 Biochemical Transformations in the Sewer System.............19
            2.1.1.5 Storage Tanks..21
        2.1.2 The Wastewater Treatment Plant...24
            2.1.2.1 Storm Tank..25
            2.1.2.2 Primary Clarification...27
            2.1.2.3 The Activated Sludge Process...29
            2.1.2.4 Secondary Clarification...37
        2.1.3 Rivers...43
            2.1.3.1 River Flow...43
            2.1.3.2 Pollutant Transport in the River..44
            2.1.3.3 Biochemical Transformations in the River...........................46
    2.2 Impact of Storm Events on the Urban Wastewater System....................57
        2.2.1 Impacts on Sewer Systems...57
        2.2.2 Impacts on Treatment Plant Performance 59
        2.2.3 Impacts on the Receiving River...62
        2.2.4 Criteria for the Assessment of River Water Quality 68
        2.2.5 The Dilemma of Control of the Urban Wastewater System..........70
    2.3 Integrated Modelling Approaches...70

2.4 Operational Management of Wastewater Infrastructure 80
    2.4.1 General Concepts..80
    2.4.2 Real-time Control of Sewer Systems 83
    2.4.3 Development of Control Strategies – Exemplified for Sewer
          Systems..85
          2.4.3.1 Off-line Development of Strategies.................................85
          2.4.3.2 On-line Development of Strategies 89
    2.4.4 Operation of Wastewater Treatment Plants 96
    2.4.5 Real-time Control of Receiving Rivers..109
    2.4.6 Integrated Real-time Control 111
    2.4.7 Concluding Remarks 113
2.5 Mathematical Optimisation Techniques 114
    2.5.1 Definition of the Optimisation Problem..114
    2.5.2 A Review of Optimisation Methods 118
          2.5.2.1 Local Optimisation..120
          2.5.2.2 Global Optimisation..122
2.6 Conclusion...126

3. Development of the Integrated Simulation and Optimisation Tool SYNOPSIS 129

3.1 Requirements on the Simulation Tool..129
3.2 Modules Simulating the Parts of the Urban Wastewater System 133
    3.2.1 Implementation of the Sewer System Module 133
    3.2.2 Implementation of the Treatment Plant Module 135
          3.2.2.1 The Original Implementation of Lessard and Beck's
                  Treatment Plant Model..135
          3.2.2.2 Modifications of the Treatment Plant Model 143
    3.2.3 Implementation of the River Module..145
3.3 Assembling the Integrated Simulation Tool...150
    3.3.1 Integration of the Simulation Software 151
    3.3.2 Variables in SYNOPSIS 152
    3.3.3 Auxiliary Routines Necessary for Simulation...............................157
3.4 Implementation of Control in SYNOPSIS 161
3.5 Optimisation Algorithms in SYNOPSIS..163
    3.5.1 Controlled Random Search 164
    3.5.2 A Genetic Algorithm 166
    3.5.3 Powell's Local Optimisation Method..170
    3.5.4 Interfacing the Simulation Tool with the Optimisation Routines......172
3.6 Summary: Overview of the Integrated Simulation and Optimisation
    Tool SYNOPSIS..176

4. Simulation of the Urban Wastewater System Using SYNOPSIS 179

4.1 Definition of a Case Study Site...179
    4.1.1 Existing Data Sets..180

    4.1.2 Definition of the Sewer System 181
    4.1.3 Definition of the Wastewater Treatment Plant.................................. 189
    4.1.4 Definition of the River... 192
    4.1.5 Overview of the Case Study Site Defined... 196
  4.2 Simulation of Dry-weather Flow.. 198
  4.3 Simulation of a Rainfall Time Series 203
  4.4 Analysis of the Control Devices of the Urban Wastewater System 209
  4.5 Potential of Reduction in Simulation Time by Selective Simulation 218
    4.5.1 Separation of Rainfall Events 218
    4.5.2 Potential Savings in Simulation Time.. 221
    4.5.3 Selective Versus Continuous Simulation 225
    4.5.4 Conclusions 227

5. Analysis of Control Scenarios by Simulation and Optimisation 229

  5.1 Definitions and Methodology 230
  5.2 Analysis of Strategy Parameters – an Example... 233
    5.2.1 Definition of a Strategy Framework 233
    5.2.2 Exploring the Parameter Space by Gridding 235
    5.2.3 Optimisation of Strategy Parameters 238
  5.3 A Top-down Approach to the Definition of Control Strategies 243
    5.3.1 Definition of Various Frameworks 244
    5.3.2 Evaluation of the Optimisation Algorithms 249
    5.3.3 Conclusions 251
  5.4 A Bottom-up Approach to the Definition of Control Strategies 255
    5.4.1 Towards a Systematic Definition of Frameworks............................ 255
    5.4.2 Analysis of Frameworks Involving Several Controllers 258
  5.5 Integrated Versus Local Control... 262
  5.6 Further Aspects... 264
    5.6.1 Sensitivity of Solutions.. 265
    5.6.2 Multi-objective Optimisation... 268
    5.6.3 Simulation Period Required for Optimisation 270
    5.6.4 Control Potential of Various Case Study Sites 273

6. Conclusions and Further Research 277

  6.1 Summary.. 277
  6.2 Suggestions for Further Research 280

Appendix A. Overview of Existing Software 285

  A.1 Software for Simulation of Sewer Systems.. 285
  A.2 Software for Simulation of Activated Sludge Wastewater Treatment
      Plants 292
  A.3 Software for Simulation of Rivers 298

Appendix B. Parameters of the Treatment Plant Model......................307

Appendix C. Rainfall Data Used in This Study..................................311

Appendix D. Detailed Results of Optimisation Runs Presented in Chapter 5..313

References...317

Index..359

Chapter 1
Introduction

1.1 Motivation of this Book

"Everything is in motion, undergoing ceaseless changes" - this statement accredited to Heraclitus is not only true for the universe and for human life, but often also in particular for water and wastewater.

In most countries of the so-called developed world, wastewater in urban areas (after its "generation" by use of toilets and other domestic and industrial uses of water) is collected and transported in sewer systems, then, in many instances, treated (to some extent) at wastewater treatment plants and finally discharged into receiving water bodies[1], such as, for example, rivers or the sea. Early examples of sewer systems are reported from Mohenjo Daro (about 2000 BC) and from ancient Rome (300 BC) (Fuchs, 1997; Lange and Otterpohl, 1997). The first wastewater treatment plants were built considerably later - the first plant in Germany, for example, was put into operation as late as 1887 in Frankfurt-Niederrad (von der Emde, 1998).

Sewer system, treatment plant and receiving water form only part of the hydrological cycle shown in Figure 1.1 (Hammer, 1996).

These three components also represent the main constituents of that part of the "water quality system" (Beck, 1976) (*cf.* Figure 1.2) which is related to wastewater. The importance of these parts is illustrated by the fact that about 70 m^3 of wastewater, 45 kg of faeces and 500 l of urine are produced per person per year (Imhoff and Imhoff; 1990; Lange and Otterpohl, 1997), most of which is conveyed through this system.

---

[1] It is interesting to note here that the wastewaters of only about 6% of the world's population are conveyed in this way (Niemczynowicz, 1997).

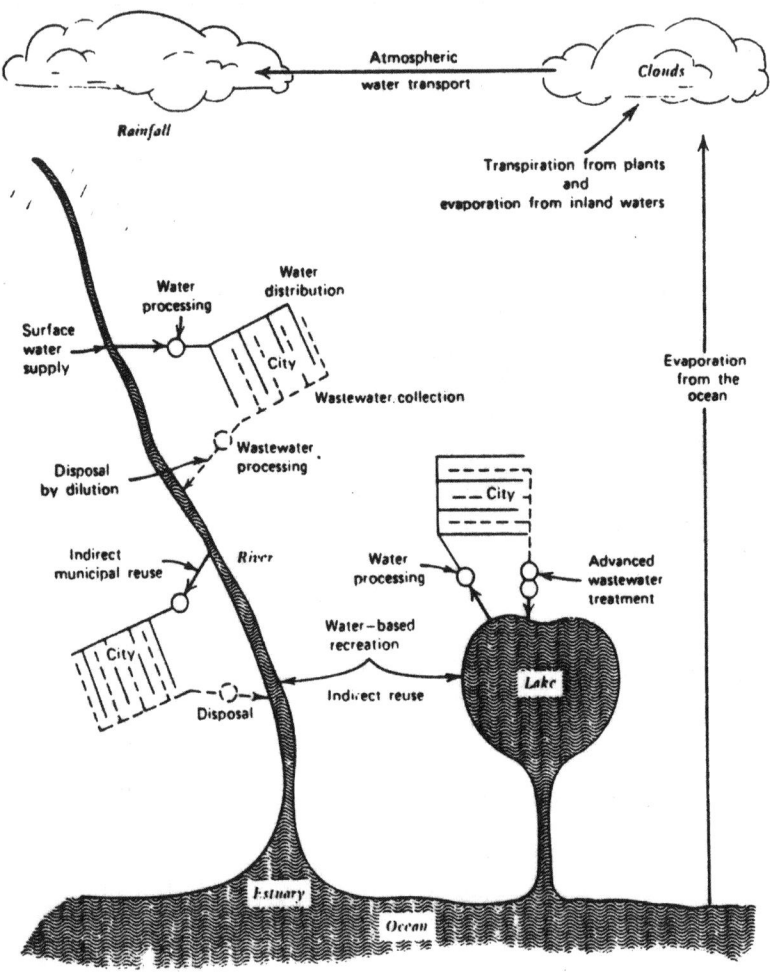

Figure 1.1. The hydrological cycle (Hammer, 1996)

The sewer system[2], treatment plant and receiving water body are considered to constitute the "urban wastewater system", which will form the scope of this book. A river will be considered as the water body receiving the discharges from combined sewer overflows (CSOs) and the treatment plant. However, although sludge treatment and disposal and groundwater-related processes (these may interact with the sewer system through infiltration and exfiltration) could be considered to be a part of this system, these are not studied here in order to limit the scope of this book. Within a river catchment, several of such urban wastewater systems may be found. However, this work will focus only on a single urban wastewater system.

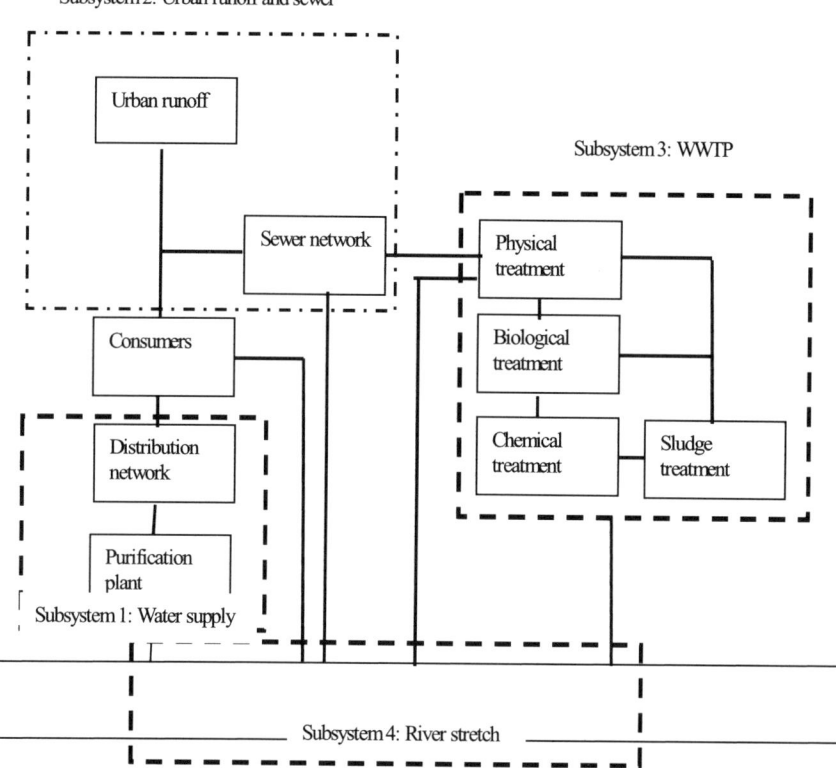

Figure 1.2. The water quality system (Reproduced from Beck (1976) with permission of Elsevier Science)

The historical development of sewer systems and treatment plants and their receiving water bodies has led to them commonly being considered as separate units with regard to the following four aspects:
- administrative responsibilities;
- standards;
- computer simulation;

---
[2] In this book, in general, only combined sewer systems are investigated.

- design and operation.

1.1.1 Administrative Responsibilities

Consideration of sewer system, treatment plant and receiving water as separate systems has led in many countries to the arrangement of separate administrative bodies being responsible for the individual parts of the urban wastewater system. An example is given by the arrangements in the Netherlands, where the water authorities, empowered by the provinces, are responsible for construction and setting of standards for wastewater treatment plants, whilst the construction and management of the sewer systems lies with the municipalities (Schilling, 1989). A similar separation of tasks may also exist within one company if responsibilities for sewer and system and treatment plant rest with different departments within that company (Pickles *et al.*, 1995).

1.1.2 Standards

Guidelines and standards commonly applied in many countries for the design and operation of sewer systems (Henderson (1995) provides a comprehensive overview) are related mainly to the sewer system itself, without detailed consideration of the other elements of the urban wastewater system. For example, design guidelines for combined sewer overflows (CSOs) in Flanders (Belgium) are based simply on a maximum number of CSO events per year. The corresponding German guideline ATV A128 (1992) uses a maximum permissible annual load of chemical oxygen demand (COD) discharged into the river as criterion. Therefore, this guideline, too, does not take into account the characteristics of the other parts of the urban wastewater system.

In a similar manner, most of the traditional standards for design and operation of wastewater treatment plants (see Jacobsen (1990) for a review of treatment plant effluent standards in various European countries) define a prescribed effluent quality ("uniform emission standards"; "end-of-pipe standard"), again, usually without detailed consideration of other parts of the urban wastewater system.

Only fairly recently, however, changes in the consideration of sewer system, treatment plant and receiving water can be observed. Notably contributions by Beck (1976) and the "Interurba" workshops held in 1992 (Lijklema *et al.*, 1993) and 2001 (to appear in *Water, Science and Technology*) constitute a significant step towards the consideration of these parts as elements of a whole, the urban wastewater system. Furthermore, a new approach to the definition of receiving water quality criteria ("Environmental Quality Objective/Environmental Quality Standard

(EQO/EQS)") (Tyson *et al.*, 1993) has been proposed, which take the use of the receiving water body and its particular characteristics into account for the definition of discharge criteria, thus motivating their definition to be based on river water quality. In this respect, the Danish criteria expressing water quality in terms of return period and duration of low DO concentrations (Spildevandskomiteen, 1984, 1985) and, in a similar way, the criteria defined for DO and ammonia in the UPM Manual (FWR, 1994) should be mentioned. These criteria may, in conjunction with appropriate simulation models, assist in the definition of site-specific treatment plant standards. Also the new EU directive on urban wastewater treatment (CEC, 1991) considers, at least in a simple way, the characteristics of the receiving water body in the definition of treatment plant effluent standards. A particular impetus towards integrated management of river basins and, thus, wastewater infrastructure is given by the recently adopted EU Water Framework Directive (CEC, 2000). This will lead to considerable efforts in research and practice of water management across Europe. This is becoming particularly apparent in the increased application of river quality models.

1.1.3   Computer Software

Also the development of simulation software for the parts of the urban wastewater system, assisting in their planning, design and management, seems to have been influenced by the historical development of considering these parts as separate units. Although the state-of-the-art in simulation of sewer system, treatment plant and receiving water as separate, individual units can be considered as being fairly advanced (despite the large number of problems yet to be solved), simulation of the urban wastewater system as an entirety is still in its infancy (details will be given in Section 2.3).

1.1.4   Design and Operation

As a direct consequence of the separate consideration of the components of the urban wastewater system in the definition of the administrative responsibilities and of the related standards, also the separation of these systems in design and operational management can be considered. However, when analysing the design or operation of the components of the urban wastewater system, it becomes clear that an option which is beneficial for one subsystem may not necessarily be beneficial to another subsystem. For example, benefits gained in the sewer system by the reduction of the number, volumes or loads of CSO discharges (these criteria representing examples of typical design or operational objectives) may well be outweighed by a deterioration in treatment plant effluent quality caused by increased inflows to the plant (Vanrolleghem *et al.*, 1996a). For proper consideration of such a conflict of interests and for proper coordination of the various discharges from sewer system and treatment plant into the receiving water body, the conventional approach to consider the sewer system, treatment plant and river as separate units in the selection of design options or in the definition of control strategies does not seem to be appropriate. It becomes apparent that optimum performance of the entire urban wastewater system may be different from optimum performance of each of its parts.

Narrowing down the thread of this discussion from the consideration of design and operation in general to the operation of the components of the urban wastewater system with regard to criteria describing the performance of the entire system in particular, it appears to be plausible from this discussion that control of the urban wastewater system under joint consideration of its components leads to improved results as compared to conventional control. However, it would be prudent to perform a detailed simulation analysis in order to analyse this supposition, before any innovative control schemes are implemented in practice. This issue, formulated as a provocative question: "Is integrated control superior to conventional control?", constitutes the driving theme and the overall objective of this book. However, several subgoals will have to be achieved before this overall objective can be approached. These are detailed in the following paragraphs.

Since available software packages and recent modelling approaches, including the UPM procedure (FWR, 1994), perform the simulation of sewer system, treatment plant and receiving river in a sequential way, a new simulation tool was developed. This tool allows an interchange of information between the modules for sewer system, treatment plant and receiving water body, which is necessary for consideration and simulation of integrated control scenarios. Therefore, this tool represents a novelty in terms of model development (*cf.* Section 2.3).

Such a tool, simulating the relevant water quantity and quality processes in all parts of the urban wastewater system, is then applied within a procedure which allows development, refinement, testing and evaluation of control strategies for the entire system. In most applications of real-time control such strategies are found either by a (often laborious) trial-and-error procedure (termed here the "strategy development loop") or by on-line application of optimisation algorithms. However, the task to consider the water flow and quality processes not only of one but of all parts of the urban wastewater system for the development and assessment of control strategies, poses new requirements on the procedures to be applied. The complexity of the processes to be considered does not support manual iteration in the strategy development loop; it furthermore resists attempts to simplify (or even linearise) the processes which are to be optimised. Control strategies are developed off-line, *i.e.*, prior to their actual application. The numerical parameters describing a control strategy are found by application of mathematical optimisation methods. These are required here to allow the consideration of complex process descriptions, in other words, these must not require strong assumptions on the objective function to be evaluated. Furthermore, due to the complexity of the processes, not much is known about the shape of the function to be optimised. Therefore, the application of algorithms focusing on local optima only may not be able to identify global solutions. In the present study, several global and (for comparison) local optimisation procedures, which use the simulation tool developed as a means to evaluate the objective function, are implemented, tested and compared with each other.

For the first time, such an off-line optimisation approach is applied to the entire urban wastewater system. For a comparative evaluation of conventional and integrated control of the urban wastewater system, a variety of control scenarios is defined. The results obtained through application of the optimisation procedures then lead to conclusions with regard to the driving question of this book.

1.2 Outline of Chapters

The present chapter (Chapter 1) introduces the background of this book and poses the driving question for this book ("Is integrated control superior to conventional control?"). Furthermore, it outlines the methodology pursued in this work and concludes with an overview of the subsequent chapters.

Chapter 2 presents a review of the state-of-the-art of those areas which are of relevance to this research. Section 2.1 provides an overview of the main processes in the parts of the urban wastewater system, *i.e.*, in sewer system, treatment plant and river. This section also describes common approaches to their mathematical representation in simulation models. Since here only rainfall events, but not, for example, failure of infrastructure (Beck, 1996), are considered as causes of transient impacts on the urban wastewater system, their impacts on the system are summarised in Section 2.2. This is followed by a review of existing software, which is provided in two parts: earlier approaches to integrated simulation of the urban wastewater system are reviewed in Section 2.3, whereas Section A.1 presents an overview of commercially available software packages for the simulation of sewer systems, treatment plants and rivers. Of primary importance for the analysis of control strategies in later chapters is Section 2.4, in which the operation of the urban wastewater system is reviewed and detailed for each of its components. This section also provides an overview of decision-finding techniques which are currently applied when taking a control decision. The definition of "integrated control", which is fundamental for the analyses described in this book, is provided in this section. Section 2.5 contains a comprehensive review of mathematical optimisation techniques and provides the background information necessary for the selection, implementation and application of optimisation routines appropriate to this research.

The topic of Chapter 3 is the development of the integrated simulation tool (called SYNOPSIS - "Software package for synchronous optimisation and simulation of the urban wastewater system"), which will be applied in the subsequent chapters. The model description commences in Section 3.1 with definitions of the requirements of a simulation tool suitable for the intended studies. The constituent submodules of SYNOPSIS are described in detail in Section 3.2. The inevitable problems encountered when linking simulation modules from various origins and their solution are the topic of Section 3.3. Section 3.4 provides information about the control devices and types of control implemented in SYNOPSIS. The optimisation algorithms implemented in SYNOPSIS as well as their interfaces to the simulation module of SYNOPSIS are detailed in Section 3.5. Finally, Section 3.6 provides an overall summary description of SYNOPSIS.

Chapter 4 is concerned with the application of the simulation routines to a case study site. Since no appropriate data from a real case study were available for use within this work, a semi-hypothetical case study site is defined in Section 4.1. In order to demonstrate the capabilities of the simulation tool and the plausibility of the results it provides, Sections 4.2 and 4.3 present and discuss simulation results obtained for dry-weather flow and rainfall input, respectively, for the case study site defined. As a preparation of later optimisation studies, Section 4.4 provides results of an analysis of the settings of the control devices implemented in SYNOPSIS. Finally, Section 4.5 discusses the feasibility of the simulation of individual rainfall events rather than of a continuous long-term time series.

Chapter 5 applies the developments and results of the previous chapters to the development and analysis of control strategies. Section 5.1 provides some definitions (including those for the terms "strategy" and "strategy framework"), which are fundamental to the discussion in this chapter, and summarises the proposed off-line optimisation approach. A simple two-dimensional example is presented in Section 5.2 as an illustration of the optimisation of strategy parameters. Section 5.3 constitutes one of the focal sections of this book, insofar as a comparative evaluation of different optimisation algorithms as well as of one of the options for the definition of strategy frameworks ("top-down") are presented. This section is complemented by Section 5.4, which defines and analyses various strategy frameworks defined by an alternative ("bottom-up") approach. Section 5.5 compares the best strategies of those obtained in the optimisation runs of the previous sections and draws conclusions about the potential of integrated control as opposed to conventional control. Finally, Section 5.6 briefly addresses some issues which are of relevance to the application of the simulation and optimisation procedure developed.

The final chapter of this book, Chapter 6, briefly evaluates the main achievements and results of the work described in this book in Section 6.1. A number of topics for further research are suggested in Section 6.2.

Chapter 2
The State of the Art

Before embarking on simulation and optimisation of the urban wastewater system as an entirety, an overview of the processes commonly observed in the individual components of the urban wastewater system is given (Section 2.1). Since its operation aims at decreasing the adverse impact of rainfall on this system, Section 2.2 provides an overview of effects caused by rainfall on the components of this system. A review of existing software for the simulation of sewer system, treatment plant and rivers is provided in Appendix A, whilst Section 2.3 lists and discusses previous approaches to integrated modelling of the urban wastewater system.

Current practice and state of the art with regard to the operation of each component of the urban wastewater system is reviewed in Section 2.4. Exemplified for real-time control of sewer systems, the different methods applied for finding the actual control decisions are presented. Section 2.5 presents an overview of mathematical optimisation methods and discusses their applicability within the context of this work. Finally, Section 2.6 develops, as a conclusion from the reviews presented in the previous sections, the methodological approach chosen for this work.

2.1 Components of the Urban Wastewater System: Basic Processes and Modelling Concepts

2.1.1 Urban Catchment Runoff and Sewer System

Water flow and quality processes within the urban catchment and the sewer system can be divided into the following categories, which will be outlined in the following subsections:

- surface runoff and washoff;

- flow within the sewer system;
- pollutant and sediment transport within the sewer system;
- biochemical processes within the sewer system;
- processes in storage tanks.

2.1.1.1 Surface Runoff in Urban Areas

The amount of rainfall on urban areas reaching the ground on impervious areas is reduced in its volume by certain losses. These include wetting losses at the start of rainfall, losses due to depression storage and losses due to evaporation (this varying with time of day and year and weather conditions). For rainfall on pervious areas, essentially the same processes are of relevance. However, additionally, infiltration into the soil takes place.

Surface runoff and sewer system simulation models consider these processes in a more or less detailed way. Wetting losses are taken into account usually by a constant subtraction from the rainfall at the start of the rainfall event. A more sophisticated implementation may be necessary for depression storage and evaporation losses, which vary with time and rainfall history (see, for example, Section 3.2.1 for references describing their implementation in the software package KOSIM). An overview of the implementation of all the different processes related to rainfall runoff in various software packages is given by Fuchs, L. (1996).

Modelling of infiltration processes for permeable areas is considered by some modellers (*e.g.,* Verworn and Kenter, 1993) as being necessary, since for larger rain events, runoff from permeable areas can contribute significantly to the total flow. One of the approaches to modelling of infiltration from pervious areas was developed by Horton. It assumes a time-variant infiltration capacity, which decreases during periods with a rain intensity greater than or equal to the infiltration capacity and increasing during other rain periods. An adaptation of this approach for use in long-term simulation is presented by Paulsen (1986). A detailed discussion of modelling approaches for surface runoff in urban areas can be found in ATV 1.2.6 (1986, 1987). Implementation details are discussed by Verworn and Kenter (1993).

After subtraction of the losses, the net rainfall is transported on the surface until it enters the sewer system. This process is often modelled either by a single linear or nonlinear reservoir, or by a cascade of linear or nonlinear reservoirs or by a unit hydrograph (ATV 1.2.6, 1987; Viessman *et al.*, 1989). These modelling approaches are also applied for simulation of flow in small sewer pipes conveying flows towards the main pipes within a subcatchment, as this is employed, for example, by the KOSIM program package.

Several approaches are available for modelling surface washoff quality. These include transport according to the Advection-Dispersion equation (*cf.* Equation (2.8)) and a first-order approach for the description of production and decay of substances. Also empirical approaches are in use (Delleur, 1996).

2.1.1.2 Flow in the Sewer System

Various types of sewer systems have been designed, which broadly fall into two categories: separate sewer systems convey stormwater runoff and wastewater from domestic and industrial sources, respectively, in separate pipe networks towards a receiving water body and to the treatment plant. The second category consists of combined sewer systems which convey all flows towards the treatment plant in a single pipe and discharge those flows which exceed the capacity of the system into the receiving water body. These discharges of combined sewage usually take place at combined overflow (CSO) structures. During some rainfall events, also surface flooding may occur. Besides rainfall runoff and industrial discharges, a major contribution to the flows in combined sewer systems consists of domestic dry-weather discharges. For these, certain diurnal patterns can be observed (details of which will be presented in Section 4.1.2). Detailed studies relating wastewater production to the use of individual domestic appliances have been carried out by Butler (1991, 1993), Butler *et al.* (1994), Butler and Gatt (1996), Friedler and Butler (1996) and Friedler *et al.* (1996).

Modelling flow in sewer systems has a long history; various approaches with various degrees of sophistication are in use. Most of the flow modelling approaches mentioned in this section are also used in river flow modelling, as will become apparent in Section 2.1.3.1. Detailed flow models allow for full consideration of backwater effects and pressurised flow by solving the full SaintVenant equations for unsteady one-dimensional flow. Examples for implementation of the full Saint Venant equations in commercial packages include HYSTEM-EXTRAN (Fuchs *et al.*, 1994b), MOUSE (DHI, 1990) and HYDROWORKS (see Appendix A.1). The Saint Venant equations (Saint Venant, 1870) consist of the continuity momentum (2.1) and the momentum (2.2) equations:

$$\frac{\partial A}{\partial t} + \frac{\partial Q}{\partial x} = 0 \qquad (2.1)$$

$$S_o - S_f = -\frac{q_l V_l}{gA} + \frac{\partial y}{\partial x} + \frac{1}{gA}\frac{\partial}{\partial x}\left(\frac{Q^2}{A}\right) + \frac{1}{gA}\frac{\partial Q}{\partial t} \qquad (2.2)$$

terms used in kinematic wave approximation

terms used in diffusive wave approximation

where

- A: cross-sectional area of channel segment [m^2]
- Q: flow rate [m^3/s]
- x: longitudinal distance [m]
- t: time [s]
- q_l: lateral inflow per unit length [m^2/(s×m)]
- S_o: bottom slope [-]
- S_f: friction slope [-]
- y: channel depth [m]
- g: acceleration due to gravity [m/s^2]
- V_l: longitudinal component of the velocity of the water inflow represented by q_l above [m/s]

The Saint Venant equations are based on the following assumptions (Havlik, 1996):

- unsteady flow in open channels is one-dimensional;
- the fluid is homogeneous and incompressible;
- the pressure distribution is hydrostatic;
- the longitudinal axis of the channel is approximated as a straight line. The channel is prismatic – *i.e.*, the channel cross-section and the channel bottom slope do not change with distance. The variations in the cross-section or bottom slope may be taken into consideration by approximating the channel as a series of several prismatic reaches;

- the average channel slope is small; the channel bed is fixed, that is, the effects of scour and deposition are assumed to be negligible (for flow routing);
- friction can be described by using the steady-state resistance laws, such as the Manning equation (Imhoff and Imhoff, 1990).

Certain assumptions allow simpler approximations of the momentum equation for modelling unsteady flow: if local and convective acceleration can be neglected, then the diffusion-analogy wave simplification is valid, obtained by omission of the final two terms of Equation 2.2. If the additional assumption can be made that backwater effects are negligible, then omission of the second term on the right-hand side of (2.2) (*i.e.*, the pressure force) is justified and the kinematic wave simplification becomes valid. Table 2.1 (after Havlik, 1996) gives an overview of some characteristics of the various approximations of the Saint Venant equations:

Table 2.1. Theoretical comparison of approximations of the Saint Venant equations

	Kinematic wave approximation	Diffusive wave approximation	Dynamic wave (Full Saint Venant equations)
Account for downstream backwater effects and flow reversal	No	Yes	Yes
Attenuation of flood waves	No	Yes	Yes
Account for flow acceleration	No	No	Yes

The solution of the system of the Saint Venant equations (or its diffusive wave approximation) is conventionally done by finite difference methods or by the method of characteristics. Finite difference methods can be divided into explicit and implicit schemes, of which the former are easier to program, but may require very small time steps for numerical stability (Courant condition) (Havlik, 1996).

Since the solution of the Saint Venant equations (or their approximations) can be computationally demanding, also simpler approaches to flow modelling can be applied, in particular for long-term simulations. One of the simplest approaches to model flow in sewer systems is to assume pure translation. By such an approach attenuation of the water wave cannot be modelled. A more sophisticated approach is the one of the Nash cascade, which models flow in subcatchments by conceptually routing it through a series of linear reservoirs, thereby achieving attenuation of the wave (ATV 1.2.6, 1987; Viessman *et al.*, 1989). Some software packages models, such as KOSIM (see Section 3.2.1), are based on this principle. Figure 2.1 depicts the concept of the reservoir cascade.

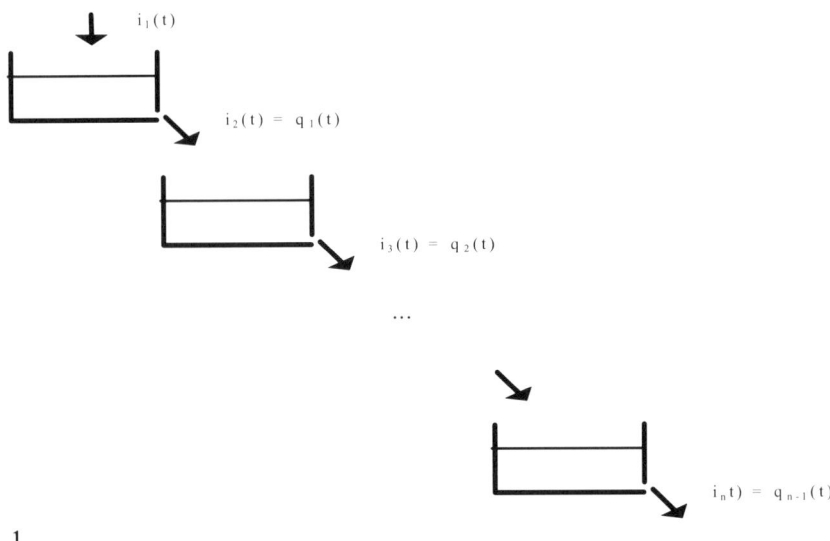

Figure 2.1. The concept of a reservoir cascade

Each of the n reservoirs in series can be described by the storage equation (describing change in stored water volume)

$$S(t) = K\, Q(t) \tag{2.3}$$

and the continuity equation (relating outflow to storage)

$$dS(t)/dt = I(t) - Q(t) \tag{2.4}$$

where

- $I(t)$: inflow at time t [m^3/s]
- $Q(t)$: outflow at time t [m^3/s]
- $S(t)$: storage at time t [m^3]
- K: storage constant [s]

A reservoir cascade of n identical reservoirs with storage constant K has the following characteristics (Harms and Kenter, 1990):

- time to maximum flow:
$$t_P = (n-1) \times K \tag{2.5}$$

- maximum ordinate (maximum flow for unit input):

$$h(t = t_p) = \frac{1}{K(n-1)!}\left(\frac{n-1}{e}\right)^{n-1} \tag{2.6}$$

- flow time of centre of gravity of the input:
$$t_L = n \times K \tag{2.7}$$

Due to its simplicity, the reservoir cascade approach allows rapid simulation; on the other hand, effects such as backwater and pressurised flows cannot be simulated - at least not directly. This constitutes a severe limitation, in particular for looped or flat networks. However, indirect consideration of backwater effects is possible by limiting a pipe capacity as a function of the downstream water level (Debebe, 1996).

Besides the linear reservoir approach, there are other, but similar, hydrological flow routing methods. These include the Muskingum–Cunge and the Kalinin–Miljukov methods (see Reda (1996) for their detailed description). Mention of these methods, originally developed for river flow modelling, is made here since these are also applied for flow routing in sewers (for example, a Kalinin–Miljukov approach is applied in the model SMUSI (Brandt *et al.*, 1989)).

2.1.1.3 Pollutant Transport in the Sewer System

Pollutant mass transport is conventionally described by advection and dispersion processes. Advection results from flow that is unidirectional and does not change the identity of the substance being transported. Diffusion refers to the movement of mass due to random water motion or mixing. It has a tendency to minimise differences in concentration by moving mass from regions of high to low concentration. Dispersion also spreads out pollutants. As opposed to diffusion, dispersion results from velocity differences in space (Chapra, 1997).

One common approach to pollutant transport modelling is given by solution of the one-dimensional Advection-Dispersion partial differential equation (ADE), which describes solute mass transport processes along the longitudinal axis of flow.

It is given here in its general form:

$$\frac{\partial(AC)}{\partial t} + \frac{\partial(AVC)}{\partial x} - \frac{\partial}{\partial x}\left(AD\frac{\partial C}{\partial x}\right) = As \qquad (2.8)$$

where

- C: concentration of a given solute at time t and position x in the longitudinal axis [g/m^3]
- A: cross-sectional area of the river [m^2]
- V: advective velocity in the x direction averaged over the cross-sectional area [m/s]
- D: coefficient of longitudinal dispersion [m^2/s]
- s: sum of solute sources and sinks not related to external water inputs [g/(m×s)]

The Advection-Dispersion equation requires a numerical solution. Some models (*e.g.,* the river model QUAL2E (Brown and Barnwell, 1987)) assume hydraulic steady-state conditions, which reduces the computational effort required to solve the ADE considerably. A simpler approach to modelling of mass transport is the representation of each conduit as a continuously stirred tank reactor (CSTR) as is done, for example, in the SWMM model (Delleur, 1996). The CSTR approach is detailed in Section 2.1.3.2.

Significant physical processes affecting pollutants in sewers are sedimentation and resuspension of pollutants on the surface and in the sewers. Figure 2.2 (ATV, A128, 1992) indicates the conditions of slope and pipe diameter under which sedimentation in sewers is likely to occur.

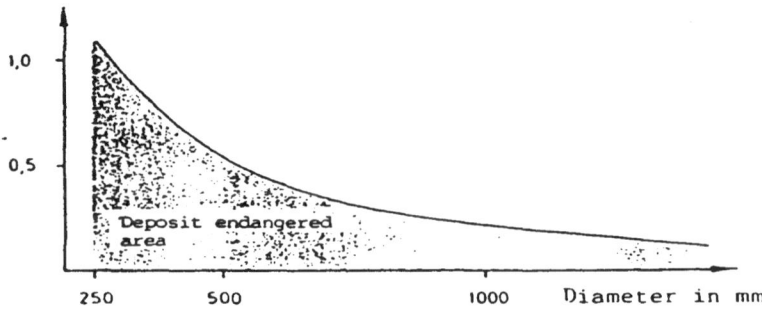

Figure 2.2. Reference values for the occurrence of sewer deposits (ATV, 1992)

At the start of rainfall, a substantial increase in the suspended solids concentration can be observed in some systems. This effect ("first flush") is discussed in more detail in Section 2.2.1. Reviews of sedimentation and washoff processes on the catchment surface and in sewers are have been given by Verbanck and Ashley (1993), Butler and Clark (1995), Delleur (1996), Ashley *et al.* (1997). The implementation in some of today's software packages (including MOSQITO, FLUPOL and KOSIM) is discussed by Bertrand-Krajewski *et al.* (1993).

Jack *et al.* (1996) state that the principal limitations of knowledge which so far have confounded attempts at accurate sewer flow quality modelling (by which, according to the context, these authors appear to mean "sediment transport") consist of:

- the inadequate modelling of gully pot performance;
- the lack of knowledge about inputs of gross solids and their interaction with 'sediments';
- the significant temporal and spatial variability of sediments and pollutants within even one sewerage network;
- the transformation mechanisms of an "active" sediment layer into a consolidated/storage layer and vice versa;
- the temporal variability of sediments and pollutants in sewerage networks in terms of short and long timescales, *i.e.*, daily and seasonally, suggests that models can only be calibrated, not verified;
- the problems of modelling sediment transport and associated pollutants;
- the important dilution of dry-weather flow pollutants associated with infiltration – for systems where this is significant.

Application of pollutant transport models in Germany is driven mainly by the ATV guideline A128 for designing storage tanks in sewer systems (ATV A128 (1992), *cf.* also ATV 1.9.3 (1992)). This guideline states that it is currently not possible to predict the (absolute) pollution of combined sewage during individual events. Therefore, guidelines as well as design practice do not attempt to simulate pollutants on an individual events basis. Instead, simple simulation of pollutant transport – assuming complete mixing – is carried out in long-term simulations.

2.1.1.4 Biochemical Transformations in the Sewer System

On their way through the sewer system, pollutants undergo biochemical processes, which can change significantly the quantity and quality of the organic matter in the wastewater (Nielsen *et al.*, 1992). More specifically, biochemical transformation processes and interactions can be observed in and between the various zones of a sewer pipe (see Figure 2.3, after Almeida, 1999a).

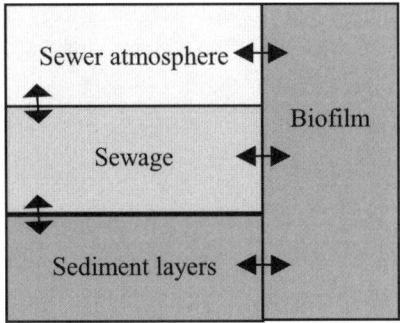

Figure 2.3. Schematic representation of zones within a sewer pipe

Processes and interactions include the following (Almeida, 1999a):

- degradation of organic matter by suspended biomass;
- growth of biofilm attached to the pipe walls (slime);
- decomposition in the sediment layers, most probably under anaerobic conditions;
- oxidation of hydrogen sulphide, resulting in the formation of sulphuric acid, in the crown of the pipes (if appropriate conditions prevail);
- reaeration of sewage;
- uptake of oxygen by the biofilm from the sewer atmosphere;
- release of hydrogen sulphide and other gases into the sewer atmosphere;
- transfer of substrate, nutrients and dissolved oxygen from the sewage into the attached biofilm;
- sedimentation and resuspension of sediments and attached pollutants;
- release of sediment interstitial water, containing dissolved pollutants;
- oxygen uptake from the sewer atmosphere by sulphur bacteria;
- sloughing of parts of the biofilm into the bulk water.

In some cases, large biochemical oxygen demand (BOD) removal rates have been observed; this ability of the sewer system for self-purification has been discussed by various authors (see references given by Nielsen *et al.*, 1992). In this context it should be noted that in many sewer systems the hydraulic retention time of the wastewater is of the same order of magnitude as in the treatment plant. Thus the composition of "young" wastewater with an age of only minutes or a few hours may be quite different from wastewater which has been under transportation for 20 hours or more. Effects of ageing on chemical oxygen demand (COD) in sewer systems is discussed in detail by Kaijun *et al.* (1995).

In-sewer processes can also have significant impacts on the other parts of the urban wastewater system: for example, sulphide, generated in the sewer system, may also affect the biological processes in the wastewater treatment plant, since it is reported to be among the causes leading to sludge bulking (Chen, 1993; Kappeler and Gujer, 1994c). Furthermore, it may also be toxic to fish in the receiving water after overflow events (Nielsen *et al.*, 1992). However, un-ionised ammonia is of much higher significance with regard to toxicity to aquatic life (see Section 2.2.3) than sulphide. According to the survey by Almeida (1999a), other influences of sewage composition on treatment plant processes include sewage septicity (reported to be associated with sludge bulking characteristics), effects of in-sewer treatment on settling characteristics (Kaijun *et al.*, 1995).

Despite the importance of these processes, in order to constrain the extent of this work, no further consideration is given to these processes.

Many of these processes are not fully understood and are the subject of extensive current research – see, for example, the papers presented at the International Conferences on "The Sewer as a Biological, Chemical and Physical Reactor" (*cf.* Hvitved-Jacobsen *et al.*, 1995; Hvitved-Jacobsen, 1998) and the work conducted by Almeida (1999a) and Huisman *et al.* (2000).

2.1.1.5 Storage Tanks

In most sewer systems storage volume is available. This can be available either in the form of storage tanks or as in-sewer storage. Real-time control of sewer systems makes deliberate use of this storage (see Section 2.4.2).

The German guidelines for the design of sewer systems (ATV A128, 1992) distinguish five different arrangements for tanks and overflow devices:

- on-line bypass tank;
- off-line bypass tank;
- on-line pass-through tank;
- off-line pass-through tank;
- simple overflow without any storage structure.

Figure 2.4 illustrates these different arrangements.

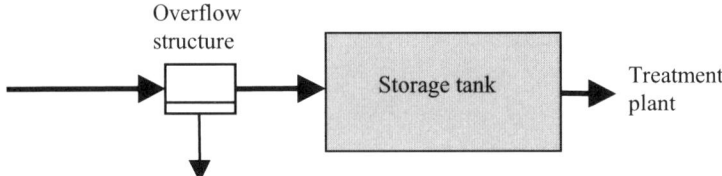

Figure 2.4a. On-line bypass tank (flow scheme) (retaining the first flush of stormwater in main stream)

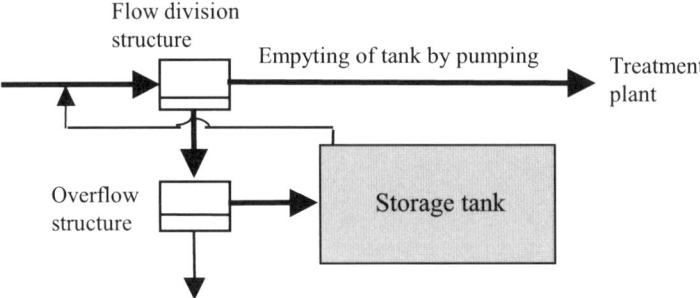

Figure 2.4b. Off-line bypass tank (flow scheme) (retaining the first flush of stormwater in bypass stream). This arrangement is called "fill-and-bypass" by Lessard and Beck (1990).

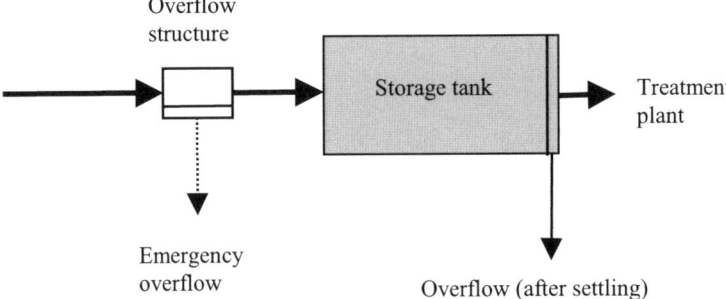

Figure 2.4c. On-line pass-through tank (flow scheme) (on-line storage tank with a maximum pass-forward capacity) (overflow for settled combined wastewater in main stream of flow)

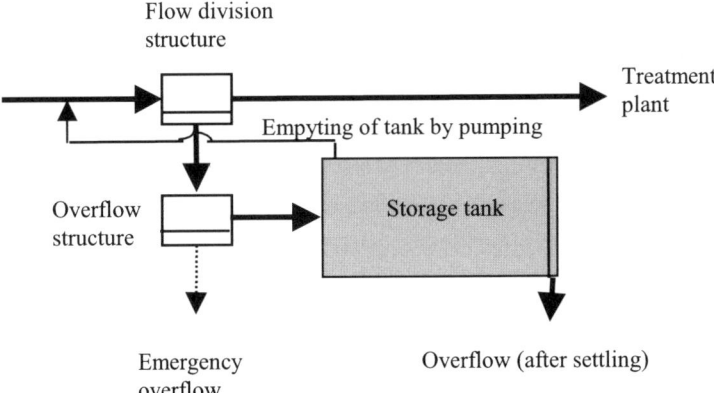

Figure 2.4d. Off-line pass-through tank (flow scheme) (overflow for settled combined wastewater in bypass stream). This arrangement is called "fill-and-treat" by Lessard and Beck (1990)

On-line tanks are characterised by the fact that sewage is always flowing through them (i.e., including during dryweather flow periods). On the other hand, off-line tanks remain dry during dryweather periods. The difference between bypass and pass-through basins lies in the location of the overflow structure of the basin: if the overflow lies upstream of the basin, the tank is called a bypass tank. All inflowing sewage is stored until the capacity of the tank is exceeded; any additional inflow then bypasses the tank and is discharged. Use of bypass tanks may help attenuate first-flush effects because the more heavily polluted sewage (occurring at the start of a storm event) is retained in the tank, whereas the diluted sewage coming in later is discharged over the overflow into the receiving water. Rauch and Renner (1996) discuss the efficiency of bypass basins. As opposed to arrangements involving bypass tanks, any overflows from pass-through tanks pass through the tank prior to being discharged. The main idea behind this is that particulate pollutants may settle while passing through the basin before they are discharged.

Simulation of tank performance is included in most sewer simulation packages. Simpler models (such as KOSIM) allow for consideration of sedimentation in on-line tanks based on settling velocity and flow velocity. A similar approach is applied by Rauch (1996b). Also more sophisticated approaches to modelling processes in storage tanks have been proposed (see, for example, the storm tank model developed by Lessard and Beck (1991b), which is also described in the next section).

In many sewer systems pumps are installed (often for filling or emptying storage tanks; or for pumping flows towards the treatment plant). Prudent operation of these pumps can increase the performance of the sewer system. These issues are discussed in more detail in Section 2.4.3.

2.1.2 The Wastewater Treatment Plant

Conventional wastewater treatment consists of three processes, including preliminary treatment (screening and grit removal), primary sedimentation and secondary (biological) treatment. The activated sludge process is applied as the secondary treatment procedure in many plants. An increasing number of treatment plants also apply methods of tertiary or advanced treatment to further reduce pollutant and nutrient content in the effluent. This section gives a brief survey of the main treatment processes relevant to this work, *i.e.*, storm sewage retention, primary sedimentation, activated sludge process and secondary sedimentation. A brief outline is also provided of modelling concepts for the various parts of an activated sludge treatment system. A good overall review of modelling concepts is given by Lessard and Beck (1991a).

Figure 2.5 provides a schematic overview of a simple activated sludge plant.

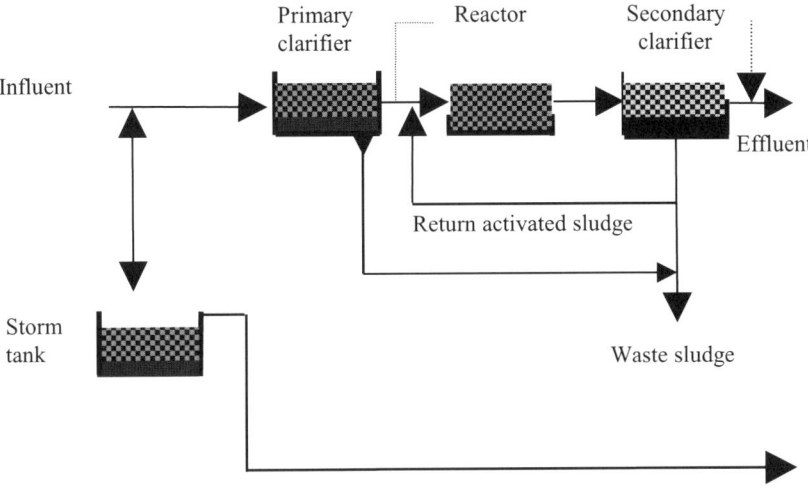

Figure 2.5. Schematic representation of an activated sludge wastewater treatment plant

2.1.2.1 Storm Tank

Process overview

Storm tanks serve the purpose of storing sewage flows induced by rainfall which are in excess of the capacity either of the treatment plant or of the interceptor sewer conveying flows to the treatment plant. In the UK, such tanks are conventionally located at the inlet of the plant. Besides temporarily storing water until hydraulic capacity is available in the treatment plant, these tanks also allow the water stored to be partially treated by sedimentation. The treatment efficiency of storage tanks depends on the flow dynamics and on the geometric properties of the tank as well as on their operation. Studies by Johansen (1985) and Durchschlag (1991) give maximum removal rates for suspended solids between 60% and 80%; for particulate COD, Michelbach and Wöhrle (1994) estimate a removal efficiency as high as 70%.

The importance of storm tanks for integrated control of the urban wastewater system results from the fact that they can be used for control of the inflow to the treatment plant and for reduction of CSO discharges. However, a careful balance has to be found between deliberate CSO discharges and prolonged increased inflows towards the treatment plant, possibly resulting in continued hydraulic overloading of the secondary clarifier (*cf.* Section 2.2.5). Thus proper operation of storm tanks is essential to the management of wastewater treatment plants.

Modelling of storm tanks

Although storm tanks are widely used and referred to as one of the most immediate means of solving problems related to CSOs, not many models exist which simulate storm tanks on a detailed event basis (Lessard and Beck, 1991b). According to these authors, many models simulating storm tanks are focused more on long-term evaluations rather than on analysis of individual events.

Goforth *et al.* (1983) relate residence time of wastewater in the tank to pollutant removal efficiency as indicated in Equation (2.9):

$$R = R_{max}(1 - e^{-kt}) \quad (2.9)$$

where

    R: pollutant removal efficiency [-]
    R_{max}: maximum pollutant removal efficiency [-]
    k: removal rate [s^{-1}]
    t: treatment (residence) time [s]

Other models were proposed by Amandes and Bedient (1980), Medina *et al.* (1981a,b) and Ferrara and Hildick-Smith (1982).

A more detailed storm tank model, distinguishing the various modes of operation which are commonly found in storm tanks, is proposed by Lessard and Beck (1991b). This model simulates suspended solids, volatile suspended solids, total COD, soluble COD, NH$_4$ and NO$_3$. The first three of these are partitioned into settleable and non-settleable fractions. Lessard and Beck's model allows the processes of sedimentation and pumping the tank's contents back to the treatment plant to be considered in more detail than is done by many other models.

More specifically, four different modes of operation are distinguished:

Fill mode: this mode is active during filling of the storm tank. The contents of the tank is assumed to be completely mixed. Allowance is made by a loss term for settling and sludge production (both as a function of the settleable solids concentration) within the tank.

Dynamic sedimentation mode: this mode closely resembles primary clarification. It is active whenever the storm tank is full and overflows are discharged into the receiving water. This part of the model is similar to Lessard and Beck's model of primary sedimentation (Lessard and Beck, 1988; see also next section).

Quiescent settling mode: this mode is active whenever the storm tank is not empty and no flows are entering or leaving the storm tank. It is assumed that all settleable matter will settle within one time step (*i.e.,* one hour) with a corresponding increase in sludge mass.

Draw mode: this mode simulates emptying the contents of the storm tank back towards the treatment plant inlet. Pumping starts as soon as the influent to the treatment plant falls below a specified value. The concentration of a fraction of the tank volume is assumed to be influenced by sludge scraping when emptying the tank.

Lessard and Beck (1991b) state that "it has not been possible to evaluate the performance of the model against field data, the results from a test simulation indicate no strong grounds for its rejection as inadequate for its present application in the development of stormwater control strategies." Simulation results by these authors suggest that the storm tank has relatively small effect on the suspended solids load that may overflow to the receiving water body. In a similar way, the return of the stored storm sewage to the main stream of the treatment plant, if scheduled appropriately, was observed to have relatively little adverse effect on the performance of primary clarification, although the same may not be true of the secondary, biological treatment processes.

2.1.2.2 Primary Clarification

Overview

The purpose of primary clarification (synonymous with "primary sedimentation") is to remove settleable solids. Typically, 50 to 70% of the total suspended solids is removed in this process (Tchobanoglous and Burton, 1991). Since some of the solids are biodegradable, biochemical oxygen demand (BOD) is typically reduced by 30 to 40% (Lester, 1990). Standard retention times are about 0.5 to 2 hours, while retention times greater than 3 hours do not improve the efficiency of the primary clarifiers (Rauch, 1996b). According to Harremoës *et al.* (1993), primary clarifiers also have an equalising effect on variations in influent wastewater concentration. Peak concentrations are damped due to a time lag of the order of the retention time. Impacts of stormwater inflow on primary clarifier performance are described in Section 2.2.2.

Modelling of primary clarifiers

Traditionally, most models of primary clarifiers employ a steady-state approach, whilst their dynamic behaviour is neglected. According to Lessard and Beck (1988), this may be feasible to some extent since the only operational control option within the primary clarifier seems to be the schedule of desludging. However, the importance of primary clarifiers is considered to arise from the fact that the efficiency of primary sedimentation affects the performance of subsequent secondary treatment units and the sludge treatment.

Problems in precise modelling of primary clarifiers are caused by the complexity of the dynamic behaviour of the sedimentation processes and include (Lessard and Beck, 1991a):

- variability of influent characteristics;
- variability of particle sizes and corresponding velocities;
- presence of complex flow patterns and density currents in the tank;
- scouring phenomena and the effect of temperature.

Nevertheless, a variety of primary clarification models has been proposed, ranging from steady-state models to lumped and distributed parameter models (see Lessard and Beck (1991a) for references). Simple models usually relate the removal efficiency to influent suspended solids concentration and/or overflow rate. An example is the model used by Rauch (1996b), which is based on an approach of Schilling and Hartwig (1988). This model relates the removal rate to inflow rate, using an exponential function. Another model is described by Alarie et al. (1980). The model is reported to give accurate results when predicting effluent suspended solids on a short-term basis. These authors also provide a detailed review of the dynamic primary clarification.models known by that time.

Lessard and Beck (1988) developed a dynamic model based on the continuously stirred tank reactor (CSTR) approach[3]. State variables modelled include suspended solids, volatile suspended solids, particulate COD, soluble COD, NH_4 and NO_3, the first three of which are partitioned into settleable and non-settleable fractions. Depending on the nature of the influent (raw sewage; raw and storm sewage; raw sewage mixed with sludge liquors returned form sludge treatment) different values are assumed for settling velocities and fractions of settleable material of SS, VSS and particulate COD. Settling and scouring from the bed of the clarifier of particulate matter are considered in the model. Furthermore, gain of NH_4 caused by releases from anoxic zones of the sludge as well as entrapment of soluble COD with particles are included in the model as additional processes.

Otterpohl and Freund (1992) propose a dynamic model for primary clarifiers which is based on removal efficiency expressed as a logarithmic function of hydraulic residence time. Paraskevas et al. (1993) describe a dynamic model for primary clarifiers, which is also based on the CSTR approach. State variables modelled are fixed and volatile suspended solids, fixed and volatile dissolved solids, suspended and dissolved BOD, and ammonia. These authors include a hopper in

---

[3] A description of the CSTR modelling approach will be given in Section 2.1.3.2.

their model (simulation of hydrolysis of ammonia is restricted to this zone). From their experimental results these authors conclude that their model is more accurate than the one proposed by Lessard and Beck (1988).

According to Harremoîs *et al.* (1993), a simple model for the primary clarifier is sufficient for realistic dynamic simulations of a complete wastewater treatment plant. Similarly, Lessard and Beck (1991a) state that simple lumped-parameter models (such as those by Alarie *et al.* (1980) and by Lessard and Beck (1988)) have proven to be reliable.

2.1.2.3 The Activated Sludge Process

Overview

The activated sludge process was discovered in 1913 by Ardern and Lockett (1914) from laboratory experiments at the Davyhulme treatment plant in Manchester (Ardern and Lockett, 1914; von der Emde, 1998). Today, this process is applied in many treatment plants throughout the world. Alternative methods of biological treatment of wastewater include the use of trickling filters, rotating biological contactors, anaerobic digestion, and the use of ponds. An overview of these methods is given by Tchobanoglous and Burton (1991).

The basic principle of the activated sludge process is to keep a high concentration of a mixed culture of micro-organisms in an artificially aerated reactor. Organic waste is fed into the reactor where it is broken down by the bacterial culture. The reactor content is referred to as the ìmixed liquorî. After a specified period of time, the biomass is separated from the treated effluent by settling in sedimentation tanks (see subsequent section). Part of the separated biomass is recycled into the aeration tank, thereby maintaining a steady population of micro-organisms. Recycling the sludge allows for a separation of the hydraulic retention time and the residence time of the biomass. The average sludge age in the system is commonly referred to as the solids retention time and is controlled primarily by the amount of sludge withdrawn from the system (waste sludge). The sludge age is a frequently applied measure in the design of activated sludge systems. Table 2.2 gives an overview of sludge ages for the design of various types of treatment plants (according to the German guidelines for the design of treatment plants, ATV A131, 1991).

Table 2.2. Minimum sludge ages for the design of treatment plants

Treatment objective	Plant up to 20000 PE	Plant for 20000 PE
Treatment without nitrification	5 days	4 days
Treatment with nitrification (design temperature 10°C)	10 days	8 days
Treatment with nitrification and denitrification (design temperature 10°C) D = 0.2 D = 0.3 D = 0.4 D = 0.5	 12 days 13 days 15 days 18 days	 10 days 11 days 13 days 16 days
Treatment with nitrification, denitrification and sludge stabilisation	25 days	not recommended

D: ratio of the volume of the denitrification zone to the volume of the activated sludge tank
PE: Population equivalent

Many variations of the original activated sludge process are in use today (see Tchobanoglous and Burton, 1991). These mainly differ in the type of flow within the system (plug flow, completely mixed flow, step feed, sequencing batch reactor), the type of aeration (conventional, tapered), and in the type of aeration devices (diffused aeration, mechanical aeration, aeration with high-purity oxygen). Control of the activated sludge process is discussed in Section 2.4.4.

Processes: degradation of organic matter

The two main processes describing the degradation of carbonaceous material in the activated sludge process can be summarised by Formulae (2.10) and (2.11) (Tchobanoglous and Burton, 1991), where COHNS represents the organic matter in wastewater:

Oxidation and synthesis:

$$COHNS + O_2 + nutrients \longrightarrow CO_2 + NH_3 + C_5H_7NO_2 \qquad (2.10)$$
$$+ \text{ other end products}$$

Endogenous respiration:

$$C_5H_7NO_2 + 5O_2 \longrightarrow 5 CO_2 + 2H_2O + NH_3 + \text{energy} \qquad (2.11)$$

Processes: nitrogen removal

Since nitrogen can have a significant impact on receiving water quality (see Section 2.1.3.3), its removal from the wastewater is desirable. The main source of nitrogen in domestic wastewater is urine (75% according to Larsen and Gujer, 1996), who have accordingly labelled it as "anthropogenic nutrient solution".

Ammonia (NH_3)[4] is toxic to fish. Furthermore, it causes oxygen depletion. Therefore, one of the objectives of wastewater treatment is to transform ammonium into less harmful substances. Ammoniacal nitrogen is converted into nitrate (NO_3) by biological nitrification. Since the presence of nitrate in treatment plant effluents is also problematic, in particular in areas where water is reused for water supply, denitrification (*i.e.*, conversion of nitrate via nitrite into nitrogen gas, thus removing the nitrogen from the wastewater) is performed at many treatment plants. Nutrient removal may also prevent eutrophication processes in the receiving water body.

Biological nitrification is the process in which nitrogen compounds in wastewater are converted into nitrite and nitrate. This process is an aerobic process in which the autotrophic nitrifying bacteria (Nitrosomonas and Nitrobacter) use carbon dioxide as their carbon source for the synthesis of new cells and nitrogen compounds (mainly ammonium) as their energy source. Two stages of oxidation occur during this process, which can be represented by the following equations (Mudrack and Kunst, 1994):

$$NH_4^+ + 1.5\ O_2 \xrightarrow{\text{Nitrosomonas}} NO_2^- + 2H^+ + H_2O$$

$$NO_2^- + 0.5\ O_2 \xrightarrow{\text{Nitrobacter}} NO_3^-$$

Nitrifying bacteria are sensitive organisms. According to Tchobanoglous and Burton (1991), several factors can inhibit their activity such as high concentrations of ammonia and nitrous acid. Other conditions necessary for nitrification to occur are a pH value within a certain range (the pH range optimum for nitrification is

---

[4] The proportion of un-ionised ammonia (NH_3) to total ammonia (NH_3+NH_4) depends on the temperature and on the pH value (see Figure 2.10).

between 7.5 and 8.6), a sufficiently high temperature (a temperature below 10°C inhibits nitrification), and a sufficiently high DO level (of more than about 1 mg/l). It should be noted that transforming ammoniacal nitrogen into nitrate-nitrogen does not facilitate nitrogen removal in itself, but it does eliminate its oxygen demand. Nitrogen removal from wastewater can be achieved by further transforming nitrate to nitrogen gas under anoxic (without oxygen) conditions as is done in the denitrification process. The reactions for nitrate reduction are:

$$NO_3 \rightarrow NO_2^- \rightarrow NO \rightarrow N_2O \rightarrow N_2$$

A large number of treatment plants are designed to perform denitrification in concert with denitrification. A variety of treatment plant designs for achieving denitrification are described by Tchobanoglous and Burton (1991) and Mudrack and Kunst (1994).

Modelling

For modelling of the activated sludge process, a wide variety of approaches has been developed. Among these are static and quasi-dynamic approaches (Hartwig, 1993; Härtel et al., 1995; ATV 2.11.4, 1997), time-series models (see, for example, Novotny et al., 1991; Carstensen et al., 1994; Berthouex and Box, 1996), reduced order (Jeppson and Olsson, 1993), and detailed mechanistic dynamic models. Of these types of mode, full mechanistic models have been recommended by practitioners for understanding treatment plant performance under varying flow conditions (FWR, 1994).

Among them is the model of Jones (1978), which is based on biochemical oxygen demand (BOD) as the main state variable. This model is implemented as one of the options available in the commercial software package STOAT (*cf.* Appendix A.2). An overview of this model is given by Güven (1995). It accounts for three mechanisms, namely co-metabolism of substrate, utilisation of substrate for maintenance of cells, and biochemical activity of the non-viable cells, which can account for the small proportion of viable bacteria present in a typical activated sludge. The model gives special consideration to the non-viable bacteria, *i.e.,* those which are still capable of biochemical activity even when they are no longer capable of cell division (WRc, 1994). The UPM Manual (FWR, 1994) characterises the activated sludge model of Jones as follows: "Although the WRc [model] is simpler than the IAWQ [model], it contains the additional element that 'dead' bacteria can still have active enzymes. This has resulted in the WRc model being applicable to domestic sewage works under average flow conditions, without needing the model parameters to be recalibrated. The IAWQ model requires the growth rates to be recalibrated for large shifts in the sludge age."

In 1986, the IAWPRC (later renamed IAWQ and, finally, IWA) Activated Sludge Model No. 1 (Henze *et al.*, 1986) was published, summarising earlier attempts at modelling carbonaceous oxidation, nitrification and denitrification. This model represents the state of the art in modelling the activated sludge process and is therefore implemented in practically all commercial software packages and is widely applied. The Activated Sludge Model No. 1 contains 19 parameters (stoichiometric and kinetic coefficients). It models eight processes involving 13 state variables, including various fractions of organic matter, biomass, nitrogen components, particulates and alkalinity. Model representation is usually done by listing the processes as rows of a matrix with the state variables as column headings. Each matrix entry indicates the appropriate stoichiometric coefficient for the relationship between the variables in the individual processes. Process equations for each state variable are easily read by summing up the products of each entry of the related column with the kinetic coefficients of the processes, which are given in the right-most column of the matrix. Figure 2.6 lists the IAWPRC Model, whilst Tables 2.3 and 2.4 list the state variables and the kinetic and stoichiometric coefficients, respectively. The assumptions made in this model as well as further explanations can be found in Henze *et al.* (1986). One of the assumptions of the original model is that nitrogen, phosphorus and other inorganic nutrients are not limiting factors for the removal of organic substrate.

Table 2.3. Definition of state variables in the IAWPRC Activated Sludge Model No. 1

Component number	Component Symbol	Definition
1	S_I	Soluble inert organic matter $M(COD) L^{-3}$
2	S_S	Readily biodegradable matter $M(COD) L^{-3}$
3	X_I	Particulare inert organic matter $M(COD) L^{-3}$
4	X_S	Slowly biodegradable substrate $M(COD) L^{-3}$
5	$X_{B,H}$	Active heterotrophic biomass $M(COD) L^{-3}$
6	$X_{B,A}$	Active autotrophic biomass $M(COD) L^{-3}$
7	X_P	Products from biomass decay $M(COD) L^{-3}$
8	S_O	Dissolved oxygen $M(-COD)L^{-3}$
9	S_{NO}	Nitrate and nitrite nitrogen $M(N)L^{-3}$
10	S_{NH}	Ammonia nitrogen $M(N)L^{-3}$
11	S_{ND}	Soluble biodegradable organic nitrogen $M(N)L^{-3}$
12	X_{ND}	Particulate biodegradable organic nitrogen $M(N)L^{-3}$
13	S_{ALK}	Alkanity - Molar units

Table 2.4. Kinetic and stoichiometric coefficients of the IAWPRC Activated Sludge Model No. 1

Kinetic event	Symbols
Heterotrophic growth and decay	$\mu_M, K_S, K_{O,M}, K_{NO}, b_M$
Autotrophic growth and decay	$\mu_A, K_{NM}, K_{O,A}, b_A$
Correction factor for anoxic growth of heterotrophs	η_B
Ammonification	k_a
Hydrolysis	k_h, K_X
Correction factor for anoxic hydrolysis	η_h
Stoichiometric coefficient	
Heterotrophic yield	Y_H
Autotrophic yield	Y_A
Fraction of biomass yielding decay products	f_P
Mass N/Mass COD in biomass	i_{XB}
Mass N/Mass COD in decay products	i_{XP}

Component (i) → / Process (j) ↓	1 S_I	2 S_S	3 X_I	4 X_S	5 X_{BH}	6 X_{BA}	7 X_P	8 S_O	9 S_{NO}	10 S_{NH}	11 S_{ND}	12 X_{ND}	13 S_{ALK}	Process Rate p_j $(ML^{-3}T^{-1})$
1 Aerobic Growth of Heterotrophic Biomass		$-\frac{1}{Y_H}$			1			$-\frac{1-Y_H}{Y_H}$		$-i_{XB}$			$-\frac{i_{XB}}{14}$	$\hat{\mu}_H\left(\frac{S_S}{K_S+S_S}\right)\left(\frac{S_O}{K_{O,H}+S_O}\right)X_{BH}$
2 Anoxic Growth of Heterotrophic Biomass		$-\frac{1}{Y_H}$			1				$-\frac{1-Y_H}{2.86Y_H}$	$-i_{XB}$			$\frac{1-Y_H}{14 \cdot 2.86Y_H}-\frac{i_{XB}}{14}$	$\hat{\mu}_H\left(\frac{S_S}{K_S+S_S}\right)\left(\frac{K_{O,H}}{K_{O,H}+S_O}\right)\left(\frac{S_{NO}}{K_{NO}+S_{NO}}\right)\eta_g X_{BH}$
3 Aerobic Growth of Autotrophic Biomass						1		$-\frac{4.57-Y_A}{Y_A}$	$\frac{1}{Y_A}$	$-i_{XB}-\frac{1}{Y_A}$			$-\frac{i_{XB}}{14}-\frac{1}{7Y_A}$	$\hat{\mu}_A\left(\frac{S_{NH}}{K_{NH}+S_{NH}}\right)\left(\frac{S_O}{K_{O,A}+S_O}\right)X_{BA}$
4 Lysis of Heterotrophic Biomass				$1-f_p$	-1		f_p					$i_{XB}-f_p i_{XP}$		$b_H X_{BH}$
5 Lysis of Autotrophic Biomass				$1-f_p$		-1	f_p					$i_{XB}-f_p i_{XP}$		$b_A X_{BA}$
6 Ammonification										1	-1		$\frac{1}{14}$	$k_a S_{ND} X_{BH}$
7 Hydrolysis of Particulate Organic Compounds		1		-1										$k_h \frac{X_S/X_{BH}}{K_X+X_S/X_{BH}}\left[\left(\frac{S_O}{K_{O,H}+S_O}\right)+\eta_h\left(\frac{K_{O,H}}{K_{O,H}+S_O}\right)\left(\frac{S_{NO}}{K_{NO}+S_{NO}}\right)\right]X_{BH}$
8 Hydrolysis of Particulate Organic Nitrogen Compounds											1	-1		$p_7 \cdot (X_{ND}/X_S)$

Figure 2.6. The IAWPRC Activated Sludge Model No. 1 (reproduced from Henze *et al.*, 1986, with permission from the copyright holders, IAWQ)

The Activated Sludge Model No. 1 of Henze *et al.*, applies a consistent notation as defined by Grau *et al.* (1987). COD is used as a measure of concentration of organic material in wastewater. The use of COD (rather than, for example, BOD) provides the advantage of clear mass balances. COD is partitioned into four fractions: soluble inert (S_I) and particulate inert (X_I) material pass through the process without undergoing any changes. The particulate material will be included at a later stage in the sludge mass and concentrations. Readily biodegradable material (S_S) is affected directly by the transformation processes (growth of heterotrophs), whereas slowly biodegradable material (X_S) must be hydrolysed before being degraded.

The large number of parameters of the IAWPRC Model No. 1 leads to problems in terms of identifiability, as reported by Beck (1983, 1986) and Lessard and Beck (1991a) and as discussed in some detail for important constituent parts of this model by Beck (1989), Dochain *et al.* (1995a), Vanrolleghem *et al.* (1995a,b), Vanrolleghem and Dochain (1997) and Weijers and Vanrolleghem (1997). It should be noted that routinely sampled data are not sufficient for calibration of the model parameters. Additional sampling, usually in form of batch tests, is necessary (ATV 2.11.4, 1997). Some useful hints for determination of some of the model's parameters for a given treatment plant by measurements are given by Londong and Wachtl (1995), ATV 2.11.4 (1997), and a methodology is proposed by Spanjers and Vanrolleghem (1995).

The process of degradation of organic matter is considered in the Activated Sludge Model No. 1 as aerobic growth and decay of heterotrophs. Nitrification and denitrification processes are described as aerobic growth and decay of autotrophs and anoxic growth of heterotrophic bacteria, respectively. Furthermore, the model accounts for "hydrolysis" of entrapped organisms (converting slowly into readily biodegradable matter) and organic nitrogen as well as ammonification of soluble organic nitrogen.

Extended versions of this model were published as the Activated Sludge Model No. 2 (Henze *et al.*, 1995a; Gujer *et al.*, 1995), Model No. 2d (Henze *et al.*, 1999) and, finally, Model No. 3 (Henze *et al.*, 2000). Models No. 2 and 2d describe additionally those processes which are related to biological phosphorus removal, where Model No. 2d also considered denitrification of phosphorus-accumulating organisms, whilst the Activated Sludge Model No. 3 addresses various inadequacies of the earlier models organisms. It models 12 processes for, altogether, 13 different components.

2.1.2.4 Secondary Clarification

The secondary clarifier is an integral part of the activated sludge system. It has two main functions: it separates the biomass from the water in order to produce a good quality effluent free of settleable solids (clarification) and the biomass is thickened (thickening). Part of the biomass is then wasted as sludge and part of it is returned to the biological reactor to maintain an appropriate biomass concentration. Figure 2.7 provides a conceptual representation of a secondary clarifier.

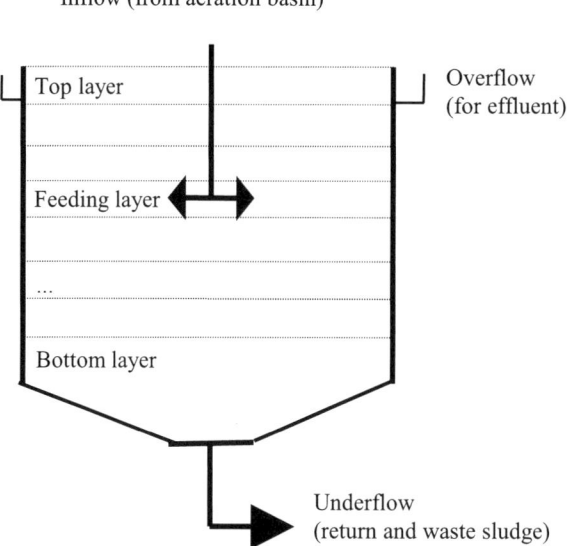

Figure 2.7. Sketch of a secondary clarifier and its conceptual partition into a number of layers for modelling purposes

Takács *et al.* (1991) describe four types of settling characteristics, all of which can normally be found in wastewater treatment plants. These types include discrete particle settling, flocculent particle settling, hindered settling and compression settling.

The operation of the secondary clarifier is crucial for the performance of the whole treatment plant (Chen, 1993). As Beck (1984) puts it "It is in the [secondary] clarifier where adverse operational problems of bulking, rising or dispersed sludge either develop or become critically apparent."

The main goal in the operation of the clarifier is to prevent excessive rise of the sludge blanket, which eventually may result in loss of sludge into the effluent. This not only increases the concentration of solids and of organic matter considerably. It also affects the performance of the activated sludge process itself, since biomass, which is necessary in the aeration tank for proper functioning of the process, is lost to the system.

Control of the activated sludge process and of the clarifier conventionally includes, besides aeration control, setting the pump rates of return activated sludge (RAS) (pumped from the secondary clarifier back to the aeration tank) and of the waste activated sludge (WAS) (usually pumped from the secondary clarifier towards sludge treatment). These two control handles are important for solids control in the conventional activated sludge process. Pumping the RAS redistributes the sludge between the secondary clarifier and the aeration tank, without affecting the total sludge mass, whereas pumping WAS removes sludge from the system.

The term "bulking sludge" refers to sludge that has poor settling characteristics and poor compactability. Causes of sludge bulking include the growth of filamentous organisms or bacterial cells swelling through the addition of water (Tchobanoglous and Burton, 1991). "Rising sludge" is caused by denitrification in the secondary clarifier. Denitrification may result in nitrogen gas becoming trapped in the sludge layer and causing the sludge to rise. Another operational problem present in the absence of filamentous organisms is "dispersed sludge", which thickens easily but gives an effluent with a high concentration of fine suspended solids.

Modelling of secondary clarification

This section gives only a brief survey of trends and milestones in secondary clarifier modelling. Detailed reviews are given by Lessard and Beck (1991a) and, more recently, by Freund *et al.* (1993), Vazquez-Sanchez (1996) and ATV 2.11.4 (1997). Parts of the following discussion are adapted from Vazquez-Sanchez (1996).

Existing models range in complexity from very simple empirical models for the clarification function of the settler (Pflanz, 1969) to some very complicated two- and three-dimensional models which consider hydrodynamic effects (Krebs, 1993, 1995, 1997; McCorquodale and Zhou, 1993; Holthausen, 1995; Ji *et al.*, 1996b). Another class of clarifier models are the so-called "quasi-dynamic models"; several of them are presented and discussed by Freund *et al.* (1993). This class of models is characterised by static formulae which are applied at every time step of the simulation. However, the dynamic layer models showed better results in a model comparison carried out by these authors.

Table 2.5 (after Krebs, 1995) gives an overview of the application areas of the various types of models. As can be seen from the Table 2.5, sophisticated layer models appear to be the most appropriate model type to be applied in the given context. Krebs (1995) states 1D layer modelling is, after careful calibration, effective for combination with an activated sludge model. However, the same author also states that 1D models cannot realistically describe dynamic processes within the sludge blanket and "there is hardly any potential for the prediction of the effluent quality".

Relevant references presenting 1D layer models include Vitasovic (1989), Takács *et al.* (1991), Dupont and Henze (1992), Härtel and Pöpel (1992), Otterpohl and Freund (1992), Chen and Beck (1993), Dupont and Dahl (1995) and Jeppson and Diehl (1995). A concise overview and comparison of most of these models is given by Grijspeerdt *et al.* (1995). Table 2.6 summarises the main milestones in the development of 1D secondary clarifier models.

Table **2.5.** Capacity of various classes of secondary clarifier models

Purpose	Conceptual model	Flux model	Sophisticated layer model	Math./Anal. 1D-modelling	Hydrodyn. 2D-modelling
Dimensioning	+	o	p	p	o
Operation[1]	-	o/-	+/p	p	p
Effluent[1]	-	-/-	o/-	o/-	p
Recycle[1]	o	+/o	+/o	+/o	+/+
Sludge blanket[1]	o	o/-	+/o	+/p	p
Flocculation	-	-	-	-	p
Combination with ASM	-	-	+	o	-
Control	-	-	+	p	o
Design (inlet/outlet)	-	-	-	-	+
Process understanding	-	o	o	+	+
Ready for applicability	+	+	+	o	-
Computer capacity	+	+	+	o	-

+: positive; o: neutral; -: negative; p: development potential
[1] Two grades are distinguished for stationary/dynamic case
ASM Activated Sludge Model

Table 2.6. Main milestones and relevant studies in 1D layered models of secondary clarifiers

Author	Main features of the model	Goals	Drawbacks
Busby and Andrews (1975)	Variable number of layers of time-varying thickness.	A dynamic model of solids-liquid separation in AS process.	A complex set of heuristics.
Vitasovic (1989)	N layers of constant thickness of five different types. Inclusion of the upward movement of the liquid in the layers above the feed point. Concept of a threshold concentration.	Inclusion of the region above the feed well. Good solids concentration profiles.	It does not include a clarification component, consequently it has low prediction capability for effluent suspended solids
Takács et al. (1991)	Extention of Vitasovic's model by modifying the settling velocity model. This allows one to take the smaller, slowly settling particles of the upper layers into account.	Unified approach to clarification and thickening function. A settling velocity model that is also valid for low solids concentrations.	It replicates fairly closely trends in MLSS and sludge blanket height but it still lacks accuracy (particularly effSS).
Härtel and Pöpel (1992)	Introduction of a correction factor in Vesilind's settling velocity. Just one layer above the influent where settling is described according to Stoke's law.	Velocity model can be applied also to transition and compression zones, leading to a more precise solids profile.	Very poor prediction of the clarification function.
Dupont and Henze (1992)	Fractioning of the solids as microflocs and macroflocs in order to predict the effluent concentration of non-settled particles under normal operation of the clarifier. The settling of the macroflocs are described according to Härtel's correction function whereas the velocity for the smaller particles is considered constant.t	Verification of the model with experimental data from a pilot plant. Easier coupling with AS Model No. 1. Improvement of Härtel's model by considering a two-component model.	Limited application to full-scale plants Limited application to full-scale plants
Otterpohl and Freund (1992)			
Dupont and Dahl (1995)	Extension of Takács' model by taking density currents and short-circuiting into account. The former is modelled by dynamically changing the inlet layer and the latter by a dilution factor.	Introduction in the Takács model of the observed effects of density current and short-circuiting within the settlers.	The constant introduced for short-circuiting is dependent on too many site-specific parameters. Further development of the model would be necessary.

Among the models proposed for secondary clarifiers is one by Lessard and Beck (1993). Its clarification part makes use of an extension of the empirical approach taken by Pflanz (1969), which was derived from data from the treatment plant of Celle/Germany. Beck and Lessard relate the SS in the effluent not just to the clarifier influent but also to the return sludge rate. Further details of this model are discussed in Section 3.2.2.

Various extensions to this model were carried out by Chen (1993), incorporating advective and dispersive transport phenomena as well as empirical expressions for approximating the effects of bulking and pin-point floc on the clarified effluent suspended solids concentration. With respect to thickening, Chen's model divides the heterotrophic biomass into floc-forming and filamentous biomass in order to simulate the commonly found sludge bulking conditions. An exponential function that takes into account the floc-forming and filamentous biomasses is included in the gravitational settling velocity equation of the suspended solids. Although this model gave slightly better results than Lessard's model, Chen concluded that the relationships among operational variables of the reactor, the components of the sludge floc and its settling velocity are too complex to be quantitatively described. Extensions to Lessard's model are also described by Vazquez-Sanchez (1996), who aimed to consider increased effluent suspended solids in clarifier situations where the sludge blanket approaches the weir crest or even overflows. These modifications are described in Section 3.2.2.2.

Kappeler and Gujer (1994a,b) propose a model for aerobic bulking, which simulates three types of micro-organisms: facultative aerobic floc-forming, obligate aerobic filamentous and nitrifying micro-organisms.

Among the secondary clarifier models most frequently referred to is that due to Takács *et al.* (1991). The model is based on the solids flux concept (Keinath, 1985, 1990; Härtel and Pöpel, 1992) and a mass balance around each layer of a one-dimensional settler. It extends the settling velocity equation used by Vesilind (1968) by using separate terms for settling characteristics of the hindered settling zone and also for zones with low solids concentration (*e.g.,* the upper layers of the clarifier). Thus this model takes a unified approach to clarification and thickening functions of settlers. According to Takács *et al.*, their settling velocity model is found to produce realistic estimates of the effluent and underflow suspended solids concentration under steady-state and dynamic conditions. However, although the trends in MLSS, suspended solids and sludge blanket height are replicated fairly closely under a variety of dynamic effects, the model could not be evaluated against field data. The pilot-scale data set used by Takács *et al.*, was not initially designed for dynamic modelling purposes and, accordingly, a full solids profile and balance was not available for the studies carried out by these authors.

Several improvements to the model of Takács *et al.* are suggested by various authors in order to enhance the prediction capability of the existing models for sludge profiles and effluent suspended solids concentration. Härtel and Pöpel (1992) propose a correction function Ω to make the settling velocity term applicable to the process of settling in the transition and compression areas. Otterpohl and Freund (1992) partition the solids into microflocs and macroflocs. Settling of the macroflocs is described according to Härtel and Pöpel's function, whereas microflocs are assumed to settle with a constant settling velocity.

Grijspeerdt *et al.* (1995) compare various 1D clarifier models using data obtained from a lab-scale treatment plant. The models investigated include those by Takács *et al.*, Otterpohl and Freund, and Dupont and Henze. Grijspeerdt *et al*, conclude that the Takács model was the most reliable of those compared. Motivated by these findings, most other studies investigating the integrated urban wastewater system (see Section 2.3) make use of the model of Takács *et al.*, for the simulation of the secondary clarifier. The main disadvantage of the Takács model is the relatively long computation time required for convergence.

Jeppsen and Diehl (1995) propose a mathematically robust model based on the approaches of Takács *et al.* Model development was designed to achieve mathematical consistency rather than the ability to predict experimental data. Their model converges if the number of layers (in most applications of Takács' model, this number is set to 10) is increased. Jeppsen and Diehl (1996) further conclude that the use of just ten layers is too coarse to explain the detailed behaviour of the settler. They state that at least 30 layers should be used in order to obtain reliable results under normal operating conditions.

As a final note in this section, it should be mentioned that Dupont and Dahl (1995) observed that most of the 1D models discussed above (Takács *et al.*; Dupont and Henze; Otterpohl and Freund) suffer from at least one of the following two problems when applied to full-scale wastewater treatment plants:
- incorrect calculation of sludge concentration profile near the effluent weirs;
- incorrect calculation of the return sludge concentration.

Dupont and Dahl succeeded in setting up a model representing both the suspended sludge concentration and the return sludge concentration by inclusion of density current and short-circuiting in the settling tank. The term "short-circuiting" refers to the observation that a substantial part of the influent to the secondary clarifier is transported straight to the return sludge outlet without effectively taking part in the actual settling process. However, this model is dependent on many site-specific parameters, and would accordingly require extensive calibration efforts.

2.1.3 Rivers

Water flow, pollutant transport and biochemical processes in sewer systems and in treatment plants take place in an engineered and thus well-defined infrastructure. These structures are deliberately designed and built to fulfil their purpose within urban wastewater management. Despite this, it is obvious they do not always perform in an optimum way.

Receiving water bodies (in this work: rivers) have various purposes, only one of them being to serve as a recipient for urban water discharges. These are usually more or less natural waters, although there are also many examples of engineered rivers (*e.g.,* Luppmen (Krejci *et al.*, 1994a)) serving as a receiving water body for urban drainage systems. Besides agricultural and industrial activities, urban drainage is considered to constitute one of the most important anthropogenic impacts on receiving waters (Lammersen, 1997b).

Since the characteristics of rivers are inherently different from those of a sewer system and treatment plant, a generalised description of the related processes and subsequent improvement of their performance might arguably be considered as more challenging than is the case for the other elements of the urban wastewater system. Furthermore, receiving rivers vary to a great extent in their type and size. ATV 2.1.1 (1989) roughly distinguishes between these six river types: streams in mountainous regions, rivers in mountainous regions, low land rivers, impounded rivers, tidal rivers and estuaries. But even within these classes rivers are quite different from each other. Thus, the extent to which the various processes discussed below will be of relevance to a particular case study will strongly depend on its individual characteristics.

2.1.3.1 *River Flow*

The most obvious state variable characterising the behaviour of a river is its flow, which is influenced by inflow from point sources (*e.g.*, discharges from sewer systems and treatment plants as well as other discharges) as well as by inflow from non-point sources (*e.g.*, from groundwater or catchment runoff). Furthermore, the river flow is influenced by rainfall runoff from the river catchment. A description of river flow for a given stretch of the river at and downstream of the location of the urban drainage and treatment system under consideration usually includes the following main components:

- base flow (flows coming from further upstream). This is considered as not being affected by the variations of discharges or flow within the river stretch given or downstream thereof. However, it is influenced by catchment runoff;
- continuous (*e.g.,* from treatment plants) and intermittent (*e.g.,* from overflow structures) discharges into this stretch of the river;
- lateral inflows from non-point sources (where applicable).

A comprehensive review of different categories of models for river flow, varying in their complexity and thus in their computational demands, is presented by Reda (1996), therefore, only a brief overview is given here. Simpler, so-called hydrologic flow routing methods are based on the continuity equation, which relates change in storage (in a given river stretch) to inflow and outflow. From this equation, a variety of modelling approaches have emerged, such as reservoir cascades, the Muskingum, Muskingum–Cunge and the Kalinin–Miljukov methods. For the hydrodynamic modelling approaches, which are based on the full Saint Venant equations for one-dimensional flow, the reader is referred to Section 2.1.1.2 for a brief overview. Details can also be found in many textbooks; a comprehensive review is also given by Havlik (1996). Higher dimensional hydrodynamic models are also available. However, for water quality modelling in rivers, generally the one-dimensional equations or approximations thereof are applied (Rauch *et al.*, 1998b).

2.1.3.2 *Pollutant Transport in the River*

Pollutant mass transport is conventionally described by advection and dispersion processes. Among the most traditional approaches to pollutant transport modelling is the solution of the one-dimenstional Advection-Dispersion equation (ADE). This equation (2.8) was discussed in Section 2.1.1.3 and is therefore not repeated here.

An alternative to the so-called Eulerian approaches to modelling of solute transport (*i.e.*, changes in hydraulic and solute state variables are observed from a fixed pont in the channel) is the Lagrangian approach. It considers the observer of dispersion as travelling with the centre of mass of a water parcel containing solute. Thus, the advective term (*i.e.*, $-V\, \partial C/\partial x$, under the assumption of the solute being fully mixed over any cross-section) can be eliminated for computation of the dispersion effects. Thus, dispersion is calculated as a correction term to the advective transport. Combination of a "pseudo-Lagrangian" model with an Eulerian flow-routing process resulted in the dynamic RATSS model, which in a comparative study was able to reasonably reproduce oscillations in streamflow and concentration for a range of situations for less computational effort than the hydrodynamic model

MIKE 11 (Norreys (1991), according to Reda (1996)). The main disadvantages of pseudo-Lagrangian schemes are the possibility of mass conservation not being fully achieved and the impossibility of simulating dead zones (Norreys, 1991; Norreys and Cluckie, 1996).

Time-series models constitute another class of mass transport models. In these models, the solute concentration in a given element is expressed by a transfer function. According to Reda (1996), further research into this method of river water quality modelling is required, despite encouraging results obtained so far. In some time-series water quality models, each channel segment is attributed a dead zone (working as a continuously mixed tank). These models consider that each such dead zone is aggregated with its respective channel segment in the main flow (Wallis *et al.*, 1989), constituting an aggregated dead zone. Thus these models are sometimes called Aggregated Dead Zone (ADZ) models (Beck *et al.*, 1989; Young, 1992), the origins of which grew out of earlier CSTR approximations (Beck and Young, 1975; Beck *et al.*, 1989). Due to their computational efficiency, ADZ models appear to be of great potential for on-line estimation and RTC purposes.

An alternative way to model pollutant mass transport is to model the stream as a series of CSTRs as done by Thomann (1963), Beck (1973), Whitehead *et al.* (1979), Beck and Finney (1987) and subsequently by Reda (1996). In terms of computational complexity, this approach lies between solution of the classical advection-dispersion equation and the algebraic equation simplicity of the ADZ model of Young (1992) (Beck and Reda, 1994). Further details of the ADZ approach can be found in Camacho (2001). As discussed by Reda, certain justifiable assumptions have to be made for application of the CSTR method. These assumptions include setting tank outflow equal to tank inflow for water quality modelling and neglecting the time derivative of weir overflows in the solute transport model. Because of its importance as one of the available options for river modelling, the concept of CSTR modelling is outlined below. The following discussion is based upon Reda (1996).

The main idea of the CSTR approach to solute transport and water flow and quality modelling is to represent each river section by a series of continuously stirred tanks. For each of these, solute mass conservation is considered as in

$$d(Vx)/dt = Q_{in} x_{in} - Q x + V r_c \qquad (2.12)$$

where

V: tank volume containing solute, calculated for each time step t [m^3]

Q_{in}:   (upstream) inflow rate into the reactor [m^3/s]

Q: outflow rate at the downstream end of the reactor [m^3/s]

x: solute concentration in the tank (assumed to be homogeneous) and at the outflow [g/m^3]

x_{in}: solute concentration in the (upstream) inflow [g/m^3]

r_c: reaction rate of that solute [1/s]

During the computation, tank volume, together with outflow and depth, is constantly updated. Whitehead *et al.* (1979) comment on the hydraulic routing procedure and note its analogy with the one derived from the Muskingum models. Thus, solute diffusion cannot be directly simulated. However, the sequence of instantaneous-mixing and gradual-release operations has proved to be useful to simulate both flood "dispersion" and actual solute dispersion.

Among the first applications of the CSTR scheme in river water quality simulation was the simulation of DO and BOD in the River Cam (Beck, 1973; Beck and Young, 1975). This model was then used for the first time to simulate flow and several, interacting, water quality determinands by Beck and Finney (1987), based on a daily measurement campaign for the Bedford-Ouse study. These authors expanded the structure of this model to five determinands (BOD, DO, Amm-N, NO$_3$-N, and chlorophyll-a, representing algae), using it to simulate algal blooms in the abundance of nitrogen compounds. A later extension to include phosphorus was reported by Bayar (1993). Reda (1996) included hourly information on flow and water quality in the River Cam, also considering treatment plant effluent discharging into this river. From the basic structure of the CSTR approach it follows that features such as backwater effects upstream of a weir and propagation of transient level differentials cannot be accommodated by this type of model.

2.1.3.3 Biochemical Transformations in the River

This section describes briefly those variables, processes and modelling approaches which are commonly considered as relevant to the description of river water quality processes. Recent and comprehensive surveys include those by Reda (1996), Rauch *et al.* (1998b), Shanahan *et al.* (1998) and Somlyódy *et al.* (1998). Detailed discussions of the various processes can also be found in most textbooks on river water quality modelling (e.g Orlob, 1983; Bowie *et al.*, 1985; James, 1993; Chapra, 1997). Therefore, only a brief overview will be given in this section. An example of a river model will be described in Section 3.2.3.

It should be noted that many biochemical processes found in rivers are similar to those occurring in wastewater treatment plants. However, temporal and spatial scales in river water quality processes are much larger than in treatment plants. Furthermore, since rivers vary significantly in their characteristics, parameters of river water quality models are expected to vary over broad domains (Somlyódy *et al.*, 1998). However, attempts have recently been made to define river models on the same basis as treatment plant models (Masliev *et al.*, 1995; Somlyódy *et al.*, 1998, see also the discussion in Section 2.3).

Figure 2.8 (after Beck and Finney, 1987) gives an example of the variety of processes affecting river water quality. The processes shown in the figure are those simulated in the Reda (1996) model.

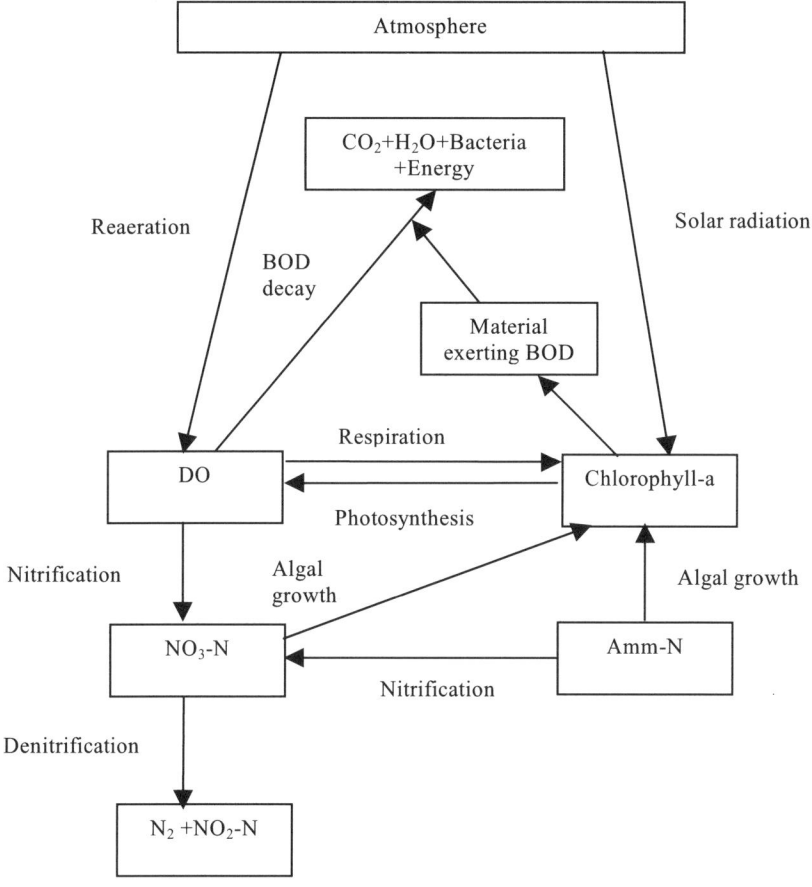

Figure 2.8. Example of in-river interactions and exchanges with the atmosphere as modelled by Reda

Oxygen

Because of its importance for aquatic life, oxygen is one of several water quality parameters which traditionally has received most attention.

Processes affecting the dissolved oxygen (DO) concentration in a river include reaeration (uptake of oxygen from the atmosphere at the water–air interface), decomposition (caused by breakdown of organic matter by organisms), sediment oxygen demand, photosynthesis, algal respiration, and nitrification.

A historic step in modelling the oxygen balance in rivers is the oxygen sag model of Streeter and Phelps (1925). The classical Streeter–Phelps equation describes the oxygen deficit caused by a point discharge of oxygen-depleting matter:

$$dD/dt = k_d \times L - k_a \times D, \qquad (2.13)$$

where

D: oxygen deficit (DO_{sat}-DO; DO_{sat} being the saturation concentration; see below) [g/m^3]

L: BOD concentration of discharge [g/m^3]

k_a: reaeration rate [day^{-1}]

k_d: decomposition rate [day^{-1}]

Most existing models of oxygen depletion in rivers are ultimately based on the Streeter–Phelps equation.

DO saturation

The saturation concentration of DO in natural water is of the order of about 10 mg/l. Of the factors affecting this value, the most important are temperature, salinity and pressure due to elevation above sea level. Chapra (1997) reports numerical relationships between DO saturation (DO_{sat}) concentration and these influencing parameters. The following two formulae making use of temperature only are widely used.

Elmore and Hayes (1960) (as reported by Bowie *et al.*, 1985):

$$DO_{sat} = 14.652 - 0.41022\,T + 0.007991\,T^2 - 0.000077774\,T^3 \qquad (2.14)$$

with T: temperature in degrees Celsius.

American Public Health Association (APHA, 1992):

$$DO_{sat} = \exp(-139.34411 + 1.575701 \times 10^5 \times T_a^{-1} - 6.642308 \times 10^7 \times T_a^{-2}$$
$$+ 1.243800 \times 10^{10} \times T_a^{-3} - 8.621949 \times 10^{11} \times T_a^{-4}) \qquad (2.15)$$

with T_a: temperature in Kelvin

However, these two formulae do not give significantly different results within the temperature range $5°C \leq T \leq 25°C$.

Water with higher salinity than fresh water has a reduced DO saturation concentration. Similarly, the saturation concentration decreases for increasing altitude above sea level. Application of an approximate formula by Zison et al. (1978) (as reported by Chapra, 1997) results in the statement that for a location of 500 m altitude above sea level, the saturation concentration decreases by approximately 4%.

Reaeration

Reaeration describes the process of exchanging oxygen across the water–atmosphere interface, until ultimately (given sufficient time) the oxygen concentration in the water reaches saturation level. In many water quality models, reaeration is described by a first-order process, involving the reaeration rate k_a (or, as in some papers, k_2):

$$dDO/dt = k_a \times (DO_{sat} - DO), \qquad (2.16)$$

where

DO: oxygen concentration [g/m^3]
DO_{sat}: DO saturation concentration [g/m^3]
k_a reaearation rate [day^{-1}]

Reaearation is influenced by many processes, such as temperature (usually considered by an Arrhenius expression – see below), wind and hydraulic characteristics (*e.g.,* flow velocity and water depth). Ideally, the value of the reaeration coefficient should be measured. Lacking data, however, empirical prediction of this coefficient based on average velocity and depth is an acceptable first approach in shallow, fluvial streams, but may be highly under-estimated in deeper waters (Reda, 1996). In the literature, many empirical formulae relating velocity and depth to reaeration can be found. A review of reaeration formulae is given by Bowie *et al.* (1985). Among the most commonly applied ones are the formulae by Owens *et al.* (1964), O'Connor and Dobbins (1958) and Churchill *et al.* (1962):

O'Connor and Dobbins: $\quad k_a = 3.93\ U^{0.5}\ H^{-1.5}$ (2.17)

Churchill $\quad k_a = 5.026\ U\ H^{-1.67}$ (2.18)

Owens et al. $\quad k_a = 5.32\ U^{0.67}\ H^{-1.85}$ (2.19)

where

k_a: reaeration coefficient [day^{-1}] at T = 20 °C

U: flow velocity [m/s]

H: water depth [m]

These formulae were derived from experiments and studies on rivers of various characteristics. Therefore, each has a limited range of applicability. According to Chapra (1997), Covar (1976) and Zison et al. (1978) suggested combining these formulae as shown in Figure 2.9.

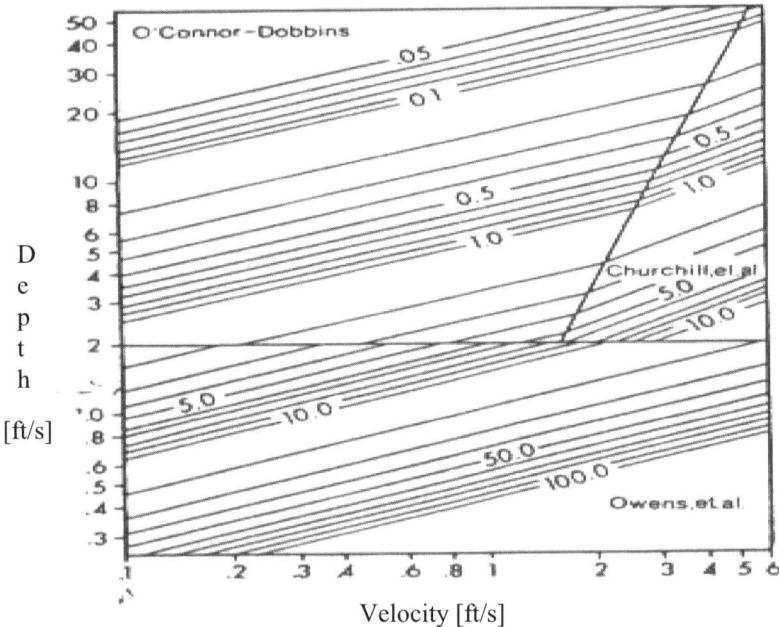

Figure 2.9. Reaearation rate [day^{-1}] versus velocity and depth (Chapra, 1997; reprinted with permission of The McGraw-Hill companies)

According to Reda (1996), Bennett and Rathbun (1972) reviewed various empirical approaches and developed their own, based on experiments in several rivers by different researchers.

They propose the following formula

$$k_a = 5.577\, U^{0.607}\, H^{-1.689} \tag{2.20}$$

with the variables defined as above.

For the range of velocities and depths of interest for the present study (see Section 4.1.4), this formula gives reaeration coefficients about 0.2 to 0.3 day^{-1} larger than those resulting from application of the formula by Owens et al. (see above).

For the river model in his studies, Rauch (1996b) uses a constant k_a value. He justifies this choice by making the statement that most of the predictive models for the k_a value are based on steady-state measurements. Thus, the results provided by those models may differ significantly from reality under fast dynamics as given by unsteady flow conditions in rivers. Lacking data which would allow a more accurate definition of the reaeration rate, Rauch sets k_a in his model to a value of 3 day^{-1}, motivated by typical reaeration coefficients as were compiled by him and which are shown in the following table.

Table 2.7. Reaeration coefficients for various water bodies

Water body	Ranges of k_a (reaeration coefficient) at 20° C
Tidal rivers and estuaries	0.1 – 0.5 day^{-1}
Deep rivers of low velocity	0.2 – 1.0 day^{-1}
Shallow rivers	1.0 – 5.0 day^{-1}
Shallow rivers of high velocity with rapids and waterfalls	> 5.0 day^{-1}

Deoxygenation

Organic matter in river water is broken down by bacteria and organisms under consumption of oxygen. These decay processes reduce oxygen and BOD concentrations in the river. In many oxygen models, these processes are described by a first-order reaction with a deoxygenation rate k_d (see, for example, the description of the river model implemented in this work, which is described in Section 3.2.3). According to Chapra (1997), the deoxygenation rate decreases with increasing water depth up to a depth of about 2.40 m.

In addition to this immediate oxygen demand, there may exist also a "delayed oxygen demand", the importance of which, in particular in small rivers, is stressed by Harremoës (1982), Hvitved-Jacobsen (1982) and Hvitved-Jacobsen and Schaarup-Jansen (1991). It is caused by degradation of material which has settled on the bottom of the river. This material originates from wastewater particulates, from other particulates (*e.g.,* leaf litter) as well as from photosynthetically produced plant matter. After settling, this organic matter is degraded and exerts a delayed oxygen demand on the water passing by (Hvitved-Jacobsen and Schaarup-Jansen, 1991). They further state that the location of the critical (minimum) DO concentration caused by delayed oxygen demand is relatively close to the point of discharge as compared to the critical point for the immediate oxygen demand. The duration of the delayed oxygen demand is reported to be typically in the order of 12 to 24 hours (DMSC, 1995). However, Reda (1996) refers to a study of a large catchment of the Milwaukee river, where a duration of several months of a CSO impact on the oxygen balance was observed (Kreutzberger *et al.*, 1980).

Some models cater for the delayed oxygen demand by simply adding a constant in the DO equation (see, for example, Lijklema *et al.*, 1996). Others relate it to the amount of adsorbed or settled particulate matter (Hvitved-Jacobsen and Schaarup-Jensen, 1991; Rauch, 1996b). Chapra (1997) provides a detailed discussion of modelling approaches for sediment oxygen demand (SOD). He proposes a SOD model capable of reproducing an observed square-root relationship between surface sediment COD and sediment oxygen demand.

Reduction of immediate oxygen demand by sedimentation

Besides causing a delayed oxygen demand, settling of organic matter in the stream reduces (immediate) oxygen demand within the river. Some authors (Reda, 1996; Chapra, 1997) summarise deoxygenation and settling effects into one single BOD removal or decay rate.

Ultimate BOD

Since BOD (here short for: BOD_5) measures the oxygen demand exerted over a period of five days, the BOD value does not describe the total (ultimate) BOD. Assuming that the first-order assumption of the BOD decay process holds, the ultimate BOD (BOD_∞) can be obtained from Equation (2.21):

$$BOD_\infty = BOD_5 \times \frac{1}{1-e^{rk_d}} \qquad (2.21)$$

where

k_d: BOD deoxygenation rate [day^{-1}]

Lijklema *et al.* (1996) include this conversion in the oxygen equation of their oxygen model.

Nitrogen processes in rivers

Also of importance to aquatic life is ammonium. There are two forms in which ammonium exists in natural waters: Un-ionised ammonia (NH_3) is toxic to fish, whereas ammonium ion (NH_4^+) is not harmful at concentrations generally encountered in natural waters (Chapra, 1997). The proportion of un-ionised ammonia of total ammonium (NH_3+NH_4) strongly depends on pH and temperature, as shown in Figure 2.10 (after Bowie *et al.*, 1985, and Alabaster and Lloyd, 1980).

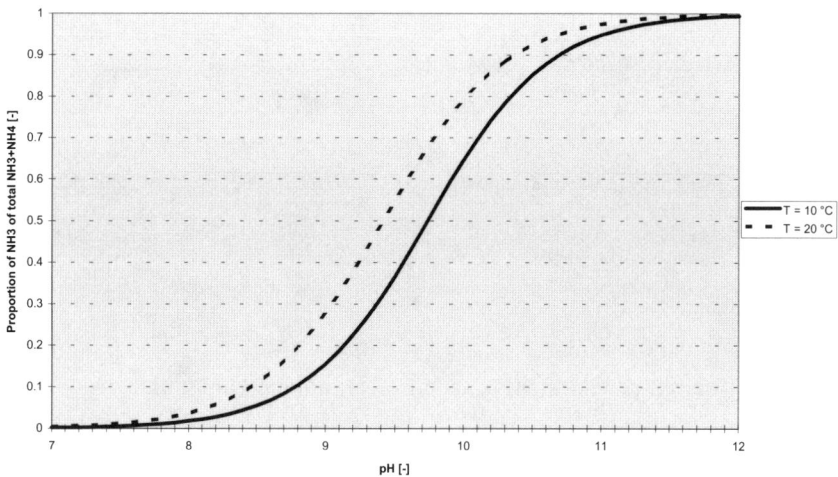

Figure 2.10. Proportion of NH_3 of total NH_3+NH_4 as a function of pH and temperature

Unless river water quality models simulate the various nitrogen compounds separately, they usually simulate the total ammonium concentration, commonly denoted by the terms "ammonium" or "ammoniacal nitrogen". Ammonium concentrations in the river are mainly affected by nitrification. Together with phosphorus, it also stimulates algal growth which may lead to eutrophication in the water body.

Figure 2.11 illustrates the nitrogen cycle in natural waters (as illustrated by Chapra, 1997). Those processes which are of particular interest to water quality modelling are briefly discussed below.

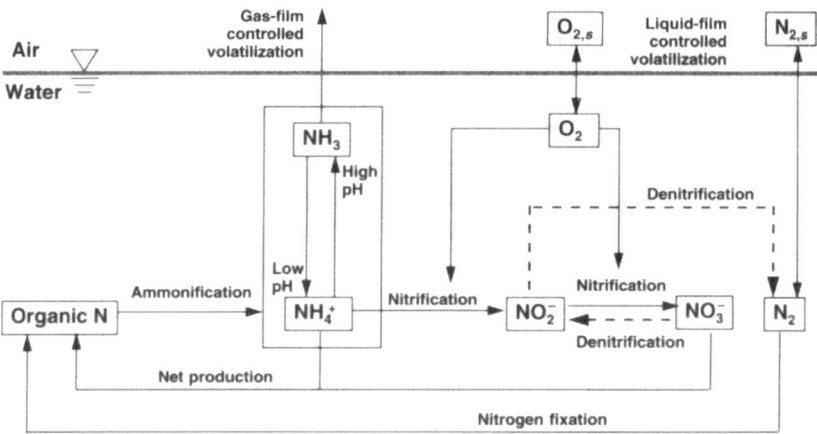

Figure 2.11. The nitrogen cycle in natural waters (Chapra, 1997; reprinted with permission of The McGraw-Hill companies)

Major transformations affecting nitrogen in rivers include the following (Reda, 1996):

- nitrification: conversion of ammonical nitrogen to nitrate, exerting an oxygen demand;
- denitrification: reduction of nitrate nitrogen to nitrogen gas, particularly under anoxic conditions. This process may be negligible in many rivers;
- uptake by algae: consumption of ammoniacal and nitrate nitrogen by algae;
- ammonification: release of ammoniacal nitrogen due to decay of organic nitrogen;
- nitrogen fixation: reduction of atmospheric N_2 by plants to organic N or ammoniated compounds.

Nitrification is dependent on sufficient oxygen levels in the river, the presence of nitrifying bacteria and an appropriate pH level. Some nitrification may also have already taken place in the treatment plant (see Section 2.1.2.3). Due to its demand for oxygen, nitrification can significantly influence the oxygen balance in the river. In river models, nitrification is frequently considered in simple source (for NO_3-N) and sink (for DO and ammoniacal nitrogen) terms. Because of the shortcomings of this approach (ignoring the presence of nitrifiers, pH level and the time-lag due to the two-step conversion from ammonium to nitrate), Chapra (1997) suggests a more detailed nitrification model, modelling organic nitrogen, ammonium nitrogen, nitrate and nitrite nitrogen as state variables. In a simpler way, some models include the limitation of nitrification due to the absence of oxygen by an appropriate Monod factor (*e.g.,* Lijklema *et al.*, 1996). Such an approach is also implemented in the simulation tool developed within this work (see Section 3.2.2). Denitrification is not always modelled, in particular when nitrate nitrogen is not considered as a state variable.

Algae: photosynthesis, respiration and decay

Important processes involving algae are photosynthesis and algal respiration, particularly in rivers or lakes with a sufficient nutrient content (having a potential for eutrophication). Photosynthesis converts inorganic nutrients into more complex organic molecules. This oxygen-generating process requires sunlight as an energy source. Therefore, it is dependent on time of the year, time of the day and, to a lesser extent, on the weather conditions. At greater water depths, the influence of light decreases (extinction). Respiration of algae, on the other hand, constitutes an oxygen-consuming process. Increased algal growth, triggered by high nutrient concentrations, may subsequently result in an increased oxygen demand through algal decay.

The relevance of these algae-related processes to a given receiving water body depends on its individual characteristics. Therefore, some river models include algae (or chlorophyll-a) as a state variable (Reda, 1996), or even several algal species (as in the EUTROF2 model described in IHE (1992)), whereas other models consider these processes simply in source/sink terms without modelling algae as such (*e.g.,* Lijklema *et al.*, 1996; Rauch, 1996b).

Influence of temperature on biological processes

Dependence of process rates on temperature is frequently expressed by a factor involving an Arrhenius constant for each individual process. Thus this influence is considered by a factor in the rate term of the form

$$\theta_P^{(T-T_0)}$$

where

θ_P: Arrhenius constant for process P [-]
T: temperature [°C]
T_0: reference temperature [°C]
(usually, $T_0 = 20°C$)

An Arrhenius constant $\theta_P > 1$ denotes that this process rate increases with increasing temperature. Typical ranges for θ_P for various processes are listed by Bowie et al. (1985).

Current trends in river water quality modelling

Somlyódy et al. (1998) state, after having reviewed the state of the art and the problems of river water quality modelling (*cf.* Rauch et al., 1998b; Shanahan et al., 1998; Somlyódy et al., 1998), that future trends in water quality modelling will be on the refinement of the description of the biochemical transformation processes, since today's models for the physical transport processes are considered to be well developed. Improvements in models should aim at establishing closed mass balances (thus suggesting the use of COD as a basis for expressing state variables, rather than BOD) and adequate consideration of sediment-related processes. These developments culminated in the publication of the *River Water Quality Model No. 1* of the IWA Task Group on River Water Quality Modelling (Reichert et al., 2001; Shanahan et al., 2001; Vanrolleghem et al., 2001; *cf.* also Reichert and Vanrolleghem, 2001). This model is described in an ASM-like matrix notation and characterised by closed mass balances based on the elementary composition of organisms.

Applications, some of which have been addressed in earlier studies (Beck and Reda, 1994), could include (Somlyódy et al., 1998):

- dynamic problems of combined stormwater overflows and nonpoint source pollution;
- impact of improved wastewater treatment plant operation and control;
- extreme and surprising pollution events;
- improved assessment of artificially influenced rivers;
- data collection;
- understanding, research, education and improved communication (*e.g.*, between wastewater treatment and river water quality experts);
- regulatory applications including catchment planning.

2.2 Impacts of Storm Events on the Urban Wastewater System

The previous section having provided a general overview of the processes in the components of the urban wastewater system, the impacts on this system caused by rainfall are the topic of this section. Since control of the urban wastewater system, as described in Chapter 1, aims at alleviation of the adverse impacts of rainfall, the related effects are summarised in this section.

2.2.1 Impacts on Sewer Systems

These can be listed as follows.

S0: Flow increase

Rainfall on the catchment area induces, after reduction through evaporation, wetting and depression storage losses, surface runoff. In urbanised areas, a large portion enters the sewer system and contributes to an increased flow in sewers. Washoff of pollutants sedimented on the surface (*e.g.,* roof areas, streets) as well as contributions from air pollution constitute the pollution load entering the sewer system due to rainfall runoff. Depending on the nature of the particular rainfall event, flows within the sewer system can suddenly rise to a multiple of the dry-weather flow (DWF). During some rainfall events, the capacity of the sewer system is exceeded, causing surcharging and ultimately surface flooding. That incoming flow rate the sewer system can cope with is conveyed towards the treatment plant. In general, flow peaks are attenuated whilst they are conveyed through the network. Depending on the nature of the particular rainfall event, the catchment characteristics and on the design and operation of the system, part of the (raw) wastewater is discharged at CSO structures directly into the river, not undergoing any treatment process (except possibly for some sedimentation and gross solids retention in storage tanks located within the sewer system or at the overflow structures or for separation at vortex separators). During rainfall, the rate of flow directed to the treatment plant is increased compared to the dry-weather situation. Common design practice within the UK aims at provision of a flow capacity towards the treatment plant of at least 6 DWF (of which 3 times DWF will be treated, and the excess stored in storm tanks at the treatment plant inlet).

S1: Impact on pollutant concentrations: Dilution of solutes

Since rainfall runoff usually has lower concentrations of organic matter and soluble material than dry-weather flow (*cf.* Tables 4.4 and 4.5), the combined sewage will be diluted with regard to these pollutants.

S2: Impact on pollutant concentrations: first flush

Frequently a substantial increase in the concentration of solids can be observed at the start of a rainfall event. This effect ("first foul flush") can be attributed to the erosion of solids which have been deposited during preceding dry-weather flow periods. The extent to which this effect can be observed is influenced by various factors, which, besides the characteristics of the rainfall event and the preceding dry-weather period, include size and slope of the sewer system, design of the basins within the system, and street and gully pot cleansing frequency (Tchobanoglous and Burton, 1991; Gupta and Saul, 1996).

The flush phenomenon of urban wet-weather discharges is presently a controversial subject (Saget *et al.*, 1996). Thornton and Saul (1986) define the first flush as the initial period of storm flow during which the concentration of pollutants is significantly higher than later during the rainfall event. Saget *et al.* give a more specific definition of the term "first flush effect". Their definition is based on pollutant load-flow volume curves; first flush is said to occur if the coefficient a in the representation (2.22) of a function relating pollutant load to flow volume conveyed in the course of a rainfall event is lower than 0.185; this corresponds to an event during which at least 80% of the pollutant load is transported within the first 30% of the flow volume.

$$Y = X^a \qquad (2.22)$$

where

X: fraction of flow volume discharged [-]
Y: corresponding fraction of pollutant load [-]
a: exponent to be found from data [-]

Carrying out a study on 197 events from fourteen French catchments, Saget *et al.* (1996) conclude that the first flush phenomenon occurs only very rarely, resulting in the fact that the first flush effect cannot be used to elaborate a treatment strategy. This conclusion is said not to be sensitive to the definition of "first flush" in that paper. In contrast to these statements, Larsen *et al.* (1998) observe first flush effects in the Aalborg catchment. The importance of first flush effects for sewer systems in the UK is stressed by Mance (1981). Attempts to related the extent of first flush effects to catchment and storm characteristics are reported by Gupta and Saul (1996) and Arthur and Ashley (1998).

2.2.2 Impacts on Treatment Plant Performance

Although the impact of storm events on wastewater treatment plants can be considered as being of only indirect nature (impact is through the changed characteristics of the treatment plant influent), these are nevertheless far from insignificant, as summarised below. More detailed reviews can be found in Lessard (1989) and Durchschlag et al. (1991). Additional useful references include Wolf (1990), Durchschlag et al. (1992), Harremoës et al. (1993), Lijklema et al. (1993), Londong (1994), Otterpohl et al. (1994a), Vazquez-Sanchez (1996).

T0: Flow increase

Most obviously, increased flow conveyed in the sewer system towards the treatment plant results in increased inflow to the plant. Mention has to be made of this since increased flow has impacts on the performance of virtually all parts of the treatment plant not only because of hydraulic effects, but also because of shorter residence times. Depending on the design of the treatment plant, any inflow in excess of the designed maximum capacity of the treatment plant will be diverted at a flow regulation structure at the inlet of the treatment plant.

T1: Increase in screenings and grits

Increased flow, resulting in wash-off of sedimented litter and floatable material from within the sewer system, leads to increases of the screenings caught at the inlet of the treatment plant. Similarly, the amount of grit to be removed increases (Tchobanoglous and Burton, 1991).

T2: Deterioration of primary clarifier effluent

As the hydraulic retention time decreases and the flow velocity in the primary clarifier increases with an increase in the influent flow rate, a reduction of the settling performance of the primary clarifier can be observed during rain periods. According to Henze (1987), the removal efficiency for settleable solids drops during rain from 90% to 60%. It should be noted that the effect of the disturbed flow continues and can be noticed for some time after the flow itself has gone back to dry-weather flow levels.

An important observation to be made concerns the time-lag between increased flows and dilution of pollutants in the primary clarifier: although the increased flow is effective immediately, the concentrations in the primary clarifier effluent remain at dry-weather flow level for about the retention time. Thus the effect of dilution with rainwater does not become apparent for some time. This explains why usually, a high pollutant load in the effluent of the primary clarifier can be observed at the start of rainfall. The resulting high ammonia load is of particular relevance to the nitrification processes in the aeration tank (see T3). The absolute amount of ammonium being led to the aeration tank depends on the time of day (*cf.* dry-weather flow pattern) and on the duration of the preceding dry-weather period (Durchschlag *et al.*, 1991).

T3: Deterioration of aeration tank performance, including nitrification

As long as the aeration capacity of the plant is sufficient to cope with the increased loads and as long as sludge handling is not disturbed, removal of organic matter is not significantly affected by additional treatment plant loads due to stormwater discharges. As Harremoës *et al.* (1993) state, physical–chemical removal mechanisms (acting primarily on particulate matter) are fast compared to the hydraulic retention time in the activated sludge tank. Thus the efficiency of these processes will normally not be influenced by the actual hour-to-hour situation. A similar statement is made for biological processes (acting on soluble matter), resulting in the statement that for low and medium loaded processes the soluble substrate concentration in the effluent is low and not influenced by the loading.

Ammonium concentrations in the aeration tank effluent, and subsequently in the treatment plant effluent, often show a peak at the start of a rain event. This effect is observed on full-scale plants (Otterpohl, 1990a) as well as in lab-scale experiments (Schwendtner and Krauth, 1992). An explanation for this effect is that an increased ammonium load applied to the aeration tank from the primary clarifier (this may also be caused by the effects described under T2) leads to faster growth of the autotrophic biomass in the aeration tank. However, the efficiency of the metabolism of the autotrophic biomass is limited (it is essentially dependent on the NH_4 concentration in the reactor (Monod kinetics), if it is below about 1 mg/l and oxygen not a limiting factor). For this reason, a rise in NH_4 concentrations in the beginning of combined water flows results in faster growth of autotrophs at a higher effluent concentration level (Harremoës *et al.*, 1993). Londong (1994) states that effluent concentration peaks of ammonium are caused by influent load peaks, thus measures should be taken to buffer these load peaks in the inflow zone, *e.g.*, in tanks, or to bypass the primary clarifier for diluted sewage at the start of rainfall.

T4: Drop in temperature

An additional cause of deterioration in treatment plant performance due to stormwater flow is a drop in temperature, since typically rainwater has a lower temperature than dry-weather flow. Decreased temperature leads to lower rates of the bacterial growth processes, most of which occur in the activated sludge tank.

T5: Washout of solids to the secondary clarifier

Due to increased hydraulic load on the aeration tank, part of the activated sludge may be shifted from the aeration tank to the secondary clarifier. Thus the concentration of biomass in the activated sludge tank will be reduced. In extreme cases half of the activated sludge can stay in the secondary clarifier for the entire period of the combined water flow, as observed by Otterpohl *et al.* (1994a) in a field study. This can have detrimental effects since biomass balance in the aeration tank is not maintained anymore. Secondly, the potential exists for this sludge to be discharged into the effluent (see T7), resulting in increased solids concentrations.

T6: Increase in solids and organic matter discharged from the secondary clarifier

Both functions of the secondary clarifier, clarification and thickening, can be significantly disrupted during a hydraulic shock, as observed by Olsson and Chapman (1985). Small increases in flow rate will automatically increase the effluent solids, with a corresponding increase in discharged organic matter (Durchschlag *et al.*, 1991).

T7: Build-up of sludge blanket and sludge overflow from secondary clarifier

Excessive amounts of biomass discharged from the aeration tank to the secondary clarifier (*cf.* T4) will lead to a build-up of the sludge blanket. In extreme cases, the sludge blanket builds up to such an extent that it reaches the crest of the secondary clarifier and is discharged directly to the effluent. Excessive loss of biomass has not only a direct impact on the river water quality, but also a persistent decrease of treatment plant efficiency in the subsequent dry-weather period as a consequence. Biomass necessary for the activated sludge process to function properly is lost. The recovery of the plant is slow due to the slow re-establishment of biomass in the reactor (Ashley *et al.*, 1997).

T8: Increased sludge volume

Although discussion of sludge treatment and disposal is not within the scope of this work, it should be mentioned that during wet-weather, sludge discharge, in particular from the primary clarifier, may increase considerably (Durchschlag *et al.*, 1991).

2.2.3 Impacts on the Receiving River

Since water flows conveyed through the urban wastewater infrastructure will eventually be led into the receiving water body, the study of their impacts on the river is of crucial importance for the assessment of design and operation of the urban wastewater system. Detailed reviews on impacts of urban discharges on rivers as the receiving water body are given by House *et al.* (1993), Reda (1996), Schilling *et al.* (1997). Gammeter (1996), Fuchs (1997) and Lammersen (1997a,b) discuss impacts of urban discharges on aquatic life.

R0: Increase in flow

As a result of rainfall, direct as well as indirect impacts on flow in the receiving river can be observed:

(a) intermittent discharges from CSOs and storm tanks (indirect impact);
(b) increased continuous discharges from the treatment plant effluent (indirect impact);
(c) (where applicable:) increased lateral inflows into the river (direct impact);
(d) increased upstream flow (due to catchment runoff, infiltration, and discharges from upstream urban wastewater systems) (direct impact).

Increased discharges from CSOs and storm tanks (see S0) and from the treatment plant (see T0) are considered here as indirect impacts, since these flows pass at least one of the other main components of the urban wastewater system. River water quality is degraded by these discharges since the quality of the overflows and treatment plant discharges is commonly worse than the quality of the river base flow. Due to different runoff characteristics on natural and urban catchments, these effects usually do not occur at the same time, but with certain time lags. Frequently, these will be observed roughly in the order (a) to (d). Deliberate use of these time lags can be made for operational control of urban wastewater systems (see, for example, Beck and Reda, 1994). The increase in base flow (d) is strongly dependent on the catchment characteristics and varies from one rain event to another. For example, one of the rain events Reda (1996) used for his simulation studies is characterised by a total depth of 15 mm over 4 hours and results in a (temporary) twofold increase in base flow. In another case study, the small Maisenbach river in Southern Germany (with a base flow of just 20 l/s), an increase to 80 times dry-weather flow was observed during a rain event (Fuchs, S., 1996).

R1: Oxygen depletion

The decrease of oxygen levels in the river following rainfall is a consequence mainly of four factors (FWR, 1994):

- low DO levels in CSO and storm tank discharges;
- degradation of dissolved BOD;
- degradation of BOD attached to sediments;
- resuspension of polluted bed sediments exerting an additional oxygen demand.

The timing of these impacts (such as CSO discharge; DO sag of an earlier event) on oxygen depletion can have a significant effect on the extent of their impact (see, for example, Rauch, 1996b; Lammersen, 1997a). Also nitrification occurring in the river contributes to oxygen depletion. This may be of particular relevance when increased ammonium loads are discharged. On a longer timescale, low oxygen concentrations may also result as a consequence of eutrophication (Schilling *et al.*, 1997) (*cf.* R4).

With regard to oxygen depletion, fish are the most sensitive organisms (ATV 2.1.1, 1989; Lammersen, 1997a,b). Reporting on various studies, including the study carried out by Milne *et al.* (1992), Lammersen states that concentrations of oxygen above a level of 5 mg/l do not affect invertebrates; if oxygen drops only occasionally below 5 mg/l, growth of fish was not observed to be hampered, whereas concentrations of less than 3 mg/l result in significant hindrance of fish growth. Sensitivity of fish to low oxygen concentrations increases with higher temperatures, since lethal concentrations are higher at higher temperatures.

R2: Resuspension of sediments

A sudden increase in hydraulic load can cause resuspension of deposits of heavy organic and mineral particles from the river bed. Settled sediments may also act as a reservoir for E.coli and pathogenic bacteria (Parkinson, 1996). In the Innerste river, the river sediment was also observed to be a source of ammonium (Lammersen, 1997a).

R3: Effects of solids

Discharge of sewage with high concentration of suspended solids (in particular when biomass is discharged from the secondary clarifier) inhibits biological activity within the river by reducing light penetration. Furthermore, it results in the substratum being covered with an anaerobic layer of fine sediment, which, in turn, might destroy the habitat of macroinvertebrates in the river (Schilling *et al.*, 1997). Interestingly, standard water quality models are not able to predict the effects of solids on light penetration or sediment oxygen demand (Shanahan *et al.*, 1998).

R4: Acute impacts of ammonia

Since ammonia is a strong fish toxicant (see, for example, FWR, 1994; Lammersen, 1997a,b; its toxicity to Gammarus pulex is discussed by Borchardt, 1992), short duration ammonium discharges into the river are of particular interest for the assessment of the performance of the urban wastewater system (Schilling *et al.*, 1997). This is of particular relevance with regard to loss of nitrification due to ammonium peak loads to the treatment plant (see T3). The proportion of (toxic) un-ionised ammonia to total ammonium (NH_3+NH_4) strongly depends on pH and temperature, as discussed in Section 2.1.3.3 above. Toxicity of ammonia increases at low temperatures (lethal concentrations increase by a factor of 1.5 to 5, depending on the type of fish, when increasing the temperature from 10°C to 20°C). Within a pH range of 6.5 to 7.5, ammonia toxicity is reported to decrease with increasing pH (Lammersen, 1997a,b).

According to this author, phosphates and nitrate are not of concern for acute toxicity since their concentrations are generally well below the toxic values. A similar statement can be made for nitrate which is toxic to fish only at concentrations of more than 100 mg/l NO_3-N, which is significantly higher than the concentrations found in treatment plant effluents, storm discharges and surface waters. For nitrite, concentrations describing toxicity for short-term impacts (such as 24 hours) are not known.

R5: Eutrophication

Abundance of nitrogen and phosphorus in the river may lead to excessive growth of aquatic plants and algae (eutrophication). According to House *et al.* (1993), this is mainly a problem in stagnant or semi-stagnant waters. In running waters, often the residence time is too short and turbidity too high for algae to develop in high densities. Subsequently, decay of algae might result in an increased oxygen demand.

The contribution of CSOs to the annual nutrient load discharged to the river from an urban area is generally insignificant compared to the load from the treatment plant, even after nutrient removal by nitrification, denitrification and phosphorus removal. Therefore, besides agricultural runoff (usually contributing significantly to the nutrient discharges into the river), disturbances of treatment plant performance caused by wet-weather flows might have to be considered when analysing eutrophication processes.

R6: Increase in heavy metals, toxicants, bacteria and viruses

Aquatic life may also be affected by long-term exposure to toxicants, even at low levels, due to adsorption and accumulation, resulting in chronic effects to aquatic life (*e.g.,* carcinogenic or mutagenic) (Schilling *et al.*, 1997). Concentrations of heavy metals, where present in sewage as a result of trade effluent discharges, usually rise rapidly during storm events (FWR, 1994). Similarly, organic hydrocarbons are reported to have long-term impacts (ATV 2.1.1, 1989). Straightforward assessment of the impacts of many toxic pollutants is considered to be difficult because of bio-accumulation effects, synergistic effects and the production of degraded products, which many be more harmful than the original pollutant (Schilling *et al.*, 1997).

Discharges of untreated sanitary wastes and contaminated effluents contain a wide variety and frequently high numbers of pathogenic bacteria and viruses, with risks of exposure near virtually all outfalls, even at high dilution ratios (House *et al.* 1993). Sediments may act as a storage for bacteria.

R7: Mechanical impacts: erosion of organisms.

Where a sudden increase of river flow results in higher flow velocities, stress may be caused to fish and invertebrates in the river and may lead to a drift of these organisms. Comprehensive studies addressing these issues of shear stress were carried out by Borchardt (1992), Gammeter (1996) and Fuchs (1997). Among these, Borchardt studied the impact of short-term intermittent discharges on three types of organisms (Gammarus pulex, Ephemerella ignita, Salmo tutta F. Fair). A loss of their population due to drift showed a significant increase above a certain threshold of shear stress. This threshold depends on the species under consideration, on river morphology (ATV 2.1.1, 1989) and on the characteristics of the hyporheic interstitial. Fuchs (1997) points out that there are complex interactions between impacts of shear stress, oxygen and ammonia concentrations. For the case study under investigation in his project (the Maisenbach river with an average low flow of

20 l/s), hydraulic shocks appeared to be responsible for impacts on aquatic life to a considerably greater extent than biochemical processes.

A regeneration of population lost by drift is often possible; this may take weeks to years (Rauch, 1995) (see, for example, also the study of the Luppmen creek by Gammeter, where repopulation of benthic organisms is reported to have taken several months). The time period required for repopulation depends on the intensity of the impact and on reachability of the river section under question for the organisms.

R8: Mechanical impacts: morphological changes to river bed

Morphological changes to the river bed might be caused by a sudden increase of river flow. Whether these changes do occur and, if so, to what extent, depends on the characteristics of the river and its bed.

R9: Aesthetic impacts

To conclude the list of impacts of storm water runoff on the receiving water body, mention has to be made of the discharge of trash, debris and oil, which form part of aesthetic pollution noticed by the general public. A detailed survey of the factors relevant to public perception of water quality is given by House *et al.* (1993).

Summary: Impacts on the receiving river

Although the various impacts of storm events on rivers have been listed separately above, it should be noted that synergistic effects might occur. Intermittent emissions from the urban drainage system frequently cause simultaneous peak loads of several contaminants and physical properties. The interference of those can easily lead to a synergistic detrimental effect to the recipient ecosystem (Borchardt, 1992).

Furthermore, it should be recalled that the impacts of storm events listed above vary considerably in their magnitude from site to site and will be affected by many river characteristics. The relevance of the impacts for a particular case study site depends on the water use. Water uses might include water supply, recreation, irrigation, shipping, fishery, cooling, waste disposal and hydropower. The required ecological quality of a receiving water is defined by its use as such and usually does not depend on its type (Schilling *et al.*, 1997).

A classification of the impacts listed with respect to their timescale was given by Lijklema *et al.* (1989) and is illustrated in Figure 2.12.

The State of the Art   67

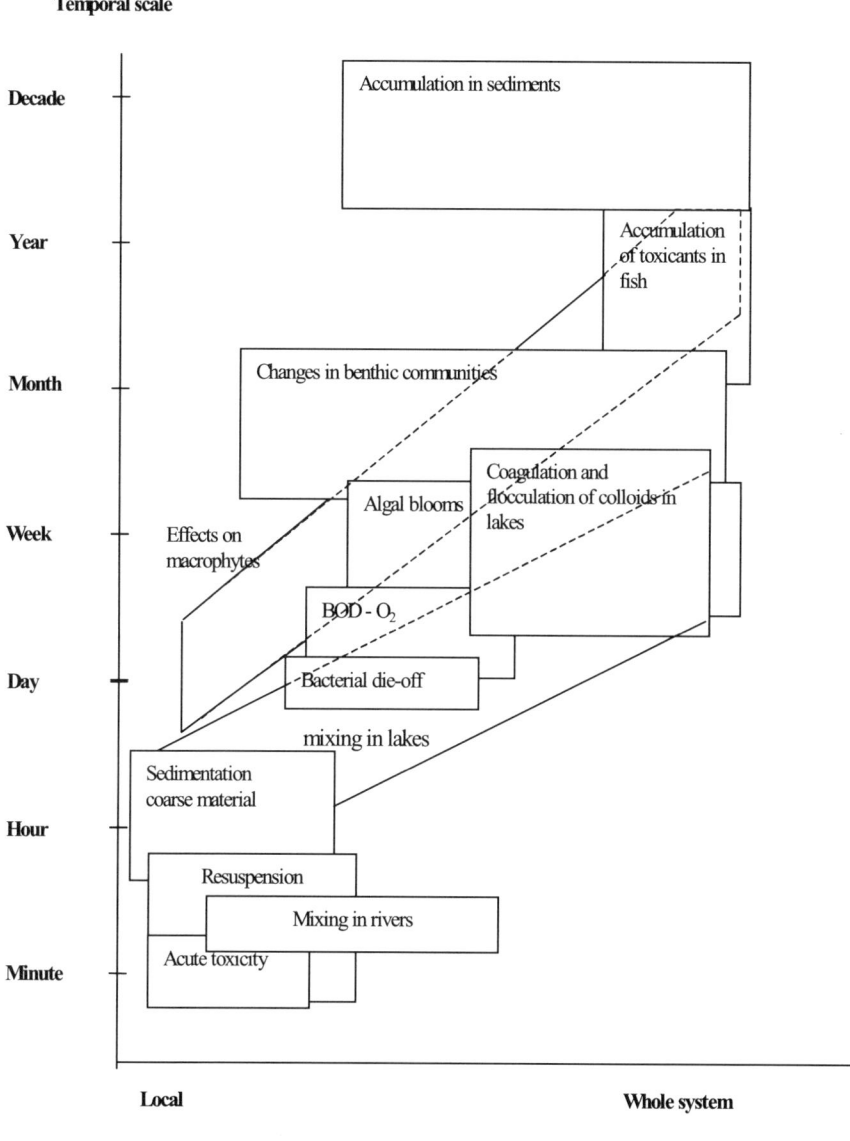

Figure 2.12. Time and spatial scales for receiving water impacts (Lijklema *et al.*, 1989)

Acute pollution denotes pollution of the receiving water body which, after a discharge event, does not have effects anymore, but only consequences (Harremoës, 1989; Rauch and Harremoës, 1996e). This type of pollution is frequently described by simple criteria such as frequency or percentage of time or by statistical criteria based on terms such as return period and exposure time. Accumulative pollution is characterised by having no immediate effects, but which gradually build up, over a period of time, to become detrimental to river water quality. Pollution with delayed impact can be classified as lying between these two classes. An example for this is delayed oxygen demand, as described above (see 2.1.3.3).

2.2.4 Criteria for the Assessment of River Water Quality

Having discussed the impacts of storm events on the urban wastewater system, the question arises how its performance can be described by a single parameter (or a set of few parameters), which can be modelled by a simulation tool. The discussion in the previous subsections suggests that DO and ammonium in the river (characterised by their concentration values, their extreme concentrations and the duration of exposure to critical values) can be considered as the two main parameters for assessing the impacts of storm events on a river (at least for those sites where the impacts of storm events are not dominated by mechanical effects (*cf.* R7, R8)).

This conclusion is supported by the Danish criteria for CSO discharges (Spildevandskomiteen, 1984, 1985) and by the criteria defined in the UPM Manual (FWR, 1994), which are based on the concepts of duration, and return period of the occurrence of extreme concentrations of these parameters. Also Schilling *et al.* (1997) suggest using criteria based on such principles, but stress also the need to perform the exact specification of criteria in a site-specific way. Tables 2.8 and 2.9 summarise the Fundamental Intermittent Standards for DO and the Derived Intermittent Standards for total ammonia of the UPM Manual (FWR, 1994). The DO and ammonium criteria are shown here, since these variables will be modelled directly by the river model implemented in this work (see Section 3.2.3).

Table 2.8. Fundamental Intermittent Standards for DO (FWR, 1994)

Return period	DO concentrations (mg/l)		
	1 h	6 h	24 h
1 month	4.0	5.0	5.5
3 months	3.5	4.5	5.0
1 year	3.0	4.0	4.5

Notes:

1. These thresholds apply when un-ionised ammonia concentrations are below 0.04 mg/l. At higher un-ionised ammonia concentrations the following correction factors apply

Un-ionised ammonia conc.	**Correction factor for DO conc.**
0.04 – 0.15 mg/l	+ 1.0 mg/l
more than 0.15 mg/l	+ 2.0 mg/l

2. The thresholds shown are different from those shown in the original research report – NRA R&D Note 123 (Milne *et al.* 1992). This is simply a reflection of a change in the way that the standards are interpreted for UPM purposes). There is no difference in the level of protection afforded.

Table 2.9. Derived Intermittent Standards for total ammonia (FWR, 1994)

pH in river after mixing	Total ammonia concentration not to be exceeded over a six hour duration more often than once a year on average (*i.e,.* one year RP) (mg/l)
7.4	15
7.6	9.7
7.8	6.2
≥ 8.0	4.0

Notes:

1. The total ammonia concentrations in the river for a particular spill event is calculated in the same way as the BOD concentration.
2. For rivers where the summer pH is less than 7.4, a Formula A overflow is unlikely to cause a breach of the Fundamental Intermittent Standards for un-ionised ammonia.

For the analyses described in this book, these criteria will be used in a simplified form (*cf.* Section 3.3.3).

2.2.5 The Dilemma of Control of the Urban Wastewater System

In summary, it follows from the discussion above that one class of impacts to river water quality are caused by intermittent discharges of relatively heavily polluted water (*i.e.,* from CSOs, storm tanks or bypasses within the treatment plant). The other class of adverse impacts is related to deterioration of the treatment plant efficiency. Thus, two main strategies for control of the urban wastewater system can be set up:

1. Treat as much water as possible in the wastewater treatment plant. This approach to control may lead to a significant reduction of treatment plant efficiency over a relatively short time. However, if sludge losses occur in the secondary clarifier, then it may take the nitrifying bacteria weeks to recover.

2. Ensure a uniform inflow to the plant by keeping water back in the sewer system and storage tanks and allowing CSO discharges when necessary.

Although this approach avoids pollutant surges to the plant, it risks more CSO discharges. Their exact number and frequency will be dependent on design and operation of the sewer system and the storage tanks. Due to the prolonged hydraulic load to the plant after the event, the plant efficiency might decrease over a period of time. The treatment plant effluent might still show increased concentrations and loads when the river flow, increased due to the same rainfall, may already be back to normal (thereby providing no further benefit of increased dilution of the treatment plant effluent in the river).

Although various approaches to integrated control can be thought of (see Section 2.4), it is expected that these two (extreme) approaches constitute the boundaries for any form of beneficial control.

2.3 Integrated Modelling Approaches

Having reviewed the fundamental processes related to water quantity and quality in the urban wastewater system and the impacts of storm events of these, the discussion now focuses on the available software packages and approaches to model these processes. A brief review of currently available software for the simulation of sewer system, treatment plant and river can be found in Appendix A. A summary of the state of the art of the modelling concepts drawn upon in integrated modelling can be found in Rauch *et al.* (2001). A discussion of previous and current approaches to simulate sewer system, treatment plant and river not only as separate units but to consider them in the simulation as integral parts of the urban wastewater system is given in this section. Here it will also be determined whether the analyses intended for this study can be based on any of the presently available tools.

One of the earliest suggestions to use simulation models of all parts of the urban wastewater system for the assessment of its performance was made by Beck (1976). In later papers Beck and Finney (1987), Beck *et al.* (1987) then focus on a "system input generator" (a time-series model with precipitation as input and river base flow as well as lateral river inflows and treatment plant inflows as output) and on dynamic models of wastewater treatment plants and rivers as the conceptual basis of integrated wastewater modelling. However, further studies by these authors described in that paper then focus on river modelling.

From the summary papers of the first INTERURBA workshop in 1992 (Lijklema *et al.*, 1993), it is clear that by that time, modelling efforts were mainly focused on the individual parts of the urban wastewater system, but more widespread interest in the integrated system was beginning to emerge. Most efforts to model integration appear to have started later. The fact that a second INTERURBA workshop was held in 2001 indicates growing interest in an integrated modelling approach. The following paragraphs summarise some recent and ongoing projects within this area.

Germany: "Total Emissions Group"

Early research focusing on integrated aspects includes the work of the "Total Emissions Study Group" in Germany. This group was established in 1988 in order to study the interactions between sewer systems and treatment plants under dynamic conditions (Durchschlag *et al.*, 1991; Härtel *et al.*, 1995). Durchschlag (1989, 1990a) and Durchschlag and Schilling (1990) linked a sewer model, which is similar to KOSIM (see Section 3.2.1), with a simple treatment plant model, based on BOD as the only state variable, represented by Monod kinetics. The secondary clarifier is modelled by an empirical function. An upgraded version of this treatment plant model constitutes today's commercial package GESIM (Kollatsch and Kenter, 1992). From simulation studies, Durchschlag concludes that increasing storage volume in the sewer system does not necessarily result in decrease of total pollution load discharged into the river.

Similar results were obtained by Schilling and Hartwig (1990) using a more detailed treatment plant model, which was calibrated against field data (Durchschlag calibrated the sewer model, but not the treatment plant model used in his study). Otterpohl *et al.* (1994a) used an improved treatment plant model, consisting of the

IAWPRC Activated Sludge Model No. 1 and primary and secondary clarifier models as described by Otterpohl and Freund (1992), for studies related to the total pollution impact of settings of the maximum inflow capacity to the treatment plant. From similar analyses for several individual events, Guderian *et al.* (1998) conclude that for every treatment plant an optimum setting of the maximum inflow capacity can be determined.

Although the authors (see, for example, Durchschlag *et al.*, 1991) stress the different characteristics of BOD and COD discharged at treatment plants and at CSOs, conclusions from these and of similar studies (*e.g.* Hansen and Pedersen, 1994) are mainly based on the total pollution loads (expressed as BOD, COD, TKN, SS, or phosphorus discharged into the receiving water) for the scenarios under question. Lacking a river model, this appears to be the best available option.

Current work in Germany on integrated modelling includes work carried out by Schütze and the SIMBA research group at ifak.

United Kingdom: Earlier work at Imperial College (University of London)

In his studies, Reda (1996) (see also Beck and Reda, 1994; Reda and Beck, 1997) developed a river water quality model (see Appendix A) and used it subsequently together with the treatment plant model of Lessard and Beck (further detailed in Section 3.2.2) for the investigation of control strategies in the treatment plant and in the river for two selected rainfall events.

In order to consider impacts of CSOs and treatment plant discharges on the river Cam, Reda defined upstream river flow series, including the CSO discharges, for three different fluvial scenarios ("dry", "extra-dry" and "high algae"). The discharges from the treatment plant used in the river model were partly computed by the treatment plant model. Particular attention had to be paid to the fact that results obtained by simulation of the Norwich treatment plant have to be adapted in order to represent the impact of the Cambridge treatment plant, which is of roughly half the size of the Norwich plant. For the concentrations of NO_3-N and DO in the treatment plant effluent, however, assumptions had to be made, since Lessard did not put particular emphasis on detailed simulation of these pollutants in the effluent. Table 2.10 summarises the interaction of variables between the treatment plant and the river models, as used in Reda's study.

Table 2.10. Parameters and their sources of information as used by Reda

(1) Parameter in the river model	(2) Storm tank overflows at the treatment plant	(3) Effluent of treatment plant	(4) Upstream flow (including CSO discharges)
Flow	Input of the river model was taken from the output of the treatment plant model. (COD concentrations were converted into BOD concentrations by application of a conversion factor of 0.45[1].)		data series defined for three scenarios ("dry", "extra dry", "high algae") and supplied externally to the river model
BOD			
Amm-N			
NO$_3$-N	constant 9 mg/l	constant 9 mg/l	
DO	constant 5 mg/l	measurement data from two events	
Chlorophyll a	constant 0 mg/l	constant 0 mg/l	

[1] The value of 0.45 was taken by Reda from Tchobanoglous (1979)

In terms of integrated modelling, Reda followed a sequential simulation approach (simulating first the treatment plant over all time steps, before embarking on river simulation) for the simulation of the two rainfall events analysed in his study. For most of the variables at the treatment plant–river interface, either assumptions had to be made or conversion factors had to be applied. Input of temperature and solar irradiation pattern was adapted from measurement data.

Belgium

A series of simulation studies considering sewer system, treatment plant and river has been carried out at the Universities of Gent and Brussels (Fronteau *et al.*, 1995, Bauwens *et al.*, 1995, 1996; Fronteau *et al.*, 1996, 1997b; Vanrolleghem *et al.*, 1996a). These studies aim at analysis of the effects of various simple variations to design and control of the sewer system and treatment plant on the receiving river. Effects were analysed in terms of total volumes and in-river concentrations of DO and ammonium. As the case study site for this study, parts of the combined sewer system of Brussels, a fictitious treatment plant and a stretch of the river Zenne were chosen.

The studies are carried out by long-term simulations performed in a sequential manner. Tools used include KOSIM for the sewer system, Lessard and Beck's (1993) model for the primary clarifier, a simplified IAWPRC Activated Sludge Model No. 1 (omitting nitrification and denitrification processes), the secondary clarifier model of Takács *et al.*, and the river model SALMON-Q (by HR

Wallingford). The results are statistically analysed and plotted as concentration–duration–frequency curves.

Due to lack of data, interactions between pollutants and sedimentation and resuspension processes are not considered in the sewer system, though sedimentation and resuspension are modelled in sewer tanks. The primary clarifier model of Lessard and Beck was extended for this study, in order to include scouring processes as well as hydrolysis of slowly biodegradable material. Since heat dynamics in the aerator were considered to be significant, these have been included in the model of the activated sludge process. For simulation of the DO balance in the river, the processes of decay and reaeration were modelled. Organic matter is simulated as BOD in the sewer system and the river (in the river, BOD is represented as dissolved and particulate fractions), whereas the treatment plant model uses COD and its fractions. BOD-COD-BOD conversion is done by conversion factors.

Long-term simulation studies carried out for the year 1986 revealed that a combination of successive non-critical CSO events can lead to a critical situation (in DO terms) in the river for nearly two weeks. Thus, Vanrolleghem *et al.* (1996a) conclude that it would not be feasible to focus on extreme overflow events only, when assessing the performance of an urban drainage system.

Later efforts in Belgium led to the development of the commercial simulator WEST++ (*cf.* Appendix A). It is currently being applied to, among other applications, integrated simulation studies aiming at the development of fast surrogate models (*cf.* Meirlaen *et al.*, 2000a, b, 2001).

Denmark–Austria

An integrated simulation model was assembled by Rauch and Harremoës and used for various studies. Parts of these studies demonstrate the importance of consideration of both CSO as well as treatment plants for a proper assessment of impacts of storm events on oxygen and ammonium balances in the receiving river (Rauch 1995, 1996b; Rauch and Harremoës, 1995, 1996a,b). Furthermore, their model was used for analyses of on-line river pollution based control of CSO structures in the sewer system (Rauch and Harremoës, 1996c,d, 1999a,b). These latter studies will be discussed in more detail in Section 2.4.3.

The model is based on conventional approaches, but does not make use of commercially available packages: rainfall–runoff processes are simulated by application of a simple approach (involving two parameters: initial loss and a time-independent runoff coefficient) and the time–area method (Viessman *et al.*, 1989). For the flow transport within the sewer system, simple translation is used. Constant concentrations of pollutants are assumed for rain water and dry-weather flow, which are assumed to be completely mixed within the sewer system. Sedimentation, resuspension and biochemical processes in sewers are neglected. A simple sedimentation model is applied for particulate COD in tanks. Simulation of the components of the wastewater treatment plant is based on modelling approaches by Schilling and Hartwig (1988) for the primary clarifier, a simplification (omission of the processes affecting oxygen) of the IAWPRC Activated Sludge Model No. 1 (Rauch, 1994) and the secondary clarifier model of Takács *et al.* (1991). In the river, water volumes and pollutant masses are routed downstream through a series of CSTRs. Simulation of the DO balance follows a Streeter–Phelps approach involving two fractions (readily and slowly biodegradable) of organic matter. Oxygen production by photosynthesis and oxygen consumption by respiration are expressed by external forcing functions. Sedimentation and degradation of sediments are modelled as a two-step process involving an intermediate state variable "settled amount of slowly biodegradable organic matter". Ammonium is considered as a conservative substance. As ammonium loading in the river is assumed to be low under dry-weather conditions, nitrification is not modelled (*cf.* Rauch and Harremoës, 1996b).

Among the particular characteristics of the Rauch model (as reported in Rauch, 1996b) is the unified set of variables, utilised in all submodules. Besides flow these include readily and slowly biodegradable organic matter with biomass (S_S and X_D, respectively), total biodegradable nitrogen (N_T), ammonia (NH_4+NH_3: S_{NH}) and, additionally in the river, DO. Within the treatment plant model, X_D is split into slowly biodegradable matter (X_S) and heterotrophic biomass (X_H). As additional variables, nitrate (S_{NO}), inert particulate matter (X_I) and autotrophic biomass (X_A) are used within the treatment plant model.

United Kingdom: UPM procedure

A significant step towards consideration of the urban wastewater system as a whole was the publication of the *Urban Pollution Management (UPM) Manual* (FWR, 1994, 1998). The UPM methodology described therein assists in the analysis of design variations of urban wastewater systems by providing a modelling and assessing procedure.

The UPM modelling procedure involves modelling of sewer systems, treatment plants and receiving water bodies (exemplified in the UPM Manual for the commercial packages MOSQITO, STOAT and MIKE11). These models are run in a sequential manner. Since detailed simulation of many events (or long time series) is fairly demanding in terms of computation time, a screening procedure is suggested for the selection of relevant events: A simple sewer system–river model (SIMPOL – see Appendix A) is to be calibrated against the detailed models and to be used subsequently for the selection of a number of events to be used for the detailed analyses. This procedure is becoming widely applied, not only in the UK (see, for example, Squibbs *et al.*, 1997), but also in other parts of Europe (Fuchs *et al.*, 1996; Dierickx *et al.*, 1998). Criticisms raised so far concern the models used for detailed modelling (Jack *et al.*, 1995, 1996), the simple screening model (Squibbs *et al.*, 1997), and the general philosophy of the UPM procedure in so far that it addresses the consequences of the pollution problem rather than its causes (Jack *et al.*, 1996).

Shortcomings of the UPM procedure with regard to the intended study on integrated simulation and control of the urban wastewater system include the computational complexity and the implementation structure (sequential processing) of the models involved and the fact that the procedure is based on the study of individual events only. On the other hand, the screening model SIMPOL is considered to be too simple for application in this study. Therefore, the UPM procedure is not applicable for this work, despite the fact that it is an important step towards integrated management of urban wastewater assets.

Simulation studies performed in a similar way include those by Hansen *et al.* (1993), Petrie and Jack (1994), Jack *et al.* (1995), Fuchs and Gerighausen (1995), Fuchs *et al.* (1996) and Holzer and Krebs (1998).

Denmark/United Kingdom: Technology Validation Project (TVP)

In 1996, a large pan-European "Technology Validation project" was launched by DHI and WRc (DHI, 1998a,b). The aim of this project, entitled "Integrated Wastewater Management", is to demonstrate that an integrated approach to the planning and management (according to the project outline, this also includes control) of wastewater facilities can now be both feasible and cost-effective. Besides pursuing objectives concerned with environmental standards for intermittent events, best practice guidelines for monitoring and consent procedures, the adaptation of sensors for water quality, adaptation of the integrated planning and management framework and enhancement of a rainfall simulator, the project aims at integration of modelling tools for planning, design and operation. Various aspects of these objectives have been investigated for six case studies in five countries.

The model integration part of this project linked the commercial packages STORMPAC (rainfall generator), MOUSE with MOUSETRAP, STOAT and MIKE11. This was done in several steps, starting from adapting the input and output formats as well as the set of determinands modelled by the individual packages.

Further improvements are concerned with streamlining the user interface ("Step 2 toolset"). The third step was aiming at linking these packages in such a way that they can be run simultaneously, thereby enabling feedback of information between the modules simulating the components of the urban wastewater system. This corresponds to the requirement of feedback for integrated simulation as it has been formulated in the early stages of this research (see Section 3.1; *cf.* also Schütze *et al.* 1996). The simulation tool at this stage is anticipated also to be used for long-term simulations; however, given the complexity of the individual submodules this perspective appears somehow to be questionable. Later steps of the model development also aim at its use for real-time control projects.

Overall, this project, which involves several well-known institutions of high reputation, addressed issues closely related to those reported in this study. Although similar concepts are used in both studies (see above), the TVP project aims at linking commercial software (resulting in the usual restrictions in terms of its accessibility).

European Cooperation: COST682 and COST624 actions

A summary of the problems encountered in integrated modelling can be found in Rauch *et al.* (1998a). This paper emerged from a workshop held as part of the European Commission's COST682 programme "Integrated Wastewater Management". Outcomes of other events of this programme focusing on this topic are summarised by Dochain *et al.* (1995b), COST682 (1996); Schilling *et al.* (1997); Vanrolleghem *et al.* (1998). The COST682 Action is currently continued as COST624 Action "Optimum Management of Wastewater Systems".

The authors (and the workshop participants) advocate that integrated wastewater management must be driven by consideration of the receiving water objectives. A methodology was proposed by Schilling *et al.* (1997). For simulation of the wastewater system, models are available and can be used. However, problems encountered include various degrees of complexity and details of the constituent models (*cf.* the different degree of fractionation of organic matter commonly found in treatment plant models on one side and in sewer and river models on the other side). Another area of concern is the different sets of state variables used in models of the components of the wastewater system. This problem can be overcome either by use of conversion factors (as, for example, by Vanrolleghem *et al.*, 1996a, and also in this study) or by definition of a common set of state variables (see below).

In the workshop summary it is stressed that, as is the case with any modelling study, the selection of models (including the necessary level of detail) to be applied in an integrated study should be done according to the needs and objectives of the particular study. Rauch *et al.* (1998a) give examples of the selection of models for three different integrated modelling studies. These authors also stress that, in some cases, model simplification might be possible due to the fact that not all water quality processes are always of relevance to a particular study. Therefore, from an integrated wastewater management perspective, existing models of the subsystem are considered as often being too complex since their development was driven by the requirement for a detailed understanding of the dynamics within that particular subsystem.

Reconciliation of different models

Driven from the observation that two of the best-known models for treatment plants (IAWPRC Activated Sludge Model No. 1; here: ASM1) and rivers (QUAL2E) are not compatible, although they roughly represent the same biochemical processes, efforts were started recently to reconcile these models (Masliev *et al.*, 1995). Inconsistencies between the models are mainly due to the different sets of state variables used. Furthermore, conventional river models lack consistency in variable and process descriptions and are characterised by incomplete mass balances. On the other hand, the ASM1 has been developed deliberately avoiding these problems. However, wastewater treatment plant and river water quality processes differ not only in the concentrations of the biomass involved, but also to a great extent in their temporal and spatial scales.

As a first step to setting up consistent models of wastewater processes, Masliev *et al.* (1995) compare processes and variables of these two models. Significant differences were found in state variables, in the representations of reactions and in the parameters. Masliev *et al.* derive an asymptotic (assuming convergence of the "fast" state variables to their equilibrium levels), reduced version of the ASM1, which is observed to be equivalent to an extended Streeter–Phelps model. This idea was taken on by the IWA Task Group on River Water Quality Modelling, who developed the River Water Quality Model No. 1 (*cf.* Section 2.1.3).

In order to provide a unified approach not only to treatment plant and river modelling but also to sewer modelling, Fronteau *et al.* (1997a,b) present results of a comparative study of a sewer system water quality model (MOUSETRAP) and the Activated Sludge Model No. 1, and make suggestions as to what elements would have to be added to a sewer model in order to ensure full compatibility with the ASM1. Significant differences are observed between these two models. These include different sets of state variables (among others, soluble and particulate BOD in the sewer model, and COD and its fractions in the treatment plant model), different sets of processes and their representations. Fronteau *et al.* (1997b) state that reconciliation of sewer models with ASM1 would essentially require modifications to the sewer model under investigation. They appear to suggest basing sewer models not solely on BOD, in order to avoid inconsistencies. Better representation of inert matter would be achieved by further integration of the water quality module with the sedimentation module. Further important extensions of sewer models would be related to biofilm processes as well as to the nitrogen and phosphorus cycles.

In summary it can be stated that a large number of approaches to integrated modelling have been pursued. Schütze (1998) provides an overview of different approaches available at that time. None of the research or commercial packages available then meets all the requirements of the work described in this book. Even nowadays, with a number of commercial tools available on the market (*cf.* Table 2.11, adapted from Rauch *et al.*, 2001), such a study as described here requires special software. The subsequent chapters of this book will describe development and application of the package SYNOPSIS.

Table 2.11. Characteristics of some of today's commercial packages for integrated simulation

Name of simulator	ICS	WEST	SIMBA
Developer	DHI, Hørsholm/DK, WRc, Swindon/GB	Ghent University/B Hemmis n.v., Kortrijk/B	ifak Barleben/D
Bi-directional interactions between submodels	Yes	Yes	Yes
Simulation of long time series feasible	Under development	Under development	Under development
Open simulation environment	No	Yes	Yes
Integrated use at a real case study reported	Yes	Semi-hypothetical	Yes

2.4 Operational Management of Wastewater Infrastructure

Previous sections have dealt mainly with the processes in the uncontrolled urban wastewater system; this section describes common approaches to actively influencing these processes by means of operational management (real-time control (RTC)) of the urban wastewater system and its subsystems. After provision of some general definitions, which will be of importance for later sections of this book, objectives and methods of real-time control are discussed for the constituent subsystems as well as for the system in its entirety.

Emphasis, however, will be less on individual examples but rather on the procedures for taking the actual control decisions. The discussion of various methods will be illustrated in Section 2.4.3 for RTC of sewer systems and extended by related remarks in the sections on treatment plant and river control. Thus, this discussion will not only summarise previously applied methods of determination of control strategies, but will also assist in the development of the methodology to be applied for the work described in later sections of this book.

2.4.1 General Concepts

A state of the art report on real-time control (Schilling, 1989) states as a definition of real-time control of urban drainage systems (UDS): "A UDS (combined or separate) is operated in real time if process data currently monitored in the system are used to operate flow regulators during the actual process."

Extending this definition to control of urban wastewater systems, Schilling *et al.* (1996) state more recently: "A wastewater system is controlled in real time if process data such as water level, flow, pollutant concentration, *etc.*, is continuously monitored in the system and, based on these measurements, regulators are operated during the actual flow and/or treatment process."

Examples of regulators used for RTC of urban wastewater systems include among others: pumps, gates, flow dividers, but also aerators and devices used for dosing chemicals (for example in the treatment plant). Real-time control attempts to use information available about the current, past and possibly also (predicted) future state of the system in order to make better use of the system's capacities aiming to achieve the operational objectives. These have to be defined for each case study site individually. Commonly these objectives include one or several of the following: prevention of flooding, reduction of overflows, optimum use of existing storage capacity available in storage tanks and in the system itself, provision of a treatment plant inflow which is optimum for treatment plant performance, maintenance of effluent standards, keeping the treatment process going, minimisation of pollution impacts on the receiving water body, minimisation of operational costs.

In general, RTC aims at optimum use of the existing transport, storage and treatment facilities. Its challenge lies in the fact that urban wastewater systems are built to operate under certain design assumptions, whilst they have to cope with a wide variety of input (storm) conditions. Therefore, RTC may help to achieve better performance of the wastewater system. Furthermore, RTC may help to defer capital expenditures (which would be necessary, for example, if new storage tanks had to be built). RTC also may (when properly designed) increase the system's ability to cope with unusual conditions (Takács *et al.*, 1995). It is interesting to note that the desire to delay capital expenditure, which otherwise would have been necessary for upgrading CSO structures, was the driving force behind the introduction of the first RTC schemes for sewer systems (Schilling, 1989).

Real-time control requires on-line measurement data, which are obtained from sensors. Use of sensors in wastewater systems imposes certain requirements on them, such as reliability, robustness and suitability for continuous recording and data transmission. (Schilling *et al.*, 1996). Sensors are currently considered as the bottleneck of RTC implementations. Details on measurement devices and regulators are discussed by Schilling (1990) for sewer systems and by Harremoës *et al.* (1993) and Vanrolleghem and Verstraete (1993) for wastewater treatment plants. Other problems frequently encountered when planning and implementing RTC systems include legal issues (who is liable if the control system does not operate as it should?) and organisational problems (often sewer system and treatment plant are operated by different institutions; *cf.* Chapter 1). These issues are addressed for example by ATV 1.2.4 (1991).

Some authors (*e.g.,* Schuurmans, 1992; Nelen, 1994a,b) make a distinction between the terms "operation" and "control". According to these authors, operation is concerned with the determination of set-points (being time-varying or constant) for the regulators in the system. Most publications on real-time control of sewer systems, for example, are concerned with operation. As opposed to the operational problem, the control problem is concerned with the regulator itself and deals with the determination of the required adjustments to the regulator which is activated by a controller in order to bring the process to its desired value (set-point) (Schilling, 1989). This problem is strongly related to the techniques of controller design, which form a part of control theory (Elgerd, 1967; Stephanopoulos, 1984). Despite this distinction, the terms "control" and "operation" will be used below as synonyms (in the sense defined above for "operation"), since this usage appears to correspond to current practice. In short, the physical device acting on the system is the regulator which requires as input a setting or a desired value of a process variable (set-point; *e.g.,* water level or concentration), which is compared with the current value of this state variable. Minimisation of the difference is then performed by a controller (*e.g.,* of PID type) which adjusts the settings of the regulator accordingly.

With regard to the information flow within a controlled system, usually a distinction between various levels of operation is made. This distinction, as presented by Schilling (1989) (see also Jørgensen *et al.*, 1995), was originally made for RTC of sewer systems. However, it can easily be extended to the other components of the urban wastewater system as well.

If no moving regulators are involved at all (*e.g.,* flows are maintained only by gravity), one can speak of *static control*. The lowest level of operation can be described as *local*. It simply involves regulators operating with constant set-points. One example would be a flow regulator which maintains flow rate or water level, using measurement information obtained in the near vicinity. Under local control, regulators are not remotely manipulated from a central control unit. Better operation in systems with two or more regulators can be achieved by *global control*. Here, regulators are operated conjunctively. A (sewer) system is said to be under global RTC if all regulators are operated in view of process measurements throughout the (sewer) system (Schilling *et al.*, 1996). Global control involves substantial use of telemetry systems and process control hardware, details of which are described by Schilling (1989, 1996c) and ATV 1.2.4 (1991).

In the literature, the term "integrated control" can be found (*e.g.,* Beck, 1996; Rauch and Harremoës, 1996c,d,e, Schilling *et al.*, 1996). However, a clear definition of this term appears still to be lacking. As used by these authors, this term seems to denote operation of one of the parts of the urban wastewater system (*e.g.,* the sewer system) taking into consideration operational <u>objectives</u> in another subsystem (*e.g.,* reduction of pollution in the receiving water).

An even further step in the scope of control employed would be to consider information about the <u>state</u> of a different subsystem when taking a control decision within one subsystem. This type of control is envisaged for example in the Aalborg RTC project (see Section 2.4.6): there, the treatment plant will provide information about its current and projected capacity for taking the control decision within the sewer system. Similarly, information about predicted flows within the sewer system (and, consequently, treatment plant loading) will be provided for the module taking the control decisions for the treatment plant.

Since no clear definition of the term "integrated control" was found, a definition is proposed here, on which the discussion in this book will be based.

Integrated control of urban wastewater systems is characterised by two aspects:

- *integration of objectives:* objectives of control within one part of the urban wastewater system may be based on criteria measured in other subsystems (*e.g.,* operation of pumps in the sewer system aiming at minimum oxygen depletion in the receiving water body);
- *integration of information:* when taking a control decision within one part of the system, information about the (present or predicted future) state of another subsystem may be used (*e.g.,* considering treatment plant effluent concentrations when performing control in the sewer system) – hence state information is transferred across subsystem boundaries.

2.4.2 Real-time Control of Sewer Systems

Extensive research was (and continues to be) carried out on real-time control of sewer systems. Comprehensive reviews, which also include a discussion of some existing implementations, are given by Schilling (1989, 1990), Nelen (1994b) and Schilling *et al.* (1996). Practical issues are discussed also by Schütze and Einfalt (1999). Thus, only a brief overview of the main concepts is given here.

Conventionally one or more of the following are chosen as objectives for RTC of sewer systems, often ranked in approximately the order shown in Figure 2.13.

- Prevention of flooding of properties and streets.
- Reduction of CSO discharges – in particular, if storage is still available at some locations within the system at the same time.
- Criteria conventionally chosen in the past to assess CSO spillages include the following: water volumes discharged at CSOs, frequency of overflow events, pollutant loads discharged at the CSOs, mean annual COD concentration of all overflow events.
- Uniform utilisation of storage capacity within the sewer system.
- Equalisation of peak discharges towards the treatment plant; optimum use of spare treatment plant capacity.
- Quick provision of storage capacity for subsequent rainfall events by emptying storage as quickly as possible at end of rainfall.
- Minimisation of operation and maintenance costs. These include, for example, energy costs incurred for the operation of pumps, which would also be affected by frequent switching of the pumps.

Figure 2.13. Objectives of real-time control in sewer systems (selection)

Usually, a combination of some of these (possibly conflicting) objectives is chosen for the operation of the drainage system (see, for example, Khelil, 1990a; Khelil *et al.*, 1990, Khelil and Schneider, 1991; Nelen, 1993; Pickles *et al.*, 1995).

Regulators in sewer systems are often used to attain desired flows within parts of the network. In most RTC applications, these devices are usually controlled by information such as water levels, flow rates upstream or downstream, rainfall information, residence time in tanks (Kollatsch and Schilling, 1990), forecast rainfall and flows (Einfalt *et al.*, 1994; Pfister *et al.*, 1994; Cluckie *et al.*, 1996; Johann and Verworn, 1996; Petruck and Sperling 1996; Petruck *et al.*, 1998). Alternatively, fixed set-points for flows and/or water levels are maintained in some cases.

Most of the earlier projects concerning RTC of sewer systems were concerned with water volumes only (see Table 2.13). The impacts of CSO discharges are assessed, for example, by total discharge volume, without taking into account performance of the treatment plant or effluent quality. However, some RTC projects (*e.g.,* Nelen, 1992, 1993) make provisions for ensuring equalised inflows towards the treatment plant. Recently, pollution aspects are also taken into account in RTC projects. This is done by computing (and minimising) the total pollution load (usually in terms of BOD or COD) discharged over the CSOs, usually, however, not considering the treatment process in detail.

Pollution-based control can distinguish between sewage of various degrees of pollution and direct the most polluted water towards the treatment plant and discharge only less polluted water (see, for example, Petruck *et al.*, 1998). Therefore, consideration of first-flush effects is possible under pollution-based RTC. As Rauch and Harremoës (1996a) observed from simulation studies, reduction of COD discharges does not closely correlate with an equal increase in water pollution. Therefore, other measures of pollution appear to be more appropriate. Rauch and Harremoës (1996c, d, 1998, 1999a) define minimisation of oxygen depletion in the river as the operational objective in some of their simulation studies. Another example of "integration of objectives" (see above) is given by Dannen (1996), Kolbinger (1996) and Petruck *et al.* (1998), who propose to use ammonium concentrations and shear stress in the river as criteria for water quality based RTC.

2.4.3 Development of Control Strategies – Exemplified for Sewer Systems

As discussed previously, RTC involves the determination of the (usually time-variant) set-points of the regulators of the urban wastewater system. Various approaches to this are outlined in this section. A more detailed overview of various methods is given by Schilling (1989) and Nelen (1994a).

A definition central to this topic and to subsequent sections of this book is the one of the term "control strategy". Schilling (1989) define a control strategy as the time sequence of the set-points of all regulators in a RTC system. Synonymously, also a set of rules which specify such a time sequence can be termed a control strategy. Control strategies can be represented and developed in a variety of ways. Several options are outlined here.

2.4.3.1 Off-line Development of Strategies

A control strategy can be determined either off-line or on-line. When it is defined off-line, it is specified prior to the actual control process and represented in a way which allows quick selection of the control action to be taken appropriate to the current state of the system. Examples of representations include (if-then) rules, scenarios and decision matrices. It is important to stress that, although in off-line approaches the control strategy is defined prior to its application and then kept fixed, it still allows for the current state of the urban wastewater system to be considered for the actual control decisions.

If-then rules relate available sensor information ("if" part of the rule), which may include forecasts where available, to corresponding control actions ("then" part of the rule). Such rule-based systems provide an easy-to-understand way of representing a control strategy (as long as the rule base is not too complex). If-then rules can be easily implemented in knowledge-based systems (these are sometimes called "expert systems"). Their rule-interpreter component processes rules, matches them to the current system state and selects the appropriate control decision. However, since at least simple rule sets can be programmed directly, implementation of rule-based systems does not necessarily have to be done within an expert system context. In some projects on RTC of sewer systems (as well as of treatment plants – *cf.* Section 2.4.4), elements of fuzzy theory (Zadeh 1965, 1979) are implemented in control rules (Yagi, 1990; Beeneken *et al.*, 1994; Fuchs *et al.* 1994a, 1997) Thus information stated in an imprecise way (such as "the water level is high" rather than "the water level is 2.41 m") can be taken into account when formulating and processing the rules. However, approaches based on the theory of fuzzy sets require careful definition of the so-called membership functions relating the (imprecise) linguistic terms to those variables which are used in the inference process.

Decision matrices constitute another way of strategy representation, which is conceptually very similar to rules: these consist of a list of all perceivable combinations of input and current state variables, relating them to the appropriate control actions. Kido and Sueishi (1990) propose a decision matrix search procedure, which allows for selection of the best control action out of a predefined set of strategies when the control decision has to be taken. Nelen (1994a) extends the rule-based method of strategy representation by including some adjustable control parameters within the rules ("control scenario"). Thus, the rules can be easily adjusted to the actual drainage conditions by a proper setting of these control parameters.

Neumann (1986), Fuchs et al. (1987), Khelil (1990b), Khelil et al. (1990, 1993a) implemented a learning algorithm in a knowledge based system. According to the definition of so-called meta rules (describing the learning process), the knowledge based system, in conjunction with a hydrodynamic sewer system model, evaluates its own performance and modifies its rule set from time to time. Cluster analysis techniques are applied for the generation of new rules (Heinemann, 1992). Although, this procedure could in principle also be applied on-line, the authors use it for off-line development of control strategies and achieve encouraging results in comparison with a linear optimisation procedure (Fuchs et al., 1987). However, as Khelil et al. (1993a) report, the number of new rules generated soon exceeds memory limitations, although the algorithm removes rules of lesser importance from the rule base. To date, no further application of this methodology has been found. Crucial for the success of this learning system is the definition of the meta rules, which requires significant experience with the system to be controlled. Hence, definition of the meta rules forms the bottleneck of this approach. Rohlfing (1993a) anticipates substantial problems when meta-rules are to be defined for very complex drainage systems.

According to Khelil (1992a), there is no well defined and recognised methodology to obtain rule sets. Details of the procedures applied in RTC projects for the development of control strategies are hardly discussed in the available literature. Notable exceptions include the description of the self-learning expert system just discussed as well as approaches described by Almeida and Schilling (1993), Fuchs and Hurlebusch (1994) and Fuchs et al. (1997). These authors derive operational strategies from optimisation runs carried out for individual events. The resulting strategies are then generalised and formulated as rules.

In most RTC projects, mainly heuristic approaches appear to have been chosen, that is, the definition of strategies is mostly driven by intuition and experience of the system developer or the staff actually operating the system. Often, testing and comparison of a variety of potential strategies is assisted by dedicated simulation tools which allow control actions to be simulated. Simulation tools used for off-line strategy development include Fitasim (Einfalt and Wolf-Schurmann, 1992; Einfalt, 1993; Einfalt and Semke, 1994), Hydroworks (Ashley *et al.*, 1995; Michas, 1995), HYSTEM-EXTRAN (Khelil, 1990b, 1992b; Khelil *et al.*, 1993a), MOUSE-ONLINE/PILOT (Nielsen *et al.*, 1994; Williams and Tidswell, 1994; Lumley *et al.*, 1995; Kjaer *et al.*, 1996) and SAMBA-Control (Petersen *et al.*, 1990; Jakobsen *et al.*, 1993b). However, the number of strategies tested by such an approach will always be finite, and there is no guarantee that an optimum strategy is found by this procedure.

Figure 2.14 illustrates this conventional way of (off-line) strategy development.

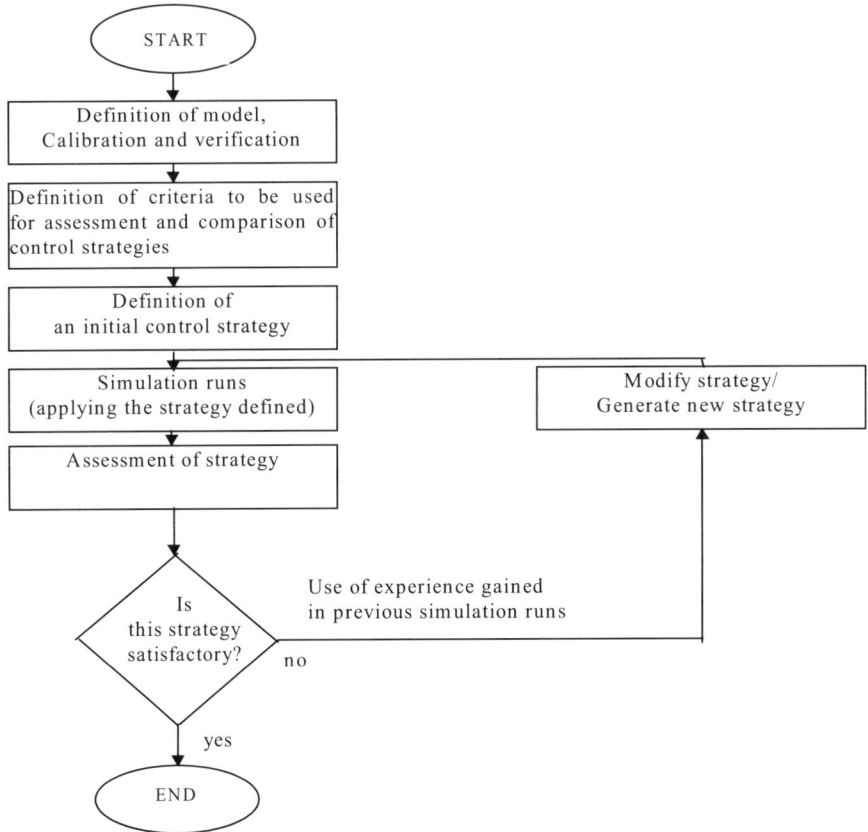

Figure 2.14. Common procedure for off-line development of control strategies

The main advantage of using rule bases, scenario definitions and decision matrices for definition of control strategies is that the implemented strategy can be executed very quickly. Addition of rules, for example in order to consider existing experience with the system or to account for emergency scenarios or structural changes to the drainage system, is (in principle) fairly easily possible (Khelil, 1992a). Compared to on-line application of optimisation methods (see below), another advantage of the off-line definition of control strategies could be described as follows. The control strategies found can be developed using process models and descriptions of control objectives of any complexity. Also, long-term effects of control can be considered when setting up the control strategies, without restrictions posed by definition of an "optimisation horizon" (as necessary in on-line optimisation approaches). Another advantage of a fixed rule set is that it simplifies the formalisation of operational procedures and their acceptance by supervising authorities (Einfalt *et al.*, 1994).

On the other hand, however, compilation of a rule base which covers the great variety of operating conditions can be a tedious task. Furthermore, for non-trivial case studies, such a rule set can quickly become quite complex. For example, the base rule set applied for RTC of the Bremen (left bank side of the Weser river) sewer system consists of just under 50 rules (including auxiliary rules), processing information from nine water level sensors and controlling the operation of two large pumping stations with seven decision variables (pumps and gates) in total (Khelil, 1990a; 1992b).

To conclude the discussion on off-line development of strategies, brief mention has to be made of linear regulators involving multivariable feedback. Design of these regulators and their application to control of water flows in sewer systems is described by Holzhausen *et al.* (1990) and by Meßmer and Papageorgiou (1992). As the latter authors state, application of a multivariable feedback controller may be an attractive alternative to other methods, as long as *a priori* consideration of nonlinearities and constraints are of lesser importance. According to Meßmer and Papageorgiou (1996), operation by linear controllers does not make strong demands on computational power, but it is considered to be generally less effective than operation based on optimisation procedures, since system restrictions can be considered only in an indirect way when designing the controller. This may also be the reason why this technique is not more widely used for global control of sewer systems.

2.4.3.2 *On-line Development of Strategies*

As an alternative to the specification of a control strategy prior to its application as described in the previous subsection, a control action appropriate to the operational objectives can be determined during the process to be controlled (*i.e.,* on-line). When such an approach is chosen, frequently optimisation methods are involved.

Application of optimisation methods requires the operational objectives to be defined as a mathematical function ("objective function"), relating the settings of the control handles to a value expressing the performance of the system. Various definitions of objective functions can be found in the literature. Usually, these include overflow volumes and/or other items of those listed in Figure 2.1.3. Sections 2.2.4 and 3.3.3 discuss the definition of potential objective functions in more detail.

Various optimisation procedures can be applied to minimisation (or maximisation) of the objective function under constraints, which are defined by initial and final states, capacity and dynamic behaviour of the system and its control devices. Characteristics of the objective function and the restrictions involved significantly influence the choice of the optimisation procedure applicable to a particular case study site. A brief overview of the application of optimisation methods to control of urban drainage systems is given here. A more detailed discussion of some of the optimisation methods can be found in Section 2.5.

Linear optimisation

Linear optimisation has been investigated in many projects on RTC of sewer systems (see Table 2.13). It requires that both the objective function as well as the constraints are expressed as linear functions. Hence, significant simplification of the description of the system and the processes involved are necessary. Linear optimisation has the advantage that (under certain conditions) the existence (and in non-degenerated cases the uniqueness) of the optimum is guaranteed and can be found by the solution algorithms widely applied. Moreover, even large optimisation problems can be solved within a short computation time. If more stringent assumptions can be made on the restrictions of the linear optimisation problem, network-flow algorithms are applicable (see, for example, Neugebauer, 1990; Neugebauer *et al.*, 1991; Nelen, 1992; Lobbrecht, 1994, 1997; Vazquez *et al.*, 1997). Network-flow algorithms result in shorter computation times. Lobbrecht (1997) reports general network-flow algorithms to be about 100 times faster than general linear optimisation codes.

In order to overcome problems associated with the system simplification necessary for the application of linear optimisation, Rohlfing (1993a) links a hydrodynamic sewer system model to a linear optimisation procedure, thus ensuring that the effects of control decisions taken are considered accurately for the optimisation process of the subsequent time step. A similar approach linking detailed simulation programs to optimisation routines is suggested by Lobbrecht (1997).

Nonlinear local optimisation

The severe restrictions imposed on the system description by linear optimisation motivates the application of nonlinear optimisation techniques.

One approach to the solution of nonlinear optimisation problems is the approach of successive linear optimisation, as discussed and applied by Nelen (1992) and Lobbrecht (1997). The nonlinear optimisation problem is approximated by a series of linear optimisation problems, which are solved applying a network-flow algorithm. After finding the optimal solution of the approximative linear problem, a new approximation to the nonlinear cost function and restrictions is determined at the optimal solution found. Then the updated linear problem is solved, and the procedure is repeated. Lobbrecht (1997) obtains convergence of this iterative procedure as long as this process is applied carefully. The main problem associated with this procedure is that approximation can be difficult and time-consuming, in particular if the processes modelled are highly nonlinear. A similar procedure is chosen by Kolbinger (1996) and Weinreich *et al.* (1996) who redefine the objective function (by varying the unit costs of overflows according to the current concentrations) at every time step in order to consider pollutants in their optimisation procedure, thus avoiding use of a nonlinear objective function.

Other nonlinear optimisation methods suggested in the literature include quadratic optimisation (Schütze, 1991), gradient based approaches (Papageorgiou and Mayr, 1988; Meßmer and Papageorgiou, 1996) and branch-and-bound algorithms (reducing the problem to the optimisation of convex subfunctions) (Knemeyer, 1992; Khelil *et al.*, 1993b).

A general problem of nonlinear local optimisation procedures is that these do not guarantee to result in a global optimum (unless, for example, certain convexity or linearity conditions are met, as in linear optimisation; see Section 2.5 for a detailed discussion of optimisation methods). Also the use of derivatives (or their approximations) may lead to problems in the case of non-smooth objective functions, as illustrated by Zunic and Seus (1993a) for an example of RTC of two sluice gates.

Global optimisation

The limitations of local optimisation techniques can be overcome by application of global optimisation methods. Their application to control of urban wastewater systems has been investigated only recently (Rauch and Harremoës, 1999a,b, 1996c,d). Although investigations described in these papers have been carried out for theoretical examples only, results seem to be encouraging. It may be argued, however, that this on-line approach of application of genetic algorithms may involve a considerable amount of computing time. Assuming, for example, a population of 50 and the number of generations required of 30 for a successful application of the algorithms described (these numbers have been derived from the papers), this might involve up to 1500 simulation runs, *i.e.*, 1500 runs (to be performed on-line) to simulate the effects of a control action whose parameters have been found by the processes of cross-over and mutation. It appears that the computational demand might contravene the need for finding a control action in a real-time control application within sufficiently short time.

However, since most global optimisation procedures make only weak assumptions on the objective function and the variable restrictions, they allow a flexible description of the system characteristics and the operational objectives of the operational problem under consideration. A more detailed discussion of this class of optimisation algorithms is provided in Section 2.5.2.2.

Dynamic programming

In an overview of optimisation techniques applied to the control of wastewater systems, mention should also be made of dynamic programming (Labadie *et al.*, 1980; Yakowitz, 1982; Papageorgiou, 1983; Papageorgiou and Mayr, 1988; Bardoel, 1995).

Stated in brief, the dynamic programming methodology is comprised of testing all possible combinations of control actions and finding the optimum one. It does, in principle, not pose strong restrictions on the definition of the objective function or on the constraints of the problem. Although dynamic programming guarantees the optimality of the solution found, its main disadvantage is that it is demanding in terms of computation time and computer storage (exponential increase with the number of state variables). Modified approaches (*e.g.*, Yakowitz, 1982; Papageorgiou, 1983) reduce this problem; however, these do not ensure convergence and optimality of the solution (Rohlfing, 1993a). Labadie *et al.* (1980) integrate the simulation of unsteady flow into a dynamic programming problem for control of a part of San Francisco's sewer network. Although dynamic programming was applied successfully to this case study, its application to more complex systems does not appear to be feasible due to its exponentially increasing computational demand.

Table 2.12, adapted from Lobbrecht (1997), summarises the characteristics of various optimisation methods with regard to their application to RTC of sewer systems.

Table 2.12. General characteristics of various optimisation procedures

	Solvable problem size	Solution speed	Model accuracy	Model complexity	Global optimum
Network optimisation	Very large	Very fast	Low	Very low	Yes
Linear optimisation	Large	Fast	Low to moderate	Moderate	Yes
Successive linear optimisation	Large	Moderate to fast	Moderate to high	Moderate	Yes, for each iteration
Nonlinear optimisation	Moderate	Slow	High	High to very high	For convex functions
Global optimisation	Moderate	Slow	Very high	Very high	Yes[5]
Dynamic programming	Very small	Very slow	High	High to very high	Yes

General advantages of mathematical optimisation performed in real time (on-line) as a means for finding control decisions as opposed to the heuristic off-line development of control strategies include the following:

Optimisation procedures can be applied without requiring substantial past experience with the system to be controlled.

Under certain assumptions and depending on the characteristics of the optimisation problem (see Section 2.5.2.1) a solution provided by an optimisation procedure is guaranteed to represent a control decision which is optimum under the assumptions made and restrictions given.

On the other hand, application of optimisation procedures suffers from the following problems:

For many solution procedures, significant simplifications of the objective function as well as of the system are required. Usually, the system description is

---

[5] In the general case, no absolute guarantee can be given for the optimality of a solution found by a global optimisation procedure.

incorporated in the restrictions on the variables. When applying a control decision found by optimisation of a simplified system, it has to be ensured that this solution is still optimum for the real system.

When applying optimisation procedures on-line, their runtime performance becomes crucial. Although this may not be a problem for solutions using linear optimisation procedures, solution of large nonlinear problems and those for which the impact of control actions over a large time period are to be considered (large optimisation horizon) may be of computational complexity beyond feasibility for runtime applications.

A summary of the discussion of this section is provided in Table 2.13. It shows methods of strategy development and operational objectives applied in earlier RTC projects for sewer systems.

Table 2.13a. Real-time control of sewer systems: methods and objectives applied in various studies. The references are detailed in Table 2.13b.

Objectives of RTC / Method of strategy development	Optimisation of flows **Flow-based RTC** (*e.g.*, minimisation of CSO volumes or frequencies, equalisation of plant influent)	Minimisation of pollution discharges **Pollution-based RTC** (*e.g.*, minimisation of CSO pollutant load discharged into the receiving water body)	Minimisation of adverse impacts in the receiving water - measured in receiving water terms **Water Quality based RTC**
Off-line development of strategies			
Heuristics (by intuition, experience; possibly aided by simulation tools): Enumeration of a finite number of potential strategies which are evaluated and compared against each other.	Aalborg [14] Aberdeen [6] Bolton [54; 57] Bradford [33] Bremen [16; 17; 18; 20] Copenhagen [4; 5] Flensburg [12; 13] San Francisco [8]	Malmö [40; 50]: Control strategy based on performance of the treatment plant Pfrimm Valley [52]	
Learning expert system	Bremen [11; 21]		
Optimisation for off-line development of strategies	Flensburg [10; 12; 13] Lisbon [2]		Presented in this work
linear regulators, controller design, multivariable feedback	Frankfurt (Main) [15; 30; 31]		
On-line determination of control decisions			
On-line linear optimisation	Aalborg [39; 50] Bremen [48; 49; 51] Fehraltorf [1; 26; 37; 38] San Francisco [8] Saverne [55]	Oslo [56]: Gelsenkirchen/Herten : [9; 27; 43]	Gelsenkirchen/Herten: [9; 27; 43]
On-line nonlinear local optimisation	Gradient based:[31; 59] Quadratic opt. [53] Other nonlinear: [19; 23; 24; 32] Branch-and-bound [22; 25]	Solution of NLP problem by sequentially solving LP problems: [29; 34; 35; 36] The Hague: [3]	
On-line global optimisation	Gen. Algorithm: [45; 46; 47; 60]		Genetic Algorithm: [44; 45;47]
Dynamic programming and related techniques	[7; 28; 41; 42; 58]		

Table 2.13 is meant to give examples of various RTC approaches. By no means is it considered to constitute a complete list of RTC projects.

Table 2.13b. References used in Table 2.13a

[1]:Almeida (1994)
[2]:Almeida and Schilling (1993)
[3]:Allit and Nelen (1994)
[4]:Andersen *et al.* (1996)
[5]:Andersen and Sørensen (1994)
[6]:Ashley *et al.* (1995)
[7]:Bardoel (1995)
[8]:Bradford (1977)
[9]:Dannen (1996)
[10]:Fuchs and Hurlebusch (1994)
[11]:Fuchs *et al.* (1987)
[12]:Fuchs *et al.* (1994a)
[13]:Fuchs *et al.* (1997)
[14]:Harremoës *et al.* (1994)
[15]:Holzhausen *et al.* (1990)
[16]:Khelil (1990a)
[17]:Khelil (1990b)
[18]:Khelil (1992b)
[19]:Khelil (1998)
[20]:Khelil *et al.* (1990)
[21]:Khelil *et al.* (1993a)
[22]:Khelil *et al.* (1993b)
[23]:Khelil *et al.* (1995)
[24]:Khelil *et al.* (1997)
[25]:Knemeyer (1992)
[26]:Knollmann (1993)
[27]:Kolbinger (1996)
[28]:Labadie *et al.* (1980)
[29]:Lobbrecht (1997)
[30]:Meßmer and Papageorgiou (1992)
[31]:Meßmer and Papageorgiou (1996)
[32]:Méthot and Pleau (1997)[33]: Michas (1995)
[34]:Nelen (1990)
[35]:Nelen (1992)
[36]:Nelen (1993)
[37]:Neugebauer (1990)
[38]:Neugebauer *et al.* (1991)
[39]:Nielsen *et al.* (1994)
[40]:Nyberg *et al.* (1996)
[41]:Papageorgiou (1983)
[42]:Papageorgiou and Mayr (1988)
[43]:Petruck *et al.* (1998)
[44]:Rauch and Harremoës (1996c)
[45]:Rauch and Harremoës (1996d)
[46]:Rauch and Harremoës (1998)
[47]:Rauch and Harremoës (1999a)
[48]:Rohlfing (1993a)
[49]:Rohlfing (1993b)
[50]:Schilling *et al.* (1996)
[51]:Schilling and Petersen (1987)
[52]:Schmitt (1996)
[53]:Schütze (1991)
[54]:Taylor (1992)
[55]:Vazquez *et al.* (1997)
[56]:Weinreich *et al.* (1996)
[57]:Williams and Tidswell (1994)
[58]:Yakowitz (1982)
[59]:Zunic and Seus (1993a)
[60]:Zunic and Seus (1993b)

2.4.4 Operation of Wastewater Treatment Plants

This section provides a brief overview of operational principles of wastewater treatment plants as they are either commonly applied or suggested in the literature. Emphasis will be not on the design of individual controllers for real-time control of treatment plants (details about these are given for example by Andrews, 1994; Vanrolleghem, 1994), but on higher-level principles. This discussion will prepare the definition and analysis of control strategies for the treatment plant and for the integrated wastewater system in Chapter 5. Since this discussion focuses on the type of treatment plant analysed in this work (nitrifying conventional activated sludge plant without denitrification and phosphorus removal processes), only those operational principles which are of relevance to this type of plant will be discussed here. Detailed reviews of treatment plant control include those by Andrews (1974), Beck (1986), Novotny et al. (1992), Chen (1993), Olsson and Jeppson (1994), Vanrolleghem (1994, 1995), Gülen (1995), Olsson and Newell (1999) and Liu (2000).

Operation of treatment plants is becoming increasingly important since today many are used not only for carbon removal (for which less operational effort is often sufficient (Kroiss, 1994)). Also increasingly stringent standards motivate the control of wastewater treatment processes (Vanrolleghem, 1994).

Berthouex and Fan (1986) state that even well attended plants are 'out of spec' (not meeting the effluent quality standards) for 8 to 9% of operation time, not including short upsets lasting less than one day. Besides faulty design, overloading and inadequately trained operators, a lack of process control leading to excessive effluent quality variations, is reported to be the cause for the insufficient performance of many treatment plants (Vanrolleghem, 1994). According to Krauth and Müller (1996), treatment plants often have a higher capacity than they are designed for, in particular when the wastewater has a temperature higher than the design temperature. Due to the variations in quantity and quality of water being treated in most treatment plants, the water may be overtreated much of the time (Andrews, 1994).

The main objectives of the operation of wastewater treatment plants usually include the following (possibly ranked in approximately the given order):

- avoidance of discharge of biomass into the effluent;
- maintenance of performance of the plant processes ("keep the plant running");
- maintenance of the overflow discharge and effluent standards;
- protecting river water quality;
- minimisation of operation and maintenance costs.

Among the crucial elements for treatment plant control are sensors and regulators which can be used to influence the processes in a beneficial way. Table 2.14 (adapted from Vanrolleghem, 1994) gives an overview of available monitoring elements for wastewater treatment plant processes.

Table 2.14. On-line monitoring equipment for wastewater treatment processes

Physical measurements			Physico-chemical measurements			(Bio-)chemical measurements		
Variable	Applicability Process ↓		Variable	Applicability Process ↓		Variable	Applicability Process ↓	
Temperature	A	∀	pH	A	∀	Respiration rate	1,2	∀
Pressure	A	∀	Conductivity	A	∀	short-term BOD [2]	1,2	∀
Liquid level	A	∀	Oxygen	1,2	∀	Toxicity	1,2	∀
Flow rate (liquid)	A	∀	CO_2	1,2	∀	COD	1,2	O
Suspended Solids	A	∃	Redox	2	∀	TOC	1,2	O
Sludge blanket	3	∃	NH_4^+ (ISE[1])	2	∃	NH_4^+	2	∃
Sludge volume	3	∃	NO_3^- (ISE)	2	O	NO_3^-	2	∃
Settling velocity	3	O	(UV absorbance)	2	∃	PO_4	2	∃
UV absorption	A	∃				Volatile Fatty Acids	2	O

Applicability: ∀: State of technology; ∃: Applicable in certain cases; O: Requires development work
Process: Unit process in the wastewater treatment plant where the sensor can be implemented: 1: Activated sludge; 2: Nutrient removal; 3: Sedimentation; A: All
[1]: ISE: Ion sensitive electrode
[2]: Details about "short-term BOD" are given by Spanjers *et al.* (1993); BOD-M3: (*e.g.*, Otterpohl, 1990a)

A detailed overview of sensors and measurement techniques is provided by Harremoës *et al.* (1993). These authors consider the development of reliable sensors as the limiting factor for the introduction of RTC in treatment plants.

Besides measurement of process performance in terms of BOD, TOC (total organic carbon), and COD, which may be hard to obtain on-line, alternative measures are suggested for the determination of BOD removal – *e.g.*, specific total oxygen uptake rate (STOUR) (Vitasovic and Andrews, 1987). Recently, progress has been made in the development of "software sensors". This term denotes combinations of robust and reliable sensors (such as available for DO, pH and redox potential) with process knowledge implemented in software directly linked to the sensor. Vanrolleghem (1994) provides an overview of related methods and developments. Furthermore, respirometric measurement techniques can be applied to deduce state information about the activated sludge process. An overview of these is given by Klapwijk *et al.* (1993) and Spanjers *et al.* (1996, 1998).

In a similar fashion to the previous table, Table 2.15 gives an overview of those process variables in a treatment plant which can be directly manipulated by operational control.

Table 2.15. Variables available for manipulation in wastewater treatment plants

Manipulable variable	Process	Applicability
Bypass / Overflow	1,2	∀
Equalisation and Buffering, Storm tanks	1,2	∃
Feeding point / Step feed	1,2	∃
Aeration intensity	1,2	∀
Internal recycle flow rates (*e.g.*, for denitrification and from sludge treatment)	2	∀
External Carbon supply	2	∃
Chemical Dosage	2,3	∃
Return Activated Sludge rate	A	∀
Waste Activated Sludge rate	A	∀
Sludge storage	A	∃

Applicability: ∀: State of technology; ∃: Applicable in certain cases
n.b.: Those control variables which will be subjected to further analysis in this work are highlighted (*cf.* Section 3.4).

Traditionally (and, to a large extent, currently), operation in most treatment plants is limited to single-loop controls. An example of this type of control is the use of controllers for maintaining a fixed set-point for the DO level in the aeration tank. Control in multiple layers is described by several authors (Beck, 1984, 1986; Couillard and Zhu, 1992) and is increasingly implemented in automatic control systems (Harremoës *et al.*, 1994). Figure 2.15 sketches this concept. Here, local controllers are coordinated by a central control unit which determines their (possibly time-varying) set-points using information also from other parts of the treatment plant. Thus, this type of treatment plant operation resembles the global control mode of RTC in sewer systems (Section 2.4.1).

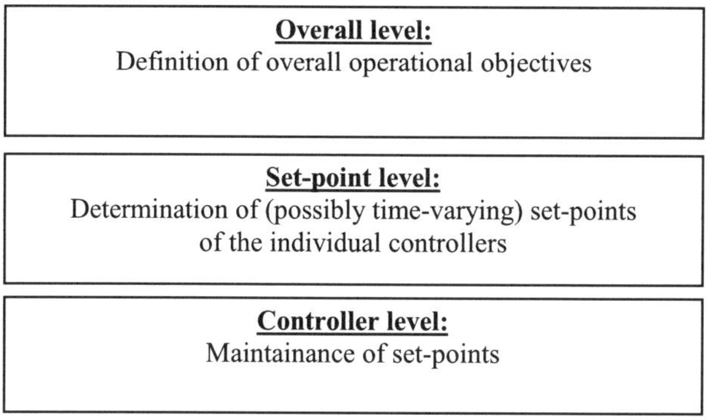

Overall level:
Definition of overall operational objectives

Set-point level:
Determination of (possibly time-varying) set-points of the individual controllers

Controller level:
Maintainance of set-points

Figure 2.15. Various levels of treatment plant operation (example)

Traditionally, on many plants, the plant manager has an important role in the upper levels of control, which are related to slower speed of process perturbation and response (Beck, 1986). A similar distinction of control levels is made by Chen (1993), who identifies the following three levels: the highest level ("Coordination of control objectives") has to deal with resolving possibly conflicting overall control objectives. The lower levels are concerned with long-term and short-term processes and, finally, with maintaining appropriate environmental conditions for the biomass.

In the following paragraphs, an overview of commonly controlled treatment plant process variables is given. With regard to the various levels of control, the discussion will not focus on the lowest level of control, which is affected by controllers maintaining set-points, which are either constant or set by an operator or at higher control levels, but on the setpoint level. The design of individual controllers is discussed in detail by Andrews (1994) and Vanrolleghem (1994) and is therefore not repeated here. In the discussion, emphasis will rather be placed on the combinations of measurement information and regulated variables, since use of these will be made in the analyses of various control options in Chapter 5.

Maximum inflow to the treatment plant

Among the measures applied in treatment plant control is the regulation of inflows to the plant and to and from storm tanks (where these exist). Inflow to treatment plants is commonly restricted in a static way (for example, in the UK commonly to three times average dry-weather flow; in Germany to two times peak dry-weather flow plus inflow from extraneous sources). Limitation of flow helps to prevent extreme hydraulic shocks which would result in further deterioration of the treatment plant performance (*cf.* Section 2.2.2). Repeatedly, the use of the traditional static threshold value has been questioned (Lessard, 1989; Wolf, 1990;

Otterpohl *et al.*, 1994a; Guderian *et al.*, 1998). Analyses by these authors suggest that a slightly larger threshold value can lead to at least equal performance of the plants studied. This observation is of particular relevance to the control of the integrated system, since increasing the treatment plant inflow capacity may reduce pollution discharges from the sewer system and storm tanks to the receiving water body. Although restriction of flow by a fixed threshold value would be considered more as a matter of treatment plant design, rather than of (dynamic) operation, it is conceivable that the influent to the treatment plant is varied according to the state of the wastewater system, thus allowing for greater flexibility in managing wastewater flows. This approach has indeed been proposed and is already implemented (Balslev *et al.*, 1993; Krauth and Müller, 1996; Schilling, 1996a). Static as well as dynamic variation of the treatment plant inflow capacity will thus be analysed in this work.

An option hardly ever used in dynamic control of treatment plants would consist of (dynamically) setting the maximum inflow rate from the primary clarifier into the aeration tank, thus controlling the overflows after the primary clarifier.

Filling and emptying of storm tanks

As already indicated in Section 2.1.2.1, storm tanks can be of great benefit to the performance of the urban wastewater system. Their prudent operation can enhance this benefit by carefully balancing between temporarily storing increased storm flows, thus reducing the inflow towards the treatment plant, and emptying them as quickly as possible in order to provide ample storage capacity for subsequent events. Storm tanks are usually filled by inflows exceeding the threshold for maximum inflows to the primary clarifier. They are emptied at or after rainfall has stopped; emptying is triggered, for example, when the inflow rate towards the treatment plant drops below a prespecified threshold value (Lessard, 1989; Lessard and Beck, 1990). It has to be taken into account that emptying the storm tanks towards the treatment plant gives increased hydraulic loading to the treatment plant, which, according to Bertrand-Krajewski (1994), can be considered to have effects similar to a second rainfall event. Timing of emptying the storm tanks is observed to be important; coincidence with the morning ammonium peak load in the influent should be avoided (Krebs *et al.*, 1996).

Lessard and Beck (1990) analyse in detail the operation of storm tanks for two rainfall events. In a similar way, when storm tanks are considered as part of the sewer system, they provide great control potential as the studies mentioned earlier about RTC of sewer systems demonstrated. Bertrand-Krajewski *et al.* (1995) and Krebs *et al.* (1996) suggest emptying the storm tank contents directly into the receiving water body at high dilution, after having allowed its contents to settle for two hours after filling. According to these authors, this operational principle results in a significant effect on the treatment plant performance only when the storage is relevant within the overall volume balance, *i.e.,* when the total rainfall is relatively small.

Aeration

Aeration is usually controlled by maintaining a set-point for the DO concentration in the activated sludge tank. A sufficient supply of oxygen is important for the removal of carbonaceous material and essential for the nitrification process. Low oxygen concentration may not only inhibit nitrification, but may also deteriorate sludge settleability, worsen effluent quality and result in predominance of filamentous bacteria (Chen, 1993). On the other hand, excessive oxygen supply results in high operational costs. The aeration costs are reported to constitute at least 25% of the operational costs (Schlegel, 1997) and about 67% of the energy costs (Hansen, 1997) of a treatment plant. Furthermore, high oxygen supply may lead to excessive nitrification and again to poorly settling sludge (Ryder, 1972). Controllers for DO in activated sludge tanks are described for example by Olsson and Andrews (1978), Schlegel and Lohmann (1981), Olsson *et al.* (1985), Haarsma and Keesman (1995), Lindberg and Carlsson (1996) and Schlegel (1997). According to Andrews (1974), the primary importance of the air flow rate is related to the costs and does not appear to have much effect on process efficiency as long as the dissolved oxygen in the reactor remains above some minimum level. Exceptions may be valid for the control of filamentous organism growth. For aeration, various schemes (including extended, high-rate, tapered and step-feed aeration) are in use (Tchobanoglous and Burton, 1991). From an operational point of view, the step-feed aeration scheme (where the wastewater is introduced to the aeration tank at several points in order to equalise the food-to-microorganism ratio) is of particular relevance. During wet-weather, the influent stream towards the aerator can be diverted towards the tail end, so that the sludge may be stored at the head of the aerator. Although this measure can help avoiding wash-out of sludge caused by high hydraulic load, it also may result in poorer treatment process performance (in particular with regard to nitrification) due to the shorter retention time in the aerator. Step-feed control was analysed by Thompson *et al.* (1989), Lessard (1989), Chen (1993) and Vazquez-Sanchez (1996).

Since aeration control is presently reasonably well practised in many activated sludge systems, DO is considered not to be a rate limiting factor for biomass growth in most cases. Therefore, and similarly to the comprehensive studies on treatment plant control conducted by Lessard (1989) and Chen (1993), aeration will not be analysed in this work.

Return activated sludge (RAS) rate

As outlined in Section 2.1.2.4, the settings of the return and waste activated sludge rates are important for solids control in the activated sludge process. According to Walker (1971), maintaining a constant MLSS concentration constitutes one of the most common methods of solids control, resulting in an equalised and improved effluent quality (Fujie *et al.*, 1990). Under steady-state assumptions, the RAS rate necessary to maintain a constant target MLSS concentration can be calculated from a mass balance (see Tchobanoglous and Burton, 1991). Alternatively, food-to-microorganism (F/M) ratio control is applied, which aims at maintaining a constant ratio of influent BOD load to MLSS mass in the aeration tank. Although F/M ratio control may be sufficient to maintain the solids inventory it may result in inadequate control of organism growth under dynamic conditions (Stenstrom and Andrews, 1979). According to Olsson (1992), these traditional control schemes do not consider the variation of either the concentrations in the aeration tank or of the settling conditions.

A number of contradictory schemes have been proposed for pumping the RAS. Fujie *et al.* (1990) provide an overview. At many treatment plants RAS is pumped simply at a constant rate (Schweighofer and Svardal, 1998). A frequently applied alternative method of RAS control is to relate it to the treatment plant inflow rate by a constant factor (*e.g.,* between 0.4 and 1.5 (Harremoës *et al.*, 1993)), thus the sludge rate is increased with increasing flow. However, this control principle may lead to deterioration of the plant performance by imposing an additional hydraulic load to the clarifier as well as by a hydraulic shock to the thickener caused by sharp changes in the RAS rate (Olsson and Jeppson, 1994). According to Wolf (1990), the sludge blanket contributes significantly to the self-regulation of the aerator-clarifier system if the RAS rate is kept constant. Also other authors report deterioration of effluent quality under proportional RAS rate control. Nelson and Buckeyne (1985), Vanrolleghem (1995) and ATV 2.12.1 (1997) suggest raising the RAS rate with increasing inflow only after a time lag. In Lessard's and Reda's (Lessard, 1989; Reda, 1996) simulation studies, an increase of the RAS rate by 43% just before (assuming availability of rainfall forecast) and during rainfall showed to be beneficial in terms of effluent and receiving water quality. Fujie *et al.* (1990), comparing in their study most of the RAS control principles mentioned here, did not observe significant obvious differences between those with respect to effluent concentrations of soluble organics. Therefore, these authors conclude that, rather than any sophisticated MLSS control, control of filamentous micro-organisms provoking sludge bulking is required. Control of sludge bulking has been studied in detail by Chen (1993).

Suggestions to relate the RAS rate to process variables other than the inflow rate include basing it on the sludge blanket height (Busby and Andrews, 1975; Lohmann and Schlegel, 1981; Tchobanoglous and Burton, 1991) (particular attention would have to be paid to the diurnal variation of the sludge blanket), or basing it directly on the MLSS or return sludge concentration (Lohmann and Schlegel, 1981). Olsson (1985) proposes a feedback/feedforward controller, which determines the RAS rate as a function of the solids concentrations in the RAS, the treatment plant effluent, the aeration tank and of the sludge blanket height. Andrews (1994) discusses a feedforward/feedback controller using influent flow rate and solids concentration as input variables.

Chen's study suggests that proportional control of RAS rate can lead to better results in terms of nitrification, but at the same time to deterioration of COD and SS concentrations. Relating the RAS rate to the organic load in the influent may lead to worse results. Those RAS control strategies in Chen's study which are based on feedback control are reported not to handle well transient influent shocks.

Waste activated sludge (WAS) rate

As stated above, the WAS rate determines the rate at which sludge is removed from the activated sludge system. If a constant mean cell residence time (sludge age) and a constant MLSS are to be maintained, the wastage rate under steady-state assumptions can be calculated from a simple mass balance involving the (target) MLSS concentration in the aeration tank and the MLSS concentrations in the effluent and in the underflow (recycle/wastage line) (Tchobanoglous and Burton, 1991; Gülen, 1995). Olsson and Jeppson (1994) state that sludge age control is important but slow; therefore it should be based on long-term averages (*i.e.,* covering more than one sludge age). Roper and Grady (1978) suggest not to adjust the WAS rate in response to changes to plant loading. Rather than that, the mean cell residence time should be kept at a suitable value, thereby allowing the self-controlling nature of the activated sludge process to adjust for changes in the loading.

Chen (1993) investigated various options for setting the RAS and WAS rates for three different rainfall events. Investigations include conventional (*e.g.,* RAS rate set proportional to inflow; or based on influent organic load; WAS set so as to maintain a given sludge age), feedback control (RAS, WAS rates defined as functions of effluent quality parameters), "optimal control" (on-line control with forecasting) and "advanced control" (involving step-feed and step-sludge control). As operational objectives, the avoidance of gross plant failures, maintenance of effluent standards and of equilibrium conditions for filamentous and floc-forming bacteria as well as improvement of effluent quality are defined and combined in one single objective function. The results of Chen's study suggest that for control during rain events none of these four controller types is particularly effective. However, the "optimal" and "advanced" types appear to be the more promising, provided longer and more accurate forecasts of the influent characteristics and expected process responses are available. The study appears to confirm that the activated sludge process has little capacity to deal with the full effects of storm runoff and that what has to be done to attenuate such disturbances is best done in the primary treatment section. Out of the tested control variables, the RAS rate and step-feed control were found to be the most influential; the WAS rate proved to be effective only in manipulating the longer-term performance of the activated sludge system.

Having reviewed the individual control devices available at wastewater treatment plants, a brief overview will be given here of some examples of control at the higher levels, *i.e.,* of combined control of some of the control devices discussed above.

Tong *et al.* (1980) and Beck (1984) propose a control strategy for the activated sludge process based on 20 rules, using fuzzy logic. These rules provide suggestions for changes in the DO set-point as well as in the RAS and WAS rates. These changes are to be performed dependent on the current state of seven state variables and presence/absence of sludge bulking and rising problems. Simulation studies applying this control strategy give encouraging results. Couillard and Zhu (1992) extended this system by defining PID fuzzy based controllers for the lower control level. Their control system results in good process performance when the treatment is under shock loads. Vitasovic and Andrews (1987) investigated rule-based control of the specific total oxygen uptake rate (STOUR) and conclude that their system is able to maintain a desired STOUR profile in the activated sludge process by manipulating the distribution of influent wastewater between reactors. Tsai *et al.* (1993) propose a fuzzy controller for control of solids concentration in the effluent of the secondary clarifier. Discussing various approaches to dynamic control, these authors state that fuzzy based control compares favourably to control based on solid flux theory, to conventional expert systems and to process prediction involving time series analysis in terms of computational speed and representation of system behaviour.

Hansen *et al.* (1994), Hansen (1997) implemented a rule-based fuzzy control system for control of aeration as a function of NH_4 and NO_3 concentrations in the aerator and time of the day (this being relevant for energy costs); their control system performs better than a conventional two-point controller and helps to reduce the energy costs significantly. Implementation of their control system on three nitrifying-denitrifying treatment plants with phosphorus removal results in good performance. Hansen (1997) concludes that future control should involve both fuzzy controllers and conventional controllers in order to combine the advantages of each of these controller types.

Besides the applications of rule based systems for the operation of treatment plants, such systems have also been proposed for diagnosis of operational problems (Gall and Patry, 1989; Beck *et al.*, 1990; Langendörfer *et al.*, 1990; Kayser *et al.*, 1992; Ladiges and Kayser, 1994; Tomei *et al.*, 1996).

Table 2.16 summarises the above discussion by showing combinations of measured information and regulated variables as implemented in practice and/or discussed in the literature. Some of these combinations will be simulated and analysed in later sections of this book.

Table 2.16. Some combinations of regulators and sensor information as employed in various studies on treatment plant operation (continued on next page)

Control device	State variables information used for control	Remarks, references
Maximum inflow rate into the plant (primary clarifier)	None (constant)	Common design practice; Variations of threshold value investigated by Lessard (1989), Lessard and Beck (1990), Otterpohl et al. (1994a), Guderian et al. (1998)
	Influent flow rate	Decrease at start of event and when storm tanks are emptied; increase otherwise (ATV2.12.1, 1997)
	Influent pollutant load/concentration	Schilling (1996a)
	Influent flow rate, Influent ammonium load	Londong (1994), Schweighofer and Svardal (1998)
	Influent ammonium concentration, Sludge blanket height	Maximum inflow rate determined as minimum value of two controllers (Krauth and Müller, 1996)
	Solids concentration in return sludge	Sludge volume loading control (Nielsen et al., 1996)
	Ammonium in aeration basin	Schweighofer and Svardal (1998)
	Filling degree of storm tank	Increase permissible inflow to TP when tank is full (Lessard, 1989)
Filling and emptying of storm tanks	Filling determined by maximum inflow rate into the treatment plant	Common design practice
	Emptying triggered by inflow rate into the plant	Constant threshold value (Lessard, 1989)
	Influent ammonium load/ concentration	Avoiding coincidence of morning NH_4 peak with storm tank returns (Londong, 1994; Krebs et al., 1996)
	Ammonium load in influent and in storm tank	Ensuring constant NH_4 load in influent (ATV2.12.1, 1997)
	Organic load in influent	Dilution of influent with storm tank returns (Maršalek et al., 1993)
Empying of storm tanks directly into the receiving water body	Time span passed for settling in the tank, dilution in the river	Bertrand-Krajewski et al. (1995); Krebs et al. (1996)

Table 2.16. Some combinations of regulators and sensor information as employed in various studies on treatment plant operation (continued)

RAS rate	Constant	(see discussion above)
	As a ratio of inflow rate	(see discussion above)
	(non-proportional) increase of RAS rate at higher inflow rates	Lessard (1989), ATV 2.12.1 (1997), Weijers et al. (1997)
	Organic load in influent	F/M control (Chen, 1993; Gülen, 1995)
	MLSS, solids concentration in return sludge	MLSS control (Walker, 1971), Lohmann and Schlegel, 1981), Gülen (1995)
	Solids concentration in aeration tank, return sludge and effluent, Sludge blanket height	Olsson (1985)
	Sludge blanket height	Busby and Andrews (1975), Lohmann and Schlegel (1981), Tchobanoglous and Burton, 1991, Gülen (1995)
	MLSS, inflow rate	Andrews (1994)
	COD, SS, NH_3 in effluent, Sludge blanket height	Chen (1993)
WAS rate	None (constant)	
	MLSS, solids concentration in return sludge	MCRT control Lohmann and Schlegel (1981), Vitasovic and Andrews (1987), Chen (1993), Weijers et al. (1997)
	Sludge blanket height	Busby and Andrews (1975), Gülen (1995), Kroiss (1994)
	Effluent COD, SS, NH_3, Sludge blanket height	Chen (1993)
Maximum inflow rate into the plant, RAS rate	MLSS, SVI	Keinath (1985, 1990), Balslev et al. (1993)
DO set-point, RAS rate, WAS rate	BOD, SS, NH_3-N effluent concentrations MLSS Solids concentration in return sludge DO set-point WAS rate Sludge performance	Tong et al. (1980), Beck (1984), Couillard and Zhu (1992)

MCRT: Mean cell residence time MLSS: Mixed liquor suspended solids
OUR: Oxygen uptake rate SVI: Sludge volume index

At many plants, only fairly simple forms of control are applied. Often, an individual regulator is set according to the value of one measured variable (single input–single output; SISO). An explanation why this type of control appears to have been fairly successful may lie in the different timescales of the processes in an activated sludge plant. Olsson and Jeppson (1994) distinguish four timescales within the activated sludge process, ranging from very fast variations (faster than minutes; *e.g.,* changes of air or liquid flow rates) over dynamics of minute to hour timescale (*e.g.,* hydraulic phenomena caused by flow rate changes; effects of aeration on the DO concentration) and dynamics of hours to days timescale (*e.g.,* those related to hydraulic retention times) to slow dynamics (with a timescale longer than days, such as cell growth and decay processes). Multiple input–multiple output (MIMO) control would consider interactions between the various regulators. One might expect that application of more advanced control (compared to classical PID and cascade control) might give advantages over classical control. At the moment, thorough comparative studies have not been undertaken to address this question, since reliable models are considered to be required for such a study (Weijers *et al.*, 1995).

A methodology of control not discussed so far, but nevertheless of relevance for the present study, is termed "Model based predictive control (MBPC)" or "Model predictive control (MPC)". It involves on-line use of an optimisation procedure whenever a control decision is to be taken. The optimisation evaluates various control alternatives by computing predictions of future system states through the use of a process model. The solution found is then suggested as the control action to be taken in this time step, after which the procedure is repeated in the next time step (Vanrolleghem, 1995). A more detailed description of the MPC approach is given by Weijers *et al.* (1997). One of the advantages of MPC is that constraints to the controllers can be considered explicitly, something which causes problems in conventional controller design. When applying MPC, it has to be ensured that the process model employed represents the processes in a sufficiently accurate way. Furthermore, it is necessary to keep the computational demand of the optimisation and selection process low enough to enable the decision-finding procedure to take a control decision at every time step. Since often purely deterministic models are characterised by runtimes beyond feasibility for on-line modelling, Harremoës *et al.* (1993) and Carstensen *et al.* (1996) suggest the use of 'grey-box' models, which constitute a combination of deterministic and stochastic modelling approaches (see Beck *et al.* (1989) and Reichert (1994b) for a definition of terms related to modelling) for this purpose.

Examples of MPC applied to the operation of wastewater treatment plants are given by Weijers *et al.* (1995), Ayesa *et al.* (1998), Chang and Chen (1997), Jumar *et al.* (2000) and Jumar and Tschepetzki (2001). The application of MPC to the control of flows within a sewer system–treatment plant–river system, applying a genetic algorithm as the optimisation procedure, is demonstrated by Rauch and Harremoës (1999a) (see also Section 2.4.3). The use of on-line models for control purposes is not restricted to them being part of a MPC procedure as described above. They can be applied for prediction of future states used in other types of control (*cf.* Carstensen *et al.*, 1996) and in control of sewer systems (Petruck *et al.*, 1998).

Furthermore, as new as the methodological approach of MPC might appear when applied to the operation of treatment plants, the underlying idea of calling an on-line optimisation procedure to determine the optimum control decision at the current time step, using a (usually simplified) process model for the evaluation of the objective function, is one of the common approaches in the area of RTC of sewer systems (see Section 2.4.3).

A different and novel methodology of control is proposed by Wilcox *et al.* (1995) and is mentioned here, since it contributes a new option to the list of how control decisions in urban wastewater systems are determined (*cf.* Table 2.13). These authors describe the application of a neural network to the control of anaerobic digestion. The neural network relates the sensor information input directly to an appropriate control action (here: dosing of bicarbonate alkalinity) as output. The network is trained prior to its application by supervised learning.

2.4.5 Real-time Control of Receiving Rivers

Although many descriptions of operational management of sewer systems and treatment plants can be found in the literature (see the previous sections), there do not appear to be many references to control performed in rivers within an urban wastewater context. Obviously, there are numerous examples of rivers which are controlled in one way or another, *e.g.,* for flood protection, for generation of hydropower or maintaining sufficient conditions for water-based transport. Some cases even lead to political crises (take as just one example, the Gabčikovo-Nagymaros Danube dam project (Dosztányi, 1987), which has significant impacts on the internal as well as foreign politics of the countries involved (Karacs, 1997). However, these are examples not concerned with rivers as recipients of urban wastewater discharges and are consequently not discussed here.

Among the potential options of river control potentially relevant to urban wastewater systems is artificial aeration, which has been studied by Whitehead (1979) and Beck (1981) and indeed performed in the River Cam and the River Ruhr in order to improve the DO balance of these water bodies. Aeration by automatic feedback-controller systems is performed immediately downstream of discharge locations. Another example of what could be considered as control action on a river, is the control of flows from retention lagoons containing treatment plant effluent into the river, thus providing the option of influencing the river water quality (*e.g.,* by prudent timing of these discharges). Beck (1976, 1981) and Whitehead (1979) suggest this as an option (alternative to provision of costly tertiary treatment) to deal with high nitrogen loads from treatment plant effluents and agricultural runoff in the Bedford-Ouse river. Another type of control is the "Thames Bubbler" – a boat in London which blows oxygen into the river to improve its oxygen balance.

Beck and Reda (1994) and Reda (1996) provide a detailed study of gate operation in the River Cam and its impacts on river water quality. Their analysis, carried out for two rainfall events combined with various river base flow scenarios, demonstrates that beneficial operation of these gates (taking rainfall information into account) can lead to a significantly improved water quality balance in the river as compared to non- or even bad control of the gates. Improvements in water quality are achieved by coordination of increased discharges from overflows and from the treatment with increased flows caused by opening of the gate structures.

Hence, as control options on receiving rivers, the following could be listed:

- artificial aeration;
- control of in-river flow (*e.g.,* upstream of urban wastewater discharges);
- control of discharges into the river from CSOs and treatment plants.

The latter type of control is not necessarily restricted to simple dilution control. Both the flow and quality characteristics of the discharges as well as those of the river could be considered here. Furthermore, this type of control could aim at minimisation of hydraulic stress to the aquatic habitat. As discussed in Section 2.2.3, for some rivers hydraulic stress has been observed as the most serious impact of urban discharges. However, numerical modelling, assessment and minimisation of these stresses appear to be difficult, although efforts have already been undertaken in this direction (Dannen, 1996).

As methods of finding the control decision used in these examples, controller design (as in the artificial aeration study) and evaluation of a number of intuitively found different options (as in the gate operation study) are applied. Therefore, no new methods can be added to the list of strategy development methods from this section.

The control options listed above as well as those which are taken within the sewer system and the treatment plant could potentially benefit from the availability of on-line river flow and water quality sensors. It may be argued, however, that the use of this sensor information may be of lesser relevance for the current control action at the actual control time step due to the time-lag of the processes in the receiving water as compared to those in sewer system and treatment plant. In-river sensors would indicate impacts of control actions taken potentially hours, if not longer periods, ago. On the other hand, given that models of the components of the urban wastewater system are now available, a study of the potential of using in-river information for control of the urban wastewater system, possibly aided by predictive river models, would be of interest (*cf.* also Beck and Finney, 1987). Within this work, however, analyses will focus on control actions performed for the sewer system and the treatment plant as well as on their impact on all components of the urban wastewater system. For taking control decisions, use will be made of sensor information assumed to be available from sewer system, treatment plant and from a location in the river upstream from the urban discharges (*cf.* also Section 3.4). However, the methodology which will be developed in this book nevertheless allows for the inclusion of more in-river sensor information at a later stage.

2.4.6 Integrated Real-time Control

As can be seen from the discussion in the previous sections, most of the approaches to control of the constituent parts of the urban wastewater system are mainly concerned with the optimisation of the performance of one particular subsystem. A review of control in urban wastewater systems published in 1992 (Zielke *et al.*, 1992) states that until then impacts of, for example, sewer control on treatment plants and rivers could not be appropriately considered. Only recently have attempts been made to consider also other subsystems when determining control actions.

Among the studies taking a wider perspective are those by Nelen (1992, 1993), taking treatment plant inflow equalisation into account when optimising flows in the sewer system. Another example is the simulation study by Rauch and Harremoës (1996d) optimising flows whilst considering impacts of overflows and treatment plant discharges on the biochemical processes in the receiving river.

An example of "integrated control" (as defined in Section 2.4.1), at least of sewer system and treatment plant, is reported from Aalborg (Lynggaard-Jensen and Harremoës, 1993; Balslev *et al.*, 1993; Harremoës *et al.*, 1994; Nielsen *et al.*, 1994;

Schilling *et al.*, 1996). Sewer system and treatment plant control are closely interwoven with each other by exchange of information across the subsystem boundaries when taking the control decision according to the overall strategy, which can be described as follows (Balslev *et al.*, 1993):

1. Under normal operating conditions, the RAS rate in the treatment plant is set to a value calculated according to Vesilind's flux theory (Keinath *et al.*, 1985, 1990).
2. The maximum inflow to the treatment plant is limited to a value which is computed from the system state of the treatment plant, ensuring that the inflow will not lead to hydraulic overload of the secondary clarifier. RTC within the sewer system is applied to the control of flow rates.
3. If the control as described under 2 is insufficient, the aerator is used as a sedimentation basin. This is achieved by switching off aeration and stirring. Hence, the solid flux to the clarifiers will be reduced. RAS rate control is performed as in 1.
4. If control as described under 1 to 3 fails, any excess flows are bypassed directly into the receiving river.

Integrated control is planned in the sense that the treatment plant will provide information about its current and projected capacity, whilst the sewer system module provides forecasts of treatment plant inflow rates expected (Nielsen *et al.*, 1994).

Another example of practical implementation of integrated control is given by the sewer system and treatment plant in Malmö-Klagshamn (Nyberg *et al.*, 1996; Schilling *et al.*, 1996). Again, emphasis of control is directed at equalisation of flows into the treatment plant, thereby enhancing its performance. Although many state variables are measured on-line (suspended solids in inflow, mixed liquor and return sludge, DO in the aeration tank, sludge blanket height), currently only the DO concentration is used within automatic control loops. As future developments, on-line pollution load control, the inclusion of treatment plant operation and of ecological constraints on the discharge sites are to be considered (Schilling *et al.*, 1996). The envisaged overall control strategy is described as follows:

The following three different control actions are activated in consecutive order:

1. Utilisation of in-line storage:

   For inflows up to approximately 2 DWF, the pump rates are increased step-wise to ensure flow is as uniform as possible. So far, only this strategy has been implemented in practice (Schilling *et al.*, 1996).

2. Step-feed operation of the treatment plant.
3. Bypass of the aerator.

If step-feed control is not sufficient, bypass of the aerator is initiated. However, due to the stringent phosphorus requirements on the effluent, bypassing is seen only as the last resort.

2.4.7 Concluding Remarks

Summarising the discussion of the various methods applied for the development of control strategies (*cf.* Table 2.13), the following list can be established:

- heuristics, intuition;
- self-learning expert system;
- off-line optimisation;
- on-line optimisation, model based predictive control;
- application of control theory;
- neural network.

Since the objective of this work is the development of control strategies which consider the urban wastewater system in its entirety, an approach which requires significant simplification of the process description (such as is required for linear and related optimisation approaches) does not appear to be feasible. Significant problems can be expected when attempting to simplify the description of water flow and quality processes within sewer system, treatment plant and receiving water body (*cf.* Sections 2.1 and 2.3) since even the simplified description of flows in sewer systems has proven to be very challenging. Furthermore, pursuing operational objectives which are directly related to river water quality as suggested in this book (see also Section 2.2.4), requires that models of sufficient complexity to describe the river water quality objectives must be applied to the development of control strategies. In this work, an optimisation approach is chosen, since it allows one to determine strategies which are optimum with regard to the chosen criteria (*cf.* Sections 2.2.4 and 3.3.3). In order to be able to consider river water quality based objectives in an appropriate way, it is proposed to apply a sufficiently detailed simulation model as a tool for the evaluation of the objective function in the optimisation procedure.

Due to the complexity of the objective function, no strong assumptions on its characteristics should be made by the optimisation procedure to be applied. Furthermore, optimisation within this work should aim not only at finding optimum solutions, but also at other "good" control strategies (*i.e.*, providing any results which are better than currently applied operation, without necessarily being optimum), as discussed by Ho (1994).

In order to be able to consider long-term impacts of control actions on river water quality and due to the computational demand of the optimisation procedures, it is proposed to apply them off-line. Thus, control strategies will be developed prior to their actual execution (*cf.* Section 2.4.3.1) by application of optimisation techniques.

Hence, the procedure proposed for and applied within this work combines various features of both the off-line and the on-line approach. The last section of this chapter will therefore provide an overview of various optimisation techniques and conclude with a selection of optimisation procedures appropriate to this work.

2.5 Mathematical Optimisation Techniques

As discussed in the previous section, an off-line optimisation approach will be applied in this book for the optimisation of control strategies. This will be done in a two-step procedure consisting of the definition of frameworks for strategies ("parametrisation of strategies") and the subsequent determination of the optimum settings of the strategy parameters involved. This procedure will be detailed in Section 5.1. For the review of optimisation techniques in this section it is sufficient to assume that the parameters of a given optimisation problem may be subject to upper and lower bounds.

This section provides a formal definition of the optimisation problem to be solved, followed by an overview of various optimisation methods. This includes an attempt to classify optimisation procedures (which, at this level of comprehensiveness, could not be found in the literature). Finally, conclusions are drawn as to which optimisation procedures are most appropriate for application within this work. The selected algorithms will be presented in more detail in Section 3.5. The discussion in this section assumes that the objective function of the optimisation problem is to be minimised (rather than maximised), unless stated otherwise. This assumption can be made here without loss of generality.

2.5.1 Definition of the Optimisation Problem

The objective of the application of an optimisation procedure in this work is to find values of the parameters describing the control strategy so that minimum possible adverse impacts of the strategy are applied to the environment. Solutions resulting not only in minimum impacts but also those leading to less detrimental impacts than the currently applied strategies will be of interest.

The simulation tool developed in the course of this work (detailed in Chapter 3) simulates the water flow and quality processes of the urban wastewater system. The control strategy, which is defined by the settings of its parameters, is applied during the simulation. The simulation tool serves as a means of computing the objective function within the optimisation procedure. It has the strategy parameters as input (besides all other input data, such as the characteristics of the case study site, the biokinetic parameters and the rainfall data) and the value of the objective function as output (see Figure 2.16; Figure 3.14 in Section 3.6 provides more details about the implementation).

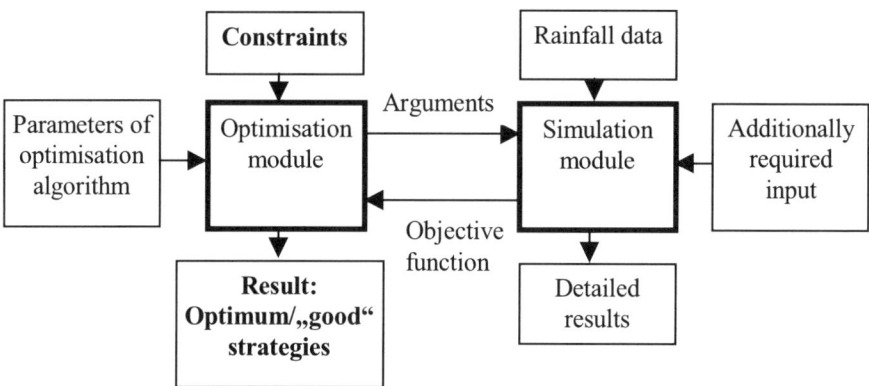

Figure 2.16. The role of the simulation tool within the optimisation procedure

Having introduced the optimisation problem, a formal definition of it is given here. For definitions of the mathematical terms used here, the reader is referred to standard mathematics textbooks; most definitions can also be found in the optimisation literature (*e.g.,* Gill *et al.*, 1981).

A function $f: \Re^n \to \Re$ and a compact subregion $X \subseteq \Re^n$ are assumed to be given. The function f represents the simulation tool (as outlined below). The set X is often called the "feasible domain" or "feasible region" and denotes the restrictions on the arguments of f. A global minimiser $\underline{a}_0 \in X$ is sought, for which $f(\underline{a}_0) \leq f(\underline{a})$ holds for all $\underline{a} \in X$. Commonly this problem is expressed as

$$f(\underline{a}) = \min_{\underline{a} \in X}! \qquad (2.23)$$

The vector $\underline{a} \in \Re^n$ represents the n ($n \in N$) parameters of the control strategy, whereas $f(\underline{a}) \in \Re^n$ denotes the value of the objective function as a function of the parameter vector $\underline{a}$. Evaluation of f is done by applying the control strategy, represented by $\underline{a}$, in a simulation run of the simulation model g, which relates the rainfall data r_i, $i=1,..., m$ (supplied as input data), and the parameter vector $\underline{a}$ to the simulation results. These results are evaluated by using them to compute the value of the objective function, which – in this notation – has been incorporated into g.

Hence, the optimisation problem can be expressed as

$$f(\underline{a}) = g(\underline{r},\underline{a}) = \min_{\underline{a} \in X} ! \qquad (2.24)$$

where:

$\underline{r} \in \Re^m$: vector of the given rain data set (finite time series), which is supplied as input to the simulation runs of the model g

$\underline{a} \in \Re^n$: parameters of the control strategy

g: $\Re^{m+n} \to \Re$ function, described by the simulation model and application of the criteria assessing the control strategy applied (*cf.* Section 3.3.3)

f: $\Re^n \to \Re$ objective function, having the parameters of the control strategy as argument

$X \in \Re^n$: feasible parameter region

For taking a control decision at any individual time step, rainfall (and any other state information) is assumed to be known for all previous, but not for any future time steps. If, however, rainfall or system state forecasts are to be included in the control decisions, a forecast module could be added to the simulation package and the state forecasts made available to the control module. This would not imply any modification of the general formulation of the optimisation problem given above.

Figure 2.17 illustrates the role of the simulation tool in the definition of the optimisation problem.

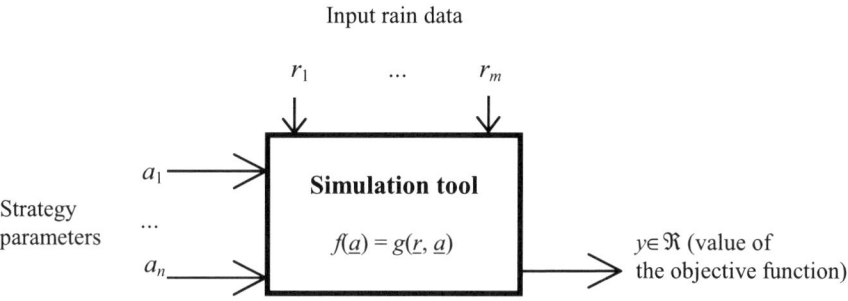

Figure 2.17. Inputs and outputs of the simulation tool in the context of (off-line) optimisation

At this stage, some remarks about the functions and the variables involved in the optimisation problem should be made. No assumption is made about the objective function f with regard to its differentiability. In particular, it is assumed that no information about its derivatives (if they exist) is available. Any attempt to express the derivatives of f analytically would involve a detailed analysis of the simulation models applied. It is argued that this task would be of too substantial complexity, if not impossible, if the various water flow and quality processes in the sewer system, the treatment plant and the receiving water body are to be taken into account. This assumption on f then has the important implication that for the optimisation procedures to be applied below, basically any set of simulation tools can be chosen, without the inherent need of simplification or analytical representation of the simulation models used. This is in contrast to most of the optimisation approaches applied so far to the operation of urban wastewater systems. These usually require some sort of simplification of the simulation model(s) involved (*cf.* Section 2.4.3).

The components of any vector $\underline{a} \in X$ (representing the parameters of a control strategy, *e.g.,* its set points) are assumed to be bound, which seems to be a realistic assumption, as in practice all set-points (*e.g.,* threshold values for water levels, concentrations, pump rates *etc.*) have values within (finite) bounds. This assumption on $\underline{a}$ ensures that the feasible region X is compact. Assuming that f has a lower bound (the objective function will be constructed accordingly – *cf.* Section 3.3.3), the following statement is valid: if f is a continuous function, then compactness of X implies the existence of a minimum of f. The compactness of X also implies, according to Brent (1973), that the constrained optimisation problem described above can be transformed into an unconstrained one. This conclusion will prove useful for the application of optimisation methods to unconstrained problems.

It should be noted, however, that even if the existence of a minimum (no statement is made about uniqueness) is formally guaranteed by these assumptions for continuous *f*, in practice one will be (or will have to be) content with any solution reasonably close to a minimum. If two different solutions are found by a minimisation routine with values of the objective function differing only slightly, then usually both solutions will be acceptable. The fact that several minima can occur in a practical application is demonstrated by Kleissen (1990), where optimisation procedures applied to parameter identification indeed resulted in two different solutions. Occurrence of several minima is not necessarily harmful (in the application being investigated within this work); it even leaves the freedom to decide, by some other criteria than the objective function, which control strategy is to be chosen. However, the optimisation method to be applied in this work should take the case of *f* having several minima into account.

It is interesting to note that the nature of this (strategy parameter) optimisation problem is very close to the parameter identifcation problem in modelling, where parameters of a model are to be found so that an error function, describing the deviations of the model results from the corresponding measurement data, is minimised. Both problems have in common, for example, that not many assumptions on the objective function of the optimisation can be made. A conclusion which can be drawn from this (and other) observation of the parallelism of these two problems is that similar solution algorithms can potentially be used (Schütze, 1996c).

2.5.2 A Review of Optimisation Methods

This section provides a review of various search and optimisation techniques described in the literature. Detailed introductions to optimisation theory and related algorithms are given by Künzi *et al.* (1979), Gill *et al.* (1981), Zhigljavski (1991), Press *et al.* (1994) and Pintér (1996). The discussion of this section will serve as a preparation for the decision on which type of optimisation techniques are most appropriate for use within this work.

Figure 2.18 attempts a classification of methods. However, no clear classification could be found in the literature. It might indeed not be possible to come up with a strict taxonomy, since there are many interrelations between approaches which are represented by different branches in the figure: for example, most global search methods include at some stage a local procedure in order to refine an approximation of a solution which has been found by the global procedure. Therefore, no clear distinction can be made between "global" and "local" methods.

The State of the Art 119

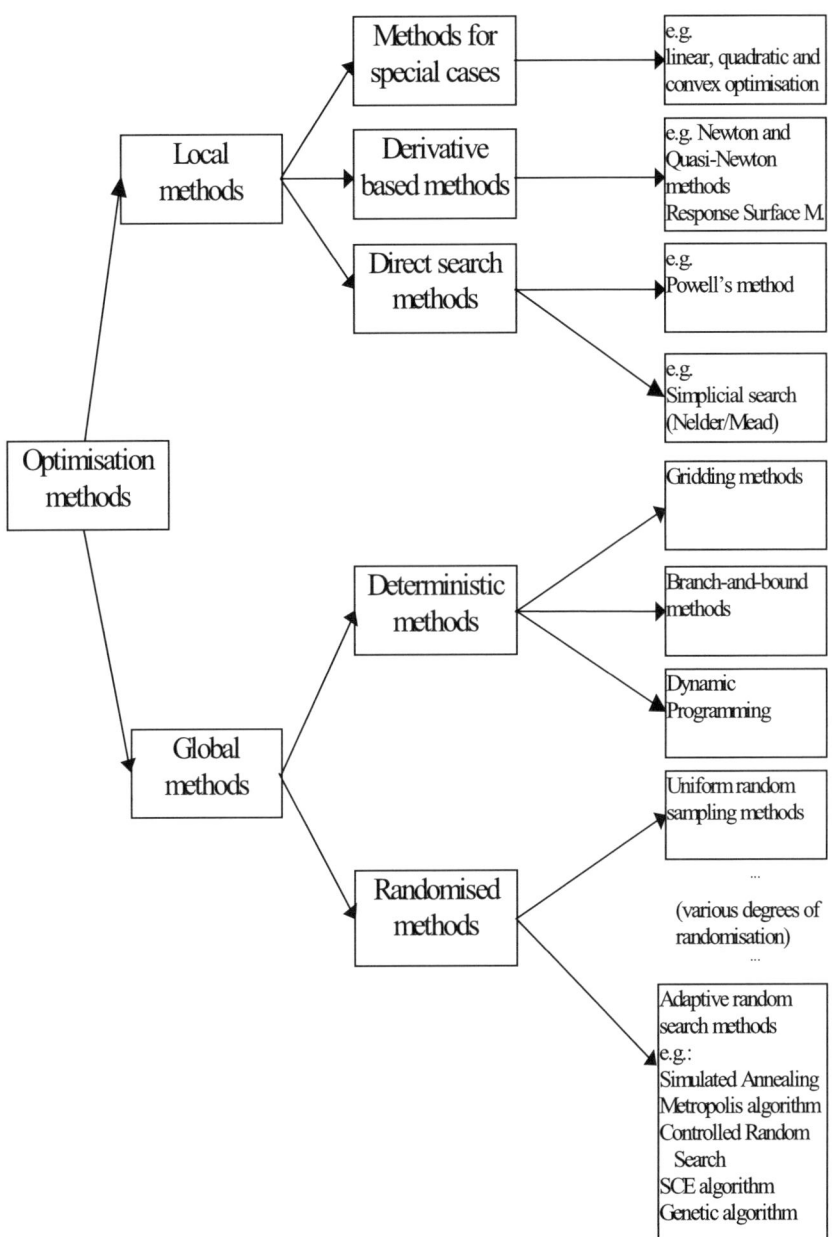

Figure 2.18. Classification of search and optimisation methods

2.5.2.1 Local Optimisation

Local optimisation methods suffer from the problem that finding a local minimum does not ensure in general that a global minimum, too, is found. However, if certain assumptions can be made on f and on the feasible domain (such as convexity or linearity), any local minimum constitutes a global minimum as well. A wide range of optimisation procedures has been developed for these cases, in particular for linear optimisation (Dantzig, 1963; Collatz and Wetterling, 1971; Neugebauer, 1990), quadratic optimisation (see Schütze (1991) for a review), and convex optimisation (Gill *et al.*, 1981). As discussed earlier, use of these methods making such specific assumptions on the optimisation problem does not appear to be appropriate within this work.

Many general methods for locating a local minimum make use of information about derivatives of the objective function. This information is either assumed to be directly available or is computed by numerical approximation ("Quasi-Newton" methods). Also the so-called "steepest-descent" and "conjugate gradient" approaches fall into these classes of derivative-based optimisation methods. Reviews of these methods can be found in the given references and in Gill and Murray (1974), Deyda (1991), Press *et al.* (1994), and detailed issues are discussed by Greenstadt (1972), Papageorgiou and Mayr (1988) and Vanrolleghem and Keesman (1996). Common to all these methods is that these either explicitly or implicitly assume the existence of derivatives and of sufficiently accurate approximations of those. Potential problems of the application of derivative-based algorithms within a RTC context are illustrated by Zunic and Seus (1993a), where a gradient procedure did not provide a solution because of non-smoothness of the objective function. Since no assumptions about the existence of derivatives are made within this work, these methods are not discussed here in further detail.

A local optimisation method which is not based on the calculation or approximation of derivatives is the method by Powell (1964), with refinements proposed by Acton (1970) and Brent (1973) (see also Press *et al.*, 1994). It is based on repeated combination of one-dimensional searches in various directions. Section 3.5.3 will present this algorithm in some detail, since it is implemented as a local optimisation method in the software package developed for this research.

Another local optimisation procedure not requiring derivative information, was proposed by Nelder and Mead (1965). For this and for related methods, several authors use the term "simplex methods". However, this term is considered to be misleading, because these methods are not related to the simplex algorithm widely used within linear optimisation. Therefore, the term "simplicial methods" is preferred here.

The idea of the simplicial methods is to define a regular simplex within the parameter space, hence the name of this approach, and to evaluate the objective function at all points defining this simplex. Starting from its initial definition, the simplex is "moved" across the search space. Whenever there are indications that it is close to a minimum, its size is reduced. Details about this method can be found in various references, including Nelder and Mead (1965) and Press *et al.* (1994). Often, simplicial search is applied in a multistart procedure, where the search is repeated with different initial settings, in order to locate several local minima (if they exist). The multistart approach is considered insufficient, although encouraging results have been achieved, as reported by Duan *et al.* (1992). However, Gan and Biftu (1997) found that for their investigations (automatic calibration of different conceptual rainfall-runoff models, including MOUSE-NAM) a local simplicial method also delivered satisfying results, when it was applied in stages: the user was involved in determining which parameters are to be calibrated in the next step, this implying an increased amount of time needed. The multistart simplicial algorithm as such was found to be highly inefficient by these authors.

Response surface methodology

Although it is hardly ever described in the optimisation literature, the "Response Surface Methodology" (RSM) can be considered as another local optimisation method. It was formally developed by Box and Wilson (1951). Descriptions of this procedure can be found in Davies (1978), Nollau (1979), Grosche *et al.* (1984), Kleijnen (1974; 1987), Khuri and Cornell (1996), a concise summary is given by Kleijnen (1982). The Response Surface Methodology has its origin in the optimisation of chemical processes. These are frequently characterised by functions, which can be evaluated only by experiments, *i.e.* for which no explicit descriptions are available. RSM essentially constitutes a gradient based optimisation procedure: The objective function is approximated locally by "meta-models", which are found for subdomains of the complete parameter domain by application of regression techniques combined with the theory of the design of experiments (see below). The substitute model obtained at each step determines the search direction, along which a move is made towards a new subdomain, where the procedure of building a meta-model is repeated. As soon as a stationary point is approached, a second-order approximation is fitted. Its transformation to canonical form reveals the nature (*e.g.*, true minimum; ridge; saddle point) of the stationary point under investigation.

Determination of the substitute models at each stage of this procedure is by regression techniques (see, for example, Draper and Smith (1981)), which evaluate the objective function at certain points of the subdomain under consideration. These points are determined according to the theory of experimental design, which states for which experimental points (parameter combinations) the variance of the resulting regression coefficients is minimum, whilst the number of function evaluations is kept low at the same time (Finney, 1963; Bandemer and Bellmann, 1976; Kleijnen, 1987; Khuri and Cornell, 1996). Application of statistical procedures (such as Analysis of Variance; F-test; t-test; see, for example, Sachs (1992)) assist in determining those arguments which are of particular significance.

In summary, the application of RSM results in a repeated series of simulations with the full simulation model, then establishing a locally valid (and often linear) substitute model, and finally approaching another (more promising) region of the parameter domain by a steepest-descent method. RSM therefore has all the benefits and drawbacks of a local optimisation technique. Although the number of function evaluations necessary for determining a minimum appears to be low at first sight due to application of the theory of experimental designs, it has to be recalled that substitute models have to be set up repeatedly during the optimisation procedure. Khuri and Cornell (1996) recommend not to implement this process of building substitute models and proceeding along the search direction found as an automatic procedure, since many intuitive decisions still have to be taken during this process. Furthermore, it should be mentioned that in many experimental designs only two values per parameter (usually, the boundary values) are used for function evaluation in the regression procedure. This results in a linear approximation of the objective function which may be too crude, unless small feasible subdomains are defined, which, in turn, would result in a larger number of function evaluations.

2.5.2.2 Global Optimisation

As all local optimisation methods suffer from the problem that (in the general case) finding a local minimum does not ensure that a global minimum has been found, global optimisation methods have been proposed, which do not make use of local properties of the objective function. Thus, the problem of getting stuck in a local minimum is circumvented. However, this advantage of global optimisation methods is usually paid for by a higher number of function evaluations necessary to obtain a solution.

Deterministic methods

Global optimisation methods can be grouped roughly into two main groups: the first consists of (purely) deterministic methods, such as gridding, dynamic programming and branch-and-bound methods.

Gridding methods evaluate the value of the objective function at a large number of points of the feasible domain. If the number of function evaluations is sufficiently high, then there might be some chance to come at least close to minima (see also Ho, 1994). Unless improved, the basic gridding method lacks efficiency, because it does not allow for any learning from function evaluation results (Swann, 1972). An obvious improvement would be to redefine the grid after a series of function evaluations, according to the results found previously.

The dynamic programming methodology was proposed by Bellman and Dreyfus (1962) for multi-stage decision problems. Brief overviews of this technique is given by Grosche *et al.* (1984), Nelen (1992) and Lobbrecht (1997). Essentially, the main procedure applied in dynamic programming consists of three steps as outlined below (Nelen, 1992):

- start at the end of n control decisions, *i.e.*, the last step of the control horizon. Determine the optimum set-points for each possible situation at this time step and store the results;
- next, time step $n-1$ is being considered. Again, determine the optimum strategy for every possible situation, but this time by taking into account the strategy as determined at $t = n$. Store the results;
- repeat this process until $t = 1$. As the initial state is known, it is possible to determine the optimum strategy for all time steps $t=1, ..., n$.

Application of dynamic programming requires that the objective function f has the Markovian property. This condition means that the value of the function f in the current time step depends only on its value at the previous time step, input at the previous step and the current system state. Most decision problems would also have a separability property, which implies the Markovian property. For the control problem within the context of this work, at least the separability condition does not seem to be met. The main advantage of the dynamic programming approach is that it does not place any restrictions on the objective function and on the constraints. However, it is very demanding in terms of computation time and computer storage.

Within the context of deterministic global search, branch-and-bound algorithms can be identified: these algorithms are based on sequentially rejecting those subsets of the feasible domain X that can be proven not to contain a global minimiser. The search then continues in the remaining subsets of X. However, these methods make certain assumptions on f in order to be able to make general statements such as that a given set cannot contain a minimiser (for example, separability of f into convex and concave subfunctions constitutes such a property of f (Knemeyer, 1992)).

Randomised methods

As soon as random decisions are involved in search methods, the term "random search" (or: "stochastic search") becomes appropriate. Zhigljavsky (1991) provides a comprehensive discussion of the theory of global random search methods. The degree of randomness varies significantly in different approaches. Uniform random search can be considered as being entirely probabilistic: here the objective function is evaluated at every point of a random sample of points of the feasible parameter domain (applying a uniform distribution). In this type of approach, no use is made of information gained in previous evaluations of the objective function. Methods making use of such information are called adaptive random search methods (Masri *et al.*, 1980; Price, 1987). The main idea here is to perform evaluations of the objective function at points centred around promising points (Duan *et al.*, 1992).

Among the earlier realisations of adaptive random search techniques is the Simulated Annealing technique (see, for example, Kirkpatrick *et al.*, 1983; Press *et al.*, 1994). A well-known example solvable by such an optimisation algorithm is the Travelling Salesman Problem, which consists of finding the shortest route for a salesman having to travel to a number of cities, visiting each of those once. The term "simulated annealing" stems from an analogy to thermodynamics where crystals attain their state of minimum energy best by being cooled down slowly. The main idea of this type of optimisation imitates this principle by not always going towards a candidate solution (thus possibly ending up in a local minimum) but allowing – occasionally – a step in a different direction. The probability of allowing for such steps is decreased slowly during the optimisation process (similar to the temperature in the annealing process). These principles of thermodynamics were first introduced into numerical calculations by Metropolis *et al.* (1953). Since a stochastic element is involved when determining new search directions, the simulated annealing procedure can be considered as a predecessor of later stochastic search methods, such as genetic algorithms and controlled random search techniques. These algorithms, which can also be termed 'evolutionary', start with an initial population of candidate solutions, sampled randomly from the feasible domain. The function values of the individuals of this population influence the process of generating new candidate solutions. Various techniques have been proposed for this process. Evaluating populations with population sizes greater than

1 enables the procedure to search for several minima at the same time. Thus such a procedure is less likely to restrict itself to a local minimum.

A Controlled Random Search (CRS) procedure was suggested by Price (1979, 1983, 1987). Its application to parameter identification problems has been reported by Klepper *et al.* (1991), Osidele (1992) and Stigter *et al.* (1997). In the CRS algorithm, new candidate solutions are generated by reflections on the centre of gravity of the current set of candidate solutions (see Section 3.5.1 for details). Although not many applications of this algorithm have been reported, it appears to be a promising procedure, as it performs relatively well for the examples discussed in the available references. Duan *et al.* (1992) developed the shuffled complex evolution (SCE) algorithm, which constitutes an extension of the CRS procedure, to which also elements of the simplicial search algorithm by Nelder and Mead (see above) and of competitive evolution (Holland, 1975) were added. Similarly to the CRS, the SCE starts with an initial set of candidate solutions. This is then split into several communities ("complexes"), which are allowed to develop separately by a combination of CRS and Nelder and Mead's simplicial search. From time to time, these complexes are mixed (shuffled), thereby passing on information about the search space gained independently by each community. Improved performance of the SCE as compared to the various versions of the CRS algorithm is reported by Duan *et al.* (1992). The SCE also avoids premature convergence as was observed in some test runs with the CRS algorithm. Kuczera (1997) observed the SCE to be robust and more efficient than a genetic algorithm procedure.

In genetic algorithms, details of which are described by Davis (1987), Goldberg (1989), Michalewicz (1992) and Beasley *et al.* (1993a,b), new generations of candidate solutions are obtained by imitation of the biological processes of crossover, mutation and selection of appropriately encoded representations (*e.g.*, bit-strings) of the elements of the feasible domain. Genetic algorithms are becoming increasingly popular in many different areas, including various water related applications (management of water reservoirs: Esat and Hall, 1994; Stanic and Avakumovic, 1996; Milutin and Bogardi, 1996; design of water distribution systems: Walters and Savic, 1996; Wong, 1996; Savic and Walters, 1997; model calibration: Wang, 1991; Wong, 1994; Savic, 1997; groundwater remediation: Rogers *et al.*, 1995; characterisation of rainfall events: Prax *et al.*, 1996; and control of urban wastewater systems: Rauch and Harremoës, 1996c,d, 1999a; *cf.* Section 2.4.3). As reported by most of these authors, the definition of the parameters of the algorithm itself (such as mutation and crossover probabilities) is crucial to the success of its application. Although substantial theoretical work has been carried out on genetic algorithms (Beasley *et al.* (1993a,b) provide a comprehensive review), there still remain many unsolved questions: Wong (1994) states: "There appears to be no accepted 'general theory' for why Genetic Algorithms possess the properties they have, though hypotheses to explain why they work have been put forward."

The following can be stated in summary: of the global optimisation methods, gridding methods and dynamic programming are computationally expensive. Branch-and-bound algorithms require certain characteristics of the objective function which cannot be presupposed in the general case. The algorithms of the randomised global methods most widely used appear to be the genetic algorithms. Therefore, an example of this class of algorithms as well as the less well-known Controlled Random Search procedure are selected for implementation and use within this work. Both algorithms will be subjected to further studies later in this book (Chapter 5). The choice of randomised methods for optimisation is also supported by Ackley's (1987) studies of various optimisation procedures for various types of objective functions. Furthermore, the procedures selected allow for restrictions on the arguments (upper and lower boundaries) to be considered in a straightforward way. In order to be able to refine a solution found by a global procedure, a local, non-derivative based local algorithm is to be implemented. For this purpose, Powell's algorithm has been chosen, since it appears to be more efficient than the simplicial search proposed by Nelder and Mead. A detailed description of the optimisation procedures which are selected for implementation and application in this work (Genetic Algorithm, Controlled Random Search and Powell's algorithm) will be given in Section 3.5.

2.6 Conclusion

Having given an overview of various areas having relevance to the present work in the previous sections, the main findings obtained are summarised in Figure 2.19. This figure also indicates the position of these findings within the framework of this work and provides an overview of the structure of the subsequent chapters.

The next chapter will describe the development of the simulation and optimisation tool, which will be applied for the analyses carried out in Chapters 4 and 5.

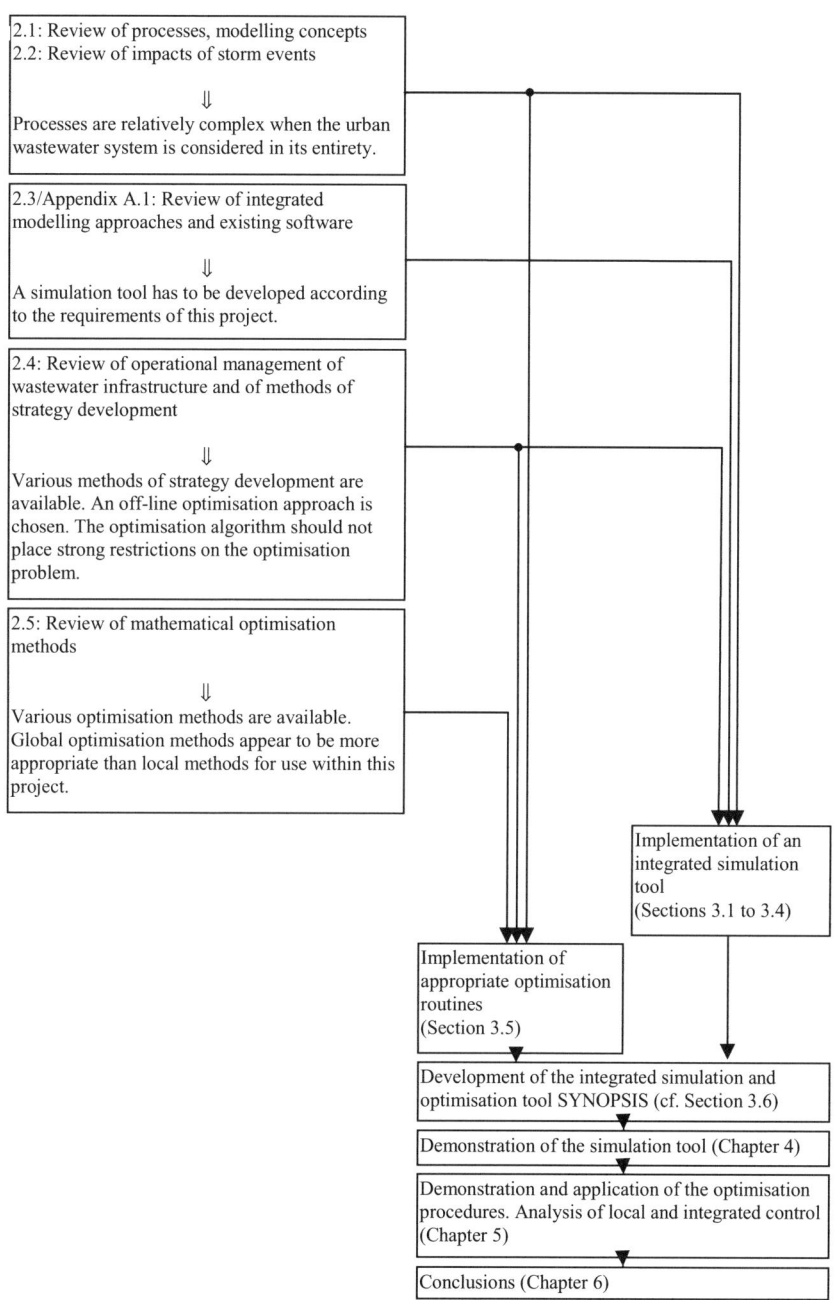

Figure 2.19. Summary of the sections of this chapter and their position within the book

Chapter 3
Development of the Integrated Simulation and Optimisation Tool SYNOPSIS

The main processes in the urban wastewater system and their representation in models have been presented in the previous chapter. The present chapter outlines the development of the integrated simulation and optimisation tool (named SYNOPSIS - "Software package for synchronous optimisation and simulation of the urban wastewater system"), which is to be used for the studies described in later chapters of this book. The development of the tool starts with the definition of a list of requirements (Section 3.1), which are based on the discussion of the previous chapter. According to these requirements, three existing software packages are selected for implementation in SYNOPSIS (Section 3.2). Section 3.3 describes their integration to an integrated simulation tool. This section also discusses the problems encountered when connecting the simulation programs. Implementation of control options is detailed in Section 3.4. Section 3.5 discusses the implementation of the optimisation procedures selected in Section 2.5 and their interfaces with the simulation tool.

Section 3.6 provides an overview of the simulation and optimisation tool developed for the particular needs of this research. Test simulations for a semi-hypothetical case study site will be presented in Chapter 4, before it is applied to the studies on the potential of integrated control in Chapter 5.

3.1 Requirements on the Simulation Tool

In Chapter 2, the main subsystems and processes in urban wastewater systems have been identified. Furthermore, the key pollutants of interest for the present work (DO, ammonium) as well as optimisation algorithms appropriate to the envisaged studies were selected. Thus, certain requirements on the simulation tool to be developed need to be derived.

The most obvious requirement is that the simulation tool should represent the impact of control on the performance of the urban wastewater system (measured by the selected criterion) in a sufficiently accurate way. Since the tool is to be used on an urban wastewater system, for which hardly any data are available (see Section 4.1), a deterministic modelling approach is chosen, since stochastic modelling would require a large set of data for model development. As operational strategies are to be analysed, neither a static nor a steady-state modelling approach is sufficient (Beck *et al.*, 1989), since the simulation tool has to cater for an appropriate representation of transient processes.

Additional requirements include the following:

Since analysis of the potential of integrated control (as defined in Section 2.4.1) is among the main objectives of this work, the simulation tool should be able to simulate this type of control. More specifically, information transfer (of sensor and control device information) between the various subsystems has to be possible. This excludes a sequential simulation approach in which the simulation of one part of the urban wastewater system (*e.g.*, the sewer system) is completed for all time steps before the simulation of the subsequent subsystems (*e.g.*, treatment plant, river) is started. Therefore, simulation of the parts of the urban wastewater system has to be done in a synchronous manner.

The simulation tool should allow the simulation of a variety of control options, including conventional as well as innovative ones. Due to the intended application of optimisation routines for determination of the parameters of control strategies, the optimisation module should have easy access to the definition of control strategies. Moreover, the simulation tool should be easily embeddable into the optimisation procedure, but, on the other hand, it should also provide the option of being run as a stand-alone module (*i.e.*, without being called by the optimisation procedure).

Because the simulation tool is to be used for the assessment of control actions on the entire urban wastewater system, it should allow for long-term simulations to be carried out. Generally, three types of simulation can be distinguished (itwh, 1995): single-event simulation, long-term series simulation, and long-term continuous simulation. In single-event simulation, only a small number of selected rainfall-runoff or pollution events are simulated, whereas in long-term series simulation all events meeting prespecified criteria (such as minimum rain depth) are simulated. In long-term continuous simulation, no events are selected at all, but a continuous time period is simulated, which includes all rainfall and dry weather periods. In many projects on RTC of sewer systems only, either single-event or long-term series simulations are performed, *i.e.*, individual rainfall events are investigated. Within an integrated context, however, the impact on a control action may extend over a long time period, thus making it increasingly difficult to define the term "event" or "event duration". This issue will be further addressed in Section 4.5.

Long-term continuous simulations are also necessary if long-term effects of pollution are to be considered (*e.g.,* accumulative pollution; see Section 2.2.3) or if statistical criteria (such as the criteria outlined in the UPM Manual, see also Section 2.2.4) are used for the assessment of pollution. Therefore, it is concluded that, even when focusing on acute pollution, which is assessed in this work in a rather simple way (see Section 3.3.3), the simulation tool should provide the option of long-term simulation. Various authors apply continuous simulation for the analysis of design and also of operational scenarios (*cf.* Gustafsson *et al.*, 1993b; Nielsen *et al.*, 1994) and also some of the integrated modelling studies presented in Section 2.3. Further support for the application of long-term simulation is given by Larsson (1995) who states the current trend in urban drainage modelling is towards continuous simulation.

Use of the simulation tool to be developed for long-term simulations implies that the components of the simulation tool should be not too detailed in order to keep the overall simulation time within feasible limits. This requirement is also reinforced by the proposed approach of using the simulation tool for the evaluation of the objective function within an optimisation procedure, which might potentially imply a large number of simulation runs (function evaluations). Therefore, the simulation tool should simulate the behaviour of the urban wastewater system with sufficient accuracy, without, however, simulating every detail if this implies substantial additional computational demand.

In order to make maximum use of previously accumulated knowledge and experience, existing simulation models are to be used for the simulation tool. It also enhances the credibility of the tool if it is based on proven and commonly accepted modelling approaches. When selecting models for the constituent parts of the simulation tool, care has to be taken that the complexity of the submodules are comparable to each other (Rauch *et al.*, 1998a). Furthermore, building the tool using a modular structure would allow straightforward extension of individual modules or the substitution of modules by updates, thus providing the option to include future developments of modelling. According to Rauch *et al.* (1998a), utilisation of existing modules for integrated modelling suffers from the problem that these often may be too complex. Following this argument, simpler models rather than complex and detailed models are chosen for use within this work.

With respect to the application of the simulation tool to a case work, its submodels (and the tool as a whole) should allow calibration and validation against field data (if not already done so), as soon as such data are available. Therefore it is

132 Modelling, Simulation and Control of Urban Wastewater Systems

desirable that any constituent model is applied elsewhere, thus supporting its applicability.

Finally, costs and availability of support are additional criteria to be considered in the process of selecting a modelling package.

In summary, the list of requirements of a simulation tool for use within this work can be stated as follows:

1. sufficiently accurate representation of the key processes and key parameters (identified as being DO and, to a lesser extent, ammonium);
2. synchronous simulation approach, thus enabling information transfer across subsystem boundaries;
3. implementation of control, including integrated control;
4. availability of two simulation modes: stand-alone simulation and simulation-optimisation
5. modest demand on computation time;
6. existing and proven software and simulation approaches are to be used wherever appropriate;
7. modular structure, allowing for future developments or alternative models to be incorporated easily;
8. affordability; availability of support.

Since existing simulation software is to be used (see item 6), the question arises whether the present study could be conducted with the modelling software discussed and suggested in the UPM Manual (FWR, 1994, 1998) (*i.e.,* MOSQITO, STOAT, MIKE11; see also Section 2.3). However, use of these software products within this work does not seem to be feasible (at least not with the present version of these models), since they do not meet requirements 2, 4, 5 and 8. Most proprietary models available on the market would suffer from problems related to 2 and 4, and possibly also with regard to 3, 5 and 8. In particular, in view of requirements 2, 3 and possibly 4, the availability of the source code of the submodules would be beneficial.

Matching these requirements with those models which were available when this work started (see Section 2.3 and Appendix A), the following set of models is selected: a research version of KOSIM (called EWSIM) for urban catchment rainfall runoff and sewer system simulation, the treatment plant model of Lessard and Beck as a model for an activated sludge treatment plant, and the DUFLOW program shell (*cf.* Appendix A) with an oxygen model of Lijklema *et al.* (1996) as a model of the biochemical processes within the receiving river. A detailed description of these constituent models will be given in the next section.

3.2 Modules Simulating the Parts of the Urban Wastewater System

The following subsections describe in detail the software packages in detail which were selected for inclusion in the simulation and optimisation tool. Some necessary modifications are also outlined in these sections. Section 3.3 will describe how these packages are connected to form an integrated simulation tool.

3.2.1 Implementation of the Sewer System Module

The sewer system module of SYNOPSIS consists of the program package EWSIM, which is an extended (research) version of the commercially available package KOSIM (itwh, 1995). Therefore, a brief description of the model KOSIM is given here, followed by some remarks about the additional features of EWSIM.

KOSIM has been developed and extended by several researchers (Paulsen, 1986; Durchschlag, 1989; Kollatsch, 1995). These references also provide further details of this model. KOSIM is used in engineering practice as well as in numerous research projects (Vanrolleghem *et al.*, 1996a) (see also Section 2.3), Kollatsch and Schilling, 1990; Schilling and Hartwig, 1990; Wittenberg, 1992; Pracejus, 1994; Barovic, 1995; Fuchs and Gerighausen, 1995). It is also officially recommended as a consent tool for design of sewer systems by the German state of Niedersachsen (NMU, 1990). In other German states, different, but conceptually very similar models are used, *e.g.,* SMUSI in the state of Hessen (Brandt *et al.*, 1989). Therefore, it can be said that the model KOSIM and its underlying modelling concepts are accepted and used widely. Flow and pollution parameters can be calibrated against field data (Durchschlag, 1989; Demuynck *et al.*, 1993).

Rainfall-runoff simulation in KOSIM distinguishes between impervious and pervious catchment areas. For impervious areas, wetting, depression (represented by a time-variant runoff coefficient) and evaporation losses are considered. Infiltration for permeable areas is simulated using Horton's approach, which has been adapted for use in long-term simulations (Paulsen, 1986). Details of these representations of the rainfall-runoff processes are given by ATV1.2.6 (1986, 1987) and by itwh (1995).

Flow routing on subcatchments in modelled by Nash cascades (see Section 2.1.1.2). Flow between the subcatchments is modelled by translation and addition of inflows from the subcatchments. From this modelling approach, it follows that backwater effects cannot be modelled by KOSIM. This might constitute a limitation for application of this program within this work. On the other hand, the above mentioned requirements of the tools to be used within this work suggest that the use of a more detailed sewer model does not seem to be feasible for reasons of computing time.

Ancillary structures such as pumps, overflows and different types of storage tanks (see also Section 2.1.1.2) can also be modelled. Surface flooding is not modelled explicitly by KOSIM; overflows occurring within subcatchments are assumed to be discharged into the river.

Pollutants are assumed to originate from two sources: domestic wastewater (described by diurnal and weekly flow and pollution patterns) and rainfall runoff (assumed to be of constant concentrations). The third component, inflow from extraneous sources (*e.g.,* infiltration into sewers), is assumed non-polluted. The pollutants originating from these three sources are routed through the system, where they are assumed to mix completely and without any interactions. Thereby, pollutants are considered as conservative. Optionally, sedimentation and resuspension on the surface and in the sewers and in basins can be modelled by KOSIM. However, according to the manual (itwh, 1995) as well as in the description of the SMUSI program (Brandt *et al.*, 1989), use of this option is discouraged, since research is still required in this area and the related model parameters are known only with great uncertainty, if at all. Another option allows sedimentation of pollutants in storage tanks to be modelled. If this option is not set, then pollutants are assumed to be completely mixed in storage tanks. In total, up to six pollutants can be modelled by KOSIM once the pollutant-specific parameters are defined.

In addition to the features of KOSIM just described, the model EWSIM allows for individual diurnal and weekly patterns of flow and each single pollutant to be considered. Also several implementations of control options are available (Kollatsch and Schilling, 1990; Pracejus, 1994).

As the source code of EWSIM was available for use within this work, EWSIM is used. Some modification of the software was carried out prior to its inclusion in SYNOPSIS. These include taking out unnecessary parts of the program (most of them being related to certain output options and the expert system based control module by Pracejus (1994), and to transferring the software from the OS/2 operating system to DOS/Windows).

In summary, it can be said that by using KOSIM/EWSIM within SYNOPSIS, a simple, though powerful program has been chosen. Some potential weaknesses are that surface flooding is not addressed in an appropriate way by this program and that backwater effects can be considered only to a limited extent. However, the discussion of the previous section supports the level of complexity chosen for modelling this subsystem.

3.2.2 Implementation of the Treatment Plant Module

Since the treatment plant model of Lessard (1989) and Lessard and Beck (1990; 1993) has been selected for implementation in SYNOPSIS (see Section 3.1), it is presented here in some detail. Some modifications and extensions, most of them related to its secondary clarifier module, were necessary in order to enable the program also to be used for situations when the sludge blanket from the secondary clarifier is discharged into the effluent. These modifications are detailed in Section 3.2.2.2.

3.2.2.1 The Original Implementation of Lessard and Beck's Treatment Plant Model

In this section, an outline of the original version of the model, as reported by Lessard (1989) and Lessard and Beck (1993) is given. Detailed descriptions of this model and its individual submodules can be found in the references given and in Güven (1995).

Lessard and Beck's model simulates a conventional activated sludge plant (including nitrification, but no denitrification). Development of the model was based on the treatment plant in Norwich-Whitlingham/UK; thus the module consists of modules for storm tanks, primary clarification, activated sludge basins and secondary clarification. Also a number of regulating devices can be simulated.

Significant extensions to this model were made by Chen (1993) in order to investigate the normal and abnormal behaviour of the system by introducing a mathematical model describing filamenteous and non-filamenteous organisms. Other extensions to Lessard and Beck's model were made by Le Dren (1996), including a model for Nitrification/Denitrification/Biological Enhanced Phosphorus Removal (NDBEPR). Lessard and Beck's model was also applied in other studies (Paracampos, 1991; Albuquerque, 1993; Côté *et al.*, 1995). For the present work, however, the less complex original version of this model is used. Some extensions which proved to be necessary were carried out and are described in the Section 3.2.2.2.

Pollutants simulated in all parts of the treatment plant include suspended solids (SS), volatile suspended solids (VSS), total chemical oxygen demand (COD_t), soluble COD (COD_s), ammonium (NH_4), and nitrate (NO_3). In various submodules, additional parameters or more detailed fractions of these parameters are modelled (see Figure 3.2 for details). The fourth-order Runge–Kutta algorithm is applied as the numerical solver for the differential equations incorporated in various parts of the treatment plant model, (see, for example, Fehlberg, 1960, 1966, 1969, 1970; Press *et al.*, 1994; Reda, 1996).

Apart from the values of the various physical (tank sizes, flow scheme *etc.*) and the biochemical parameters (kinetic and stoichiometric coefficients) and the initial conditions of the system, Lessard and Beck's model requires as input the influent flow rate and concentrations (of SS, VSS, COD_t, COD_s, NH_4, NO_3) for every time step (1 hour). Furthermore, information about the weather status (since the settling velocity in the primary clarifier is assumed to be a function of the composition of the sewage) as well as about sludge returns from the sludge processing unit into the treatment plant influent are required for every time step. Similarly, after the output module of the program has been extended for this work, the output now contains (among other information) the following for each time step: flow rates and concentrations of SS, VSS, COD_t, COD_s, NH_4 and of NO_3 of storm tank overflows, primary clarifier effluent, secondary clarifier effluent, treatment plant effluent and storm tank discharges.

Figure 3.1 provides an overview of the flow-scheme of the treatment plant modelled and of its modular structure.

The flow of variables and their transformations at the interfaces between the various submodules of Lessard and Beck's program is shown in Figure 3.2: an overview of these modules is given here.

Module 1: This module processes the input data describing the characteristics of the treatment plant influent. Within SYNOPSIS these data will be provided by the sewer system simulator.

Module 2: This module simulates a regulating device (which, in the case of the treatment plant in Norwich, is a side-weir), splitting the influent to the treatment plant into inflow towards the primary clarifier and towards the storm tank. A maximum inflow rate (set by the control module; *cf.* Section 3.4) determines how much water is admitted to the primary clarifier. All inflows in excess of this flow rate are diverted into the storm tank.

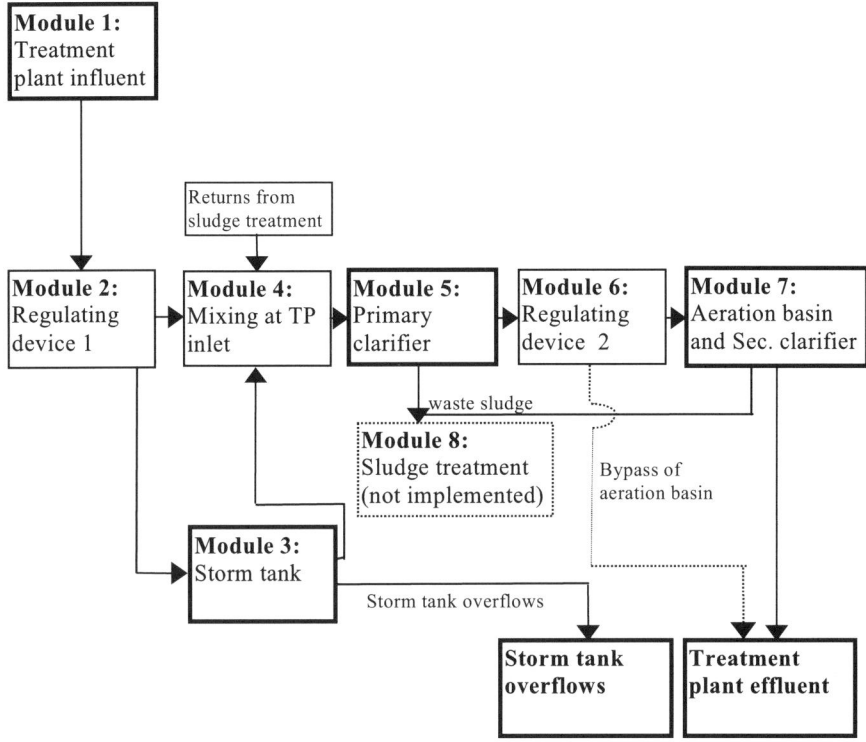

Figure 3.1. The structure of the treatment plant model of Lessard and Beck

Module 3: This submodule is an implementation of the storm tank module developed by Lessard and Beck as described in Section 2.1.2.1. As can be seen from Figure 3.2, the pollutants SS, VSS and particulate COD (COD_p) are partitioned into two fractions each (settleable and non-settleable) within the storm tank module in order to allow for an appropriate description of the settling processes.

Module 4: This module simulates complete mixing of the storm tank returns with the treatment plant influent and the returns from sludge treatment prior to this mixture entering the primary clarifier. Since module 8 (Sludge treatment) is not implemented, this module also requires input information about the returns from sludge treatment. In addition to mixing flows, Module 4 also partitions the pollutants SS, VSS and particulate COD into settleable and non-settleable fractions, since this partition is required for the primary clarifier module. Since different partitions are applied for different types of sewage (crude and storm sewage and sludge returns) and since the input of Module 4 consists not only of storm tank returns, no use can be made within this module of the partition already performed in Module 3.

Module 5: This module contains the primary clarifier model as described by Lessard and Beck (1988) (see Section 2.1.2.2).

Module 6: A flow separation device is simulated by this module, which provides the option of leading those flows which are in excess of a given threshold value directly to the effluent (aeration tank bypass). Although information about SS, VSS and soluble COD concentrations of the output of this module is not required by the aeration tank module (see Figure 3.2), it still forms a crucial part of the information about the effluent characteristics of the treatment plant.

Module 7: This module simulates the aeration tank and the secondary clarifier. It is divided into three submodules: *Module 7A* (aeration basin), *7B* (Fing) and *7C* (clarification). Although for clarity these submodules are shown as separate units in Figure 3.2, they are closely interrelated. As additional inputs, these submodules require the flow rates for return and waste activated sludges (denoted by RAS and WAS, respectively), which are set by the control module (see Section 3.4).

As a model for the activated sludge process in the aeration basin, a simplification of the IAWPRC Activated Sludge Model No. 1 was implemented by Lessard and Beck. Driven by a lack of related data, processes involving organic nitrogen, such as the ammonification of soluble organic nitrogen and hydrolysis of organic nitrogen, are not included in Lessard and Beck's model. Similarly, since no emphasis was given to the precise simulation of DO behaviour, the anoxic growth of heterotrophs was not considered in this model, resulting in denitrification not being modelled. As stated by Lessard and Beck (1993), the simpler restriction of nitrogen transformations merely to the nitrification of ammonium has been found to be adequate in other studies (*e.g.,* Gujer, 1977), including an earlier examination of the Norwich plant (Beck, 1981). Figure 3.3 provides a comparison of the IAWPRC Activated Sludge Model No. 1 with its simplification by Lessard and Beck: those processes which are modelled by Lessard and Beck are marked in the figure; differences in stoichiometric coefficients are identified by shading. However, implementation of the full IAWPRC model would appear to be easily possible if deemed necessary and as soon as appropriate data are available.

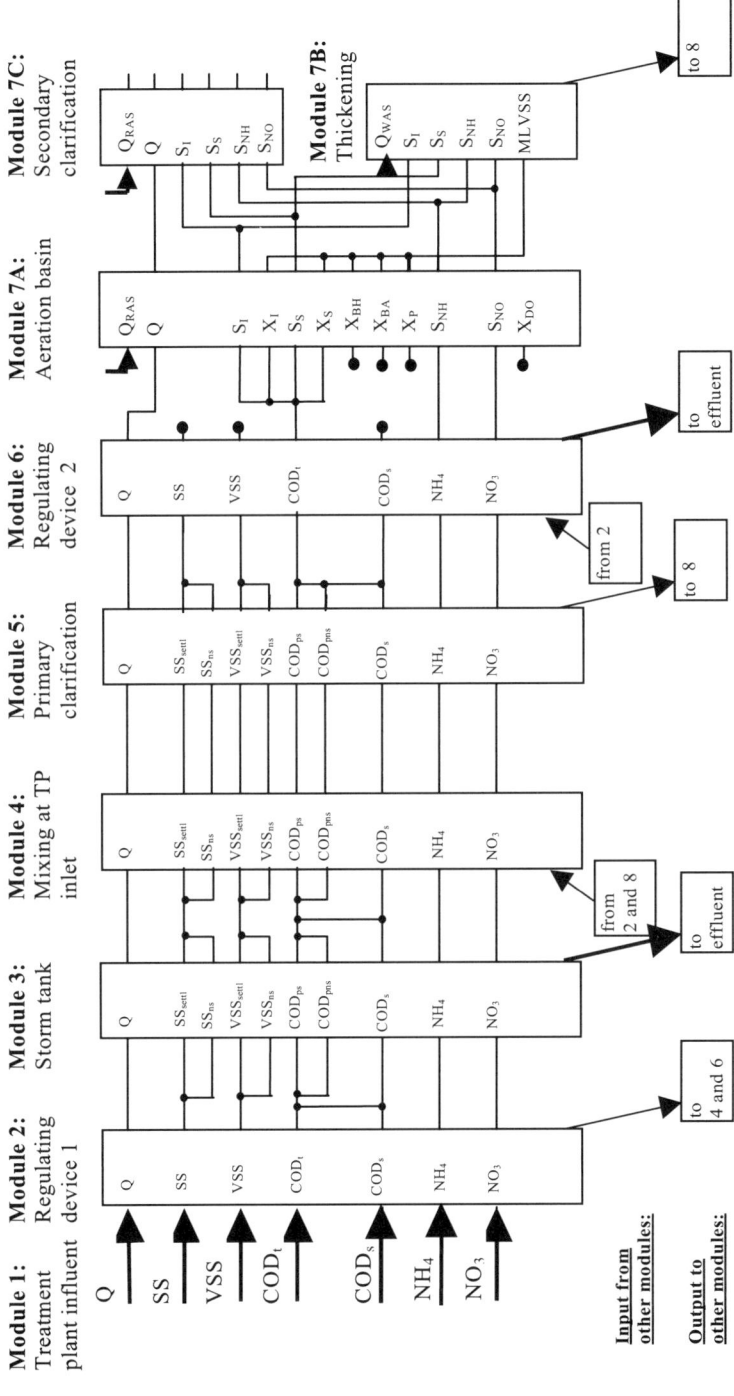

Figure 3.2a. Flow of variables in Lessard and Beck's treatment plant model. *n.b.*: Module 8 (Sludge treatment) is not implemented. Information is provided externally

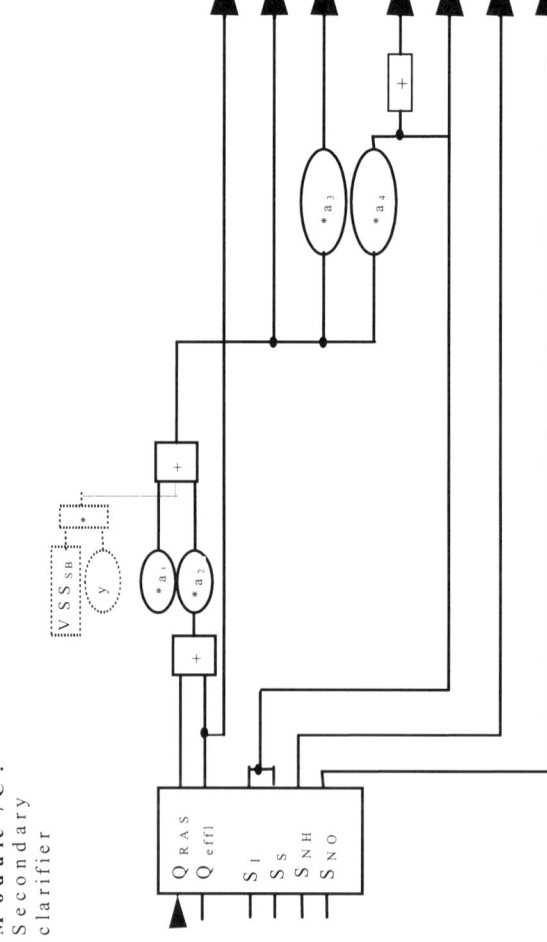

Figure 3.2b. Flow of variables in the secondary clarifier submodule of Lessard and Beck's treatment plant model. It includes the model by Pflanz (1969) for calculation of SS effluent concentration. Dotted elements indicate extensions by Vazquez-Sanchez (1996). y is defined by Equation (3.2). The factors $a3$ and $a4$ are used in Lessard and Beck's model to determine effluent VSS and COD as functions of effluent

Component (i) → Process (j) ↓	1 S_I	2 S_S	3 X_I	4 X_S	5 X_{BH}	6 X_{BA}	7 X_P	8 S_O	9 S_{NO}	10 S_{NH}	11 S_{ND}	12 X_{ND}	13 S_{ALK}	Process Rate ρ_j (ML^{-3} T^{-1})
1 Aerobic Growth of Heterotrophic Biomass		$-\frac{1}{Y_H}$			1			$-\frac{1-Y_H}{Y_H}$					$-\frac{i_{XB}}{14}$	$\hat{\mu}_H \left(\frac{S_S}{K_S+S_S}\right)\left(\frac{S_O}{K_{O,H}+S_O}\right) X_{BH}$
2 Anoxic Growth of Heterotrophic Biomass		$-\frac{1}{Y_H}$			1				$-\frac{1-Y_H}{2.86 Y_H}$				$\frac{1-Y_H}{14 \cdot 2.86 Y_H} - \frac{i_{XB}}{14}$	$\hat{\mu}_H \left(\frac{S_S}{K_S+S_S}\right)\left(\frac{K_{O,H}}{K_{O,H}+S_O}\right)\left(\frac{S_{NO}}{K_{NO}+S_{NO}}\right)\eta_g X_{BH}$
3 Aerobic Growth of Autotrophic Biomass						1		4.57	$\frac{1}{Y_A}$	$-1/Y_A$			$-\frac{i_{XB}}{14} - \frac{1}{7Y_A}$	$\hat{\mu}_A \left(\frac{S_{NH}}{K_{NH}+S_{NH}}\right)\left(\frac{S_O}{K_{O,A}+S_O}\right) X_{BA}$
4 Lysis of Heterotrophic Biomass				$1-f_P$	-1		f_P					$i_{XB}-f_P i_{XP}$		$b_H X_{BH}$
5 Lysis of Autotrophic Biomass				$1-f_P$		-1	f_P					$i_{XB}-f_P i_{XP}$		$b_A X_{BA}$
6 Ammonification										1	-1		$\frac{1}{14}$	$k_a S_{ND} X_{BH}$
7 Hydrolysis of Partikulate Organic Compounds		1		-1										
8 Hydrolysis of Partikulate Organic Nitrogen											1	-1		$\rho_7 (X_{ND}/X_S)$

Figure 3.3. Simplification of the IAWPRC Activated Sludge Model No. 1 as implemented by Lessard

As shown in Figure 3.4, the secondary clarifier is modelled in three parts (as suggested by Lessard and Beck, 1993): the *clarification* part, the *thickening* part and the *compression* part, whereby the *clarification* part is subdivided into a fixed volume *clarification* zone and a *dead* zone with variable volume. The latter serves as a buffering zone between the blanket and the fixed volume clarification zone.

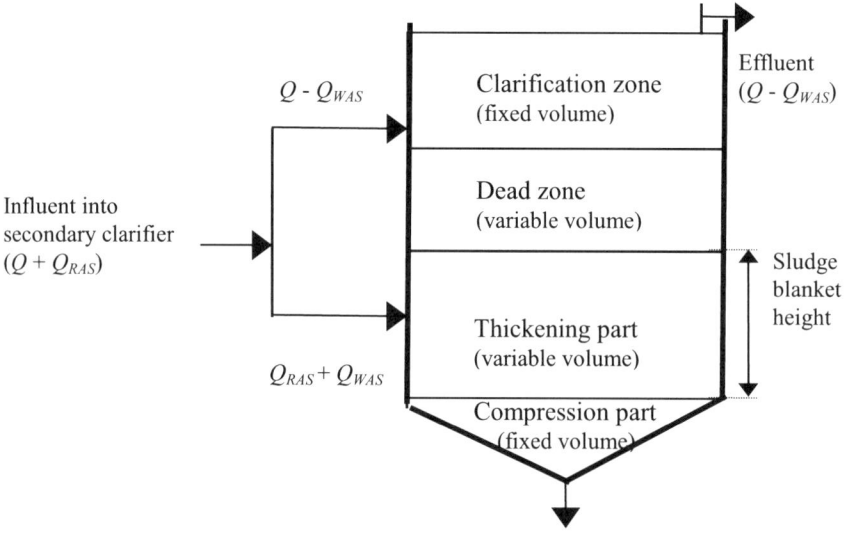

Figure 3.4. Conceptualisation of the secondary clarifier

In the original model of Lessard and Beck, the effluent suspended solids concentration is based on an empirical relationship found by Pflanz (1969) and thus determined by

$$SS_{eff} = a_1 + a_2 \times (Q + Q_{RAS}) \qquad (3.1)$$

where

SS_{eff}: effluent suspended solids concentration [g/m^3]

a_1: minimum effluent suspended solids concentration in the secondary clarifier effluent [g/m^3]

a_2: proportionality constant for the effect of flow on effluent SS [-]

Q: influent flow [m^3/s]

Q_{RAS}: return activated sludge rate [m^3/s]

It can be seen that no account is taken of the influence of the sludge blanket height on the effluent suspended solids. Under the assumption that the influent flow rate to the secondary clarifier as well as the return sludge rate are limited, it follows immediately from (3.1) that the effluent SS concentration is bound by a potentially unrealistic low upper bound. Using parameter values as applied in Lessard's original study, the SS concentration in the secondary clarifier effluent would never exceed the value of 41 mg/l. Modifications of Lessard and Beck's model in light of these observations were carried out by Vazquez-Sanchez (1996) and are also implemented in SYNOPSIS. A summary of the modifications carried out is given below.

The thickening model used by Lessard and Beck is based on the flux theory as presented by Dick and Young (1972) (see the detailed discussion by Lessard and Beck (1993)). The thickening function of the clarifier is modelled as a sequence of two completely mixed reactors: one representing the sludge blanket (variable volume) and one the compression zone (fixed volume). Five state variables are simulated by this model: four soluble substances (biodegradable and non-biodegradable soluble COD, NH_4 and NO_3), and, finally, mixed liquor volatile suspended solids (MLVSS), representing a pollutant capable of thickening.

The sludge blanket is represented by a single CSTR, although it is more usual to find ten or more such elements. However, Lessard (1989) justified this simplification by indicating that after testing the model of Tracy and Keinath (1974), which makes use of ten layers, it was demonstrated that there were no significant differences between the performances of the two representations.

3.2.2.2 Modifications of the Treatment Plant Model

The extensions by Vazquez-Sanchez (1996) allow simulations to be continued also under conditions of sludge overflows and settler failures caused by hydraulic loads. The original version of the program terminated whenever such conditions occurred.

Vazquez-Sanchez' modifications are threefold:

The fixed volume clarification zone is assigned a variable volume with a dynamically changing number of CSTR elements of about equal volume. However, since this modification resulted in occasional jumps in the effluent concentrations when the number of CSTR elements was changed, it is not included in SYNOPSIS.

The dead zone introduced by Lessard in order to describe the effluent ammonium concentration when evaluating his model based on a set of data from the Norwich treatment plant is considered to be inappropriate (Vazquez-Sanchez,

1996). According to Chen (1993), a large dilution zone results in a big increase in mixing and damping of the influent loading. Therefore, this dead zone is excluded from the model.

In order to consider the impact of the sludge blanket (when it is at high levels) on the effluent solids concentrations, Vazquez-Sanchez (1996) proposed and implemented a pragmatic extension to the clarification function by adding a term $y \times VSS_{SB}$ to the right hand side of (3.1). Here, VSS_{SB} denotes the VSS concentration in the sludge blanket, whilst y is defined by

$$y = \begin{cases} \alpha(1-SBDT)^\beta, & \text{if } 0<SBDT<1 \\ 0, & \text{otherwise} \end{cases} \quad (3.2)$$

where

$SBDT$	$SBHT_{Max} - SBHT$ [m]
$SBHT_{Max}$:	maximal sludge blanket height (at this level, sludge is discharged into the effluent) [m]
$SBHT$:	sludge blanket height [m]
α:	dimensionless parameter with $\alpha \in [0;1]$
β:	dimensionless parameter with $\beta > 1$

According to Vazquez-Sanchez (1996), α and β should be calibrated against field data. A trial and error study performed by Vazquez-Sanchez suggests values of $\alpha = 1$ and $\beta = 5$.

This extension of the clarifier function results in the SS effluent concentration being the same as described in (3.1) for low values of the sludge blanket height. The effluent SS concentration increases as soon as the sludge blanket approaches the crest of the secondary clarifier until it reaches a value of approximately VSS_{SB} when sludge is discharged into the effluent. In case of sludge spillage, the volume of the discharge is taken into account by applying a mass balance over the sludge blanket in the thickening submodel. As Vazquez-Sanchez (1996) states, although there are better 1-D conceptual models and also higher dimensional models available which give promising results, the approach presented is considered to be more beneficial within the context of this work due to its simplicity.

As in Lessard and Beck's original version of the program, as input not only the characteristics of the influent for every time step are required, but also the information about the current weather situation (influencing the nature of the sewage and thus the settling velocities in the primary clarifier) and about returns from sludge treatment (not to be confused with return activated sludge from the secondary clarifier) into the primary clarifier. This information is required by the primary clarifier submodule. Within SYNOPSIS, the weather information is passed on to the treatment plant model from the sewer system model. Returns from sludge treatment to the primary clarifier are assumed to take place once every six hours. This setting may appear to be somewhat arbitrary, but resembles in its frequency the pattern arbitrarily chosen by Lessard (1989) for his simulation exercises.

3.2.3 Implementation of the River Module

As the river simulator within SYNOPSIS, the DUFLOW shell program has been selected (see Section 3.1 and Appendix A). This package allows great flexibility since the water quality processes and the related variables and parameters can be defined freely. Hence, future developments within the area of river modelling can be easily implemented within SYNOPSIS.

Since the studies within this work focus mainly on the oxygen balance in the river, an oxygen model developed at Wageningen Agricultural University (Lijklema *et al.*, 1996) is used in a slightly simplified form as the water quality model in this work. Its main feature is that it models organic matter (expressed in terms of BOD) in two fractions (slowly and readily biodegradable). Therefore, the different characteristics in terms of biodegradability of CSO discharges and treatment plant effluents can be taken into account in the simulation. In addition to partitioning BOD in readily (x_{BODR}) and slowly (x_{BODS}) degradable fractions, also dissolved oxygen (x_{DO}) and ammonium (x_{NH4}) are simulated by this model. Algae and nitrate are not modelled in detail within SYNOPSIS due to the emphasis on oxygen within this work. Although in some case-studies algae showed to be of significant importance (*e.g.*, in the Bedford-Ouse study (Beck and Finney, 1987)), other integrated modelling approaches as well do not make use of a detailed representation of algae and nitrate (for example: Rauch, 1996b).

Naturally, the river module requires as input information about flow and concentration of the discharges from CSOs, storm tank and treatment plant effluent. Within SYNOPSIS, this information is provided by the sewer system and treatment plant modules.

The following lists the process equations used in the biokinetic model in this study. Parameters are defined in Table 3.1.

$$\frac{dx_{O2}(t)}{dt} = k_a(DO_{sat} - x_{O2}) + P + S + D + N$$

$$\frac{dx_{BODR}(t)}{dt} = \left[-v_{sedR}\frac{1}{z}(1-f_{dR}) - k_{dR}\theta_{Kd}^{(T-20)}\frac{x_{O2}}{x_{O2}+k_{O2}}\right]x_{BODR}$$

$$\frac{dx_{BODS}(t)}{dt} = \left[-v_{sedS}\frac{1}{z}(1-f_{dS}) - k_{dS}\theta_{Kd}^{(T-20)}\frac{x_{O2}}{x_{O2}+k_{O2}}\right]x_{BODS} \qquad (3.3.)$$

$$\frac{dx_{NH4}(t)}{dt} = \left[-k_N\theta_{knit}^{(T-20)}\frac{x_{O2}}{x_{O2}+k_{NO2}}\right]x_{NH4}$$

where

$DO_{sat} = 14.652 - 0.41022\,T + 0.007991\,T^2 - 0.000077774\,T^3$

DO saturation concentration according to Elmore and Hayes (1960) (as in (2.14))

$P = \alpha\beta$ (photosynthetic term, with α and β as defined below)

$S = -\theta_{SOD}^{(T-20)}\dfrac{SOD}{z}$ (Sediment oxygen demand term)

$D = -\theta_{kd}^{(T-20)}\dfrac{x_{O2}}{x_{O2}+k_{O2}}[k_{dR}BODR_\infty + k_{dS}BODS_\infty]$

(Deoxygenation term)

$N = -4.33 k_N x_{NH4}\theta_{kN}^{(T-20)}\dfrac{x_{O2}}{x_{O2}+k_{NO2}}$ (Nitrification term)[6]

$BODR_\infty = x_{BODR}\dfrac{1}{1-e^{-5k_{dR}}}$ (Ultimate readily biodegradable BOD)

$BODS_\infty = x_{BODS}\dfrac{1}{1-e^{-5k_{dS}}}$ (Ultimate slowly biodegradable BOD)

The parameters of the model are defined in Table 3.1. The values of most of the parameters have been set according to the default values as specified in Lijklema *et*

---

[6] Although the theoretical stoichiometric ratio for oxidation of ammoniacal nitrogen would be 4.57, a value of 4.33 is used here in order to allow for reduced oxygen requirement due to involvement of a small part of ammoniacal nitrogen in cell synthesis. The value of 4.33 was also used by Beck and Finney (1987) and Reda (1996).

al. (1996). The values of k_{O2}, k_N were set to values corresponding to those reported in the review by Bowie *et al.* (1985).

Table 3.1. Coefficients of the river model implemented in SYNOPSIS

Parameter	Description	Unit	Value chosen
k_a	Reaeration	1/day	(see below)
k_{dR}	BODR decomposition rate	1/day	0.60
k_{dS}	BODS decomposition rate	1/day	0.15
v_{sedR}	BODR sedimentation rate	m/day	1.0
v_{sedS}	BODS sedimentation rate	m/day	0.2
f_{dR}	Dissolved fraction BODR	-	1.0
f_{dS}	Dissolved fraction BODS	-	1.0
k_{O2}	BOD half-saturation constant	mgO_2/l	0.5
k_N	Nitrification rate	1/day	0.2
k_{NO2}	Nitrification half-saturation constant	mgO_2/l	2.0
SOD	Sediment oxygen demand	$g/(m^2 \times day)$	2.5
α	Constant weighting factor for photosynthetic activity	mgO_2/l	(see below)
β	Factor for seasonal and diurnal pattern of photosynthesis	-	(see below)
T	Water temperature	°C	(see below)
z	Water level	m	computed in flow part of DUFLOW
Temperature coefficients			
θ_{kl}	for reaeration	-	1.024
θ_{kd}	for BOD deoxygenation	-	1.047
θk_N	for nitrification	-	1.050
θ_{SOD}	for SOD	-	1.047

Although the model allows for BOD removal by settling to be considered, this feature is disabled by the present settings of the f_{dR} and f_{dS} values.

Various formulae for the calculation of the reaeration coefficient as a function of river flow velocity and depth have been presented in Section 2.1.3.3. Although the characteristics of the hypothetical river implemented (see Section 4.1.4) with a dry-weather flow velocity of 0.28 m/s and a depth of 1.58 m would suggest use of the formula by O'Connor and Dobbins (*cf.* Figure 2.9), the formula by Owens *et al.* is implemented here. However, the resulting reaeration coefficient (0.968 day^{-1}) is similar to the one which would be obtained by application of O'Connor and Dobbins' formula (1.034 day^{-1}). During simulation, the value of the reaeration coefficient is calculated at every time step from the current values of velocity and depth. Although more sophisticated modelling approaches for sediment oxygen demand do exist (see Section 2.1.3.3), a simple approach, where SOD is expressed by a constant term, which is normalised by depth, is chosen here.

As discussed in Section 2.1.3.3, the oxygen concentration in the river is also influenced by photosynthesis and temperature. Both of these lead to diurnal and seasonal variations of the DO concentration. Since no long-term data on those parameters commonly employed in descriptions of photosynthetic effects (such as light intensity, algal biomass and a coefficient describing oxygen production by photosynthesis) are available for this study, a simple approach is chosen here. The term defining the photosynthesis coefficient P in (3.3) includes two factors, α and β. The weighting factor α controls the overall extent of the diurnal variation of DO as a consequence of photosynthesis. The factor β describes the influence of sunlight on photosynthesis and is computed according to Equation 3.4.

$$\beta(t) := \begin{cases} \sin(\pi * \frac{(t - t_{sunrise})}{t_{daylength}}), & for\ t \in [t_{sunrise}; t_{sunrise} + t_{daylength}] \\ 0, & otherwise \end{cases} \quad (3.4)$$

where

t: time of the day [h],

$t_{sunrise}$: time of sunrise [h],

$t_{daylength}$: length of the day (hours between sunrise and sunset) [h]

The values of $t_{sunrise}$ (hour of sunrise) and $t_{daylength}$ depend on the date and on the geographical location; these are computed using simple astronomical formulae (Rohr, 1982).

As shown in Figure 3.5, $\beta(t)$ is a half-sine function, which is zero during the night and assumes values between 0 and 1 during the day. Its maximum is assumed at noon. Strictly speaking, maximum sunlight intensity will not be observed at 12 o'clock noon as the definition of β suggests. This is due to use of summer time in the summer months, as well as to corrections necessary resulting from Equation of Time (Rohr, 1982) and differences in longitudes of the location under consideration and the time meridian defining the time zone (for Britain: the Greenwich meridian). However, these effects are considered to be of minor significance within the context of this work.

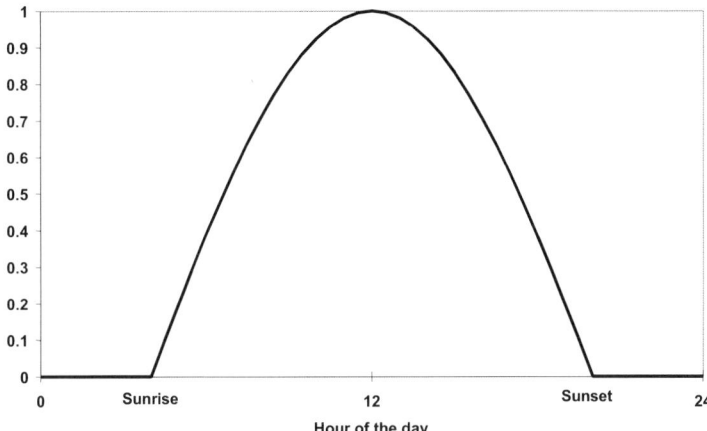

Figure 3.5. The value of β describing photosynthetic activity as a function of time of the day and of time of the year

The approach of expressing the influence of photosynthesis by a half-sine function is widespread (see, for example, Orlob, 1983; Bowie *et al.*, 1985; Rauch, 1996b).

For the river water temperature, a fixed pattern (describing for example diurnal and seasonal variations) can be defined within SYNOPSIS. Alternatively, an external time series can be supplied as an input file. As will be discussed in Section 4.1.4, temperature will, however, be assumed to be constant for the case study of a hypothetical river described in that section.

Since rainfall usually results in increased base flow in the river due to catchment runoff, the simulation tool should allow for this effect to be modelled. Within SYNOPSIS, a simple time-series approach based on Schreider *et al.* (1996) is chosen. The additional base flow is represented by an expression which involves the base flow values of earlier time steps and rainfall values of current and past time steps (see (3.5)). In the paper by Schreider *et al.* (1996), the net amount of rainfall is used for determination of catchment runoff, whereas here, for simplicity, gross rainfall is used. Within SYNOPSIS, consideration of the difference between gross and net rainfall is included in the definition of a simple constant runoff coefficient. An extension to more sophisticated approaches would be easily possible. However, since the case study site of the river is of a hypothetical nature anyway, no attempt is made to consider the catchment runoff processes in more detail.

More specifically, the additional flow in the river caused by rainfall runoff in the river catchment is modelled in SYNOPSIS according to Equations 3.5a and 3.5b

$$Q_n := A \times c_{runoff} \times U_n \qquad (3.5a)$$

with

$$U_n := \sum_{k=1}^{N} \gamma_k U_{n-k} + \sum_{k=1}^{N} \delta_k R_{n+1-k} \qquad (3.5b)$$

where

n: current time step

N: number of time steps considered for catchment rainfall runoff computation

Q_n: additional flow volume in river (generated by catchment runoff) in time step k [m³]

A: Catchment area as used for runoff calculation [ha]

c_{runoff}:   Runoff coefficient [-]

R_k: Rainfall in time step k [mm]

γ_k, δ_k: ($k=1,...,N$) dimensionless time series coefficients, describing the influence of earlier runoff and rainfall values on current runoff value [-]

(U_k, R_k, γ_k, δ_k are assumed to be zero for all k with $k \leq 0$)

3.3 Assembling the Integrated Simulation Tool

Having presented the constituent submodels for sewer system, treatment plant and receiving river in the previous section, this section describes the assembly of these parts into an integrated simulation tool. As will become apparent, merging individual programs into an integrated simulator is more complex than the mere adjustment of the source code. Similar experiences are also described by Rauch (1996a). Section 3.3.1 describes the general aspects of merging the constituent models. One of the main problems encountered when integrating different models is the differences in their sets of variables. Thus, Section 3.3.2 will discuss this issue and derive a set of conversion factors which relate BOD and COD at the various interfaces within the simulation tool. Finally, Section 3.3.3 provides an overview of some of the auxiliary programs which have been developed for pre- and post-processing of simulation runs. These programs provide options for the preparation of a series of simulation runs for sensitivity studies, for exploration of the search space, and for generating graphical output.

3.3.1 Integration of the Simulation Software

As discussed in Section 3.1, one of the key requirements for a simulation tool to be used within this work is its ability to simulate the parts of the urban wastewater system synchronously. This can be achieved in either of two ways. One option is to run the related models as parallel processes in a multitasking operating system (such as Unix or OS/2) (Jamsa, 1988; Young, 1988). Care has then to be taken with process communication and synchronisation. An alternative option is to actually merge the programs into one single package and switch between the programs at every simulation time step and make provision for information transfer between the models. Implementation of the various models as parallel processes appears to be the more elegant solution. However, attempts in this direction on an OS/2 platform led to a significant overhead for process administration and synchronisation, which resulted in poor performance of the simulation. Although it is believed that the implementation could have been improved with regard to its performance, eventually the alternative option of merging the programs was chosen. Besides shorter simulation times, it also provides the advantage of being based on a widely used hardware and operating system platform (PC under DOS/Windows).

The sewer system and the treatment plant programs (*cf.* Section 3.2) were merged and incorporated in one simulation package. However, the programs were connected only at the necessary parts (communication between them; overall control of the simulation; interfaces with the control module – see Section 3.4) and in such a way that the constituent programs still represent (almost) independent units. This provides the advantage that each of these two modules can be exchanged by other programs should this be necessary at a later stage for the inclusion of future modelling developments (*cf.* requirement on modular structure). Ideally, the river model would be incorporated into the simulation model in a similar way. For the time being, the river module is run as a sequential process after the linked simulation of sewer system and treatment plant has finished. Full inclusion of the river model would be possible as soon as the source code is available and the necessary modifications are made. Therefore, the present implementation of SYNOPSIS allows river information to be considered for control actions only to a limited extent: Since information of only that part of the river which is upstream of the sewer system and treatment plant discharges is available prior to the simulation of sewer system and treatment plant, so far only information from upstream river sections can be used for control decisions in sewer system and treatment plant (see also Section 3.4). Although this certainly represents a restriction on the information transfer required between the submodules (*cf.* Section 3.1), the implementation of sewer system, treatment plant and upstream river processes in SYNOPSIS still allows the analysis of a large number of integrated control strategies, as will be presented in subsequent chapters.

When connecting various simulation models, reconciliation of their time steps is also necessary. Each submodule should be run with a time step adequate for description of its processes. At the interfaces between the modules appropriate conversions have to be made. Since the default time step of the sewer model is 5 minutes, which appears to be appropriate for simulation of its processes (including CSO events which may not be described reasonably by time steps of significantly larger size) and since rainfall data for long-term simulation are available in 5-minute time steps, this time step is chosen for the sewer model. Lessard and Beck's treatment plant model was run in various studies (for example, Lessard, 1989; Chen, 1993; Güven, 1995; Vazquez-Sanchez, 1996) with a time step of one hour, which appeared to be an appropriate choice. Therefore, and in order not to carry out too many modifications to this model which was calibrated for a real treatment plant, this time step has also been chosen for the simulation of the treatment plant within this work. Various trial runs were carried out of the river model in order to establish the most appropriate settings of the time steps for river flow and quality simulation. A compromise had to be found between more accurate simulation (using small time steps) and faster simulation (resulting from large time steps). Results from simulations with different time steps suggest the use of a time step of 20 minutes for both, quantity and quality simulations.

Due to the different time steps of the constituent models, the results provided by the sewer model (in time steps of 5 minutes) are averaged accordingly before being passed over to the treatment plant and river models. Since the river model uses a smaller time step than the treatment plant model, input data to the river model from the treatment plant is linearly interpolated when passed on to the river model.

3.3.2 Variables in SYNOPSIS

Since existing simulation models with different sets of process variables were selected for use within this work, consideration has to be given as to how the variables used in the various submodules can be converted to related variables used in the other submodules. Clearly, such an approach appears is far from ideal and may be, in fact, one of the weakest points in integrated modelling. However, despite its inelegance, such an approach has been chosen for most projects involving integrated modelling (see Section 2.3).

Since the starting point for the development of SYNOPSIS is the treatment plant model, which is based on a real treatment plant, and since the sewer model allows any pollutant to be simulated, the set of variables of the treatment plant model will also be used within the sewer system model. Thus, flow, suspended solids (SS), volatile suspended solids (VSS), total COD (COD_t), soluble COD (COD_s), ammonium (NH4) and nitrate (NO_3) are simulated in these two submodules. Since the river model is based on the oxygen model of Lijklema *et al.* (1996) (see Section 3.2.3), its variable set, consisting of flow, ammonium (AMM), BOD in two fractions (BOD_R, BOD_S) and DO is used for the river submodel. Figure 3.6 gives an overview of the variables used in the various submodules of SYNOPSIS.

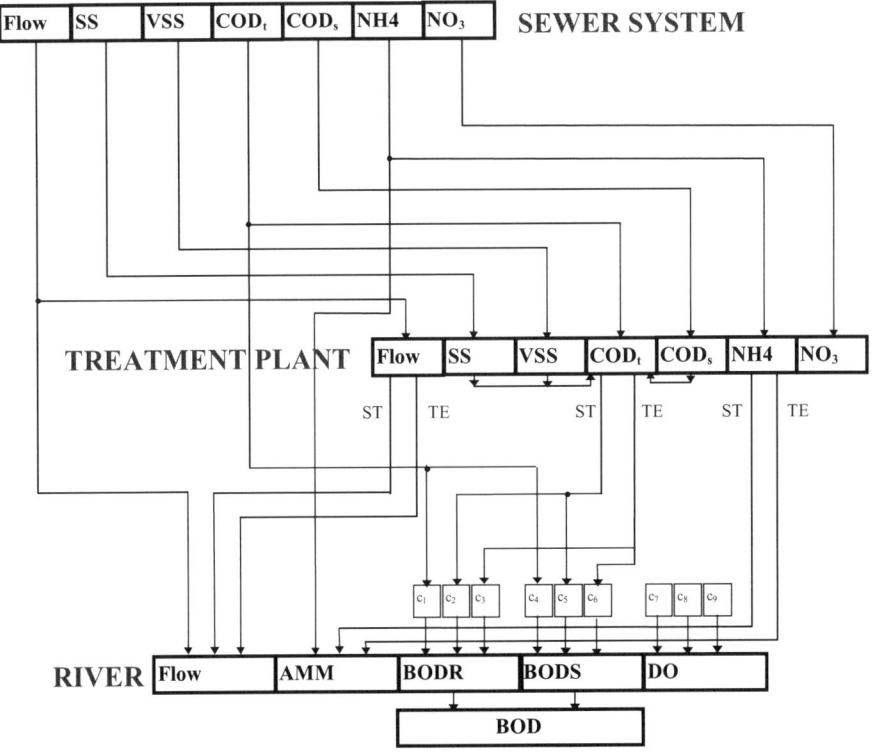

Figure 3.6. Variables in the simulation modules of SYNOPSIS and their interactions.
n.b.: Storm tank discharges are denoted by ST; the treatment plant effluent is denoted by TE

As can be seen from this figure, a number of conversions of state variables (indicated by c_1 to c_6) becomes necessary at the interfaces (see Table 3.2). The boxes marked c_7 to c_9 depict assumptions made about the DO concentration of the various discharges, which will be discussed in Section 4.1.4.

The factors c_1 to c_6 will be applied as follows for the conversion of COD into the BOD fractions.

BODR	CSO discharges:	$BODR = c_1 \times COD$
	Storm tank discharges:	$BODR = c_2 \times COD$
	Treatment plant effluent	$BODR = c_3 \times COD$
BODS	CSO discharges:	$BODS = c_4 \times COD$
	Storm tank discharges:	$BODS = c_5 \times COD$
	Treatment plant effluent	$BODS = c_6 \times COD$
BOD	Total BOD	$BOD = BODS + BODR$

The derivation of conversion factors presented here starts from values of BOD/COD ratios derived from data found in the literature. The range of values found indicates that this ratios is different for different wastewaters at different locations. Table 3.3 presents these values.

Table 3.3. BOD/COD ratios derived from the literature (after Wan (1997))

Author	Raw Sewage	Settled sewage/ Primary effluent	Final treatment plant effluent
Tebbutt (1992)	0.43	0.44	0.22
Tchobanoglous and Burton (1991)	0.44	-	-
Hall and Ellis, after Wan (1997)	0.17 - 0.36	-	-
Ranchet and Philippe, after Wan (1997)			
- Wet-weather	0.25	0.31[1]	-
- Dry weather	0.54	0.60[1]	-

[1] After 2 hours of quiescent settling

Other relationships between BOD and COD include those involving a constant term in a linear equation. Ademoroti (1986) established the equation

$$BOD = 0.70 \times COD - 6.9.$$

Since the river model applied in this work distinguishes between readily and slowly biodegradable fractions of BOD (*i.e.* BODR and BODS), more specific conversions have to be found in order to obtain BODR and BODS values from COD information available from the treatment plant model.

As a crude approximation, the soluble readily biodegradable (S_S) fraction of COD (see Section 2.1.2.3) can be considered to correspond to BODR and the sum of the particulate biodegradable (X_S) fraction and the biomass (X_H, X_A) as the BODS (*cf.* Maryns, 1996; Rauch, 1996b). Assuming that this approximation and, at the same time, the overall ratios of BOD to COD as shown in Table 3.3 are valid, the conversion factors can be derived as follows, making the following two definitions.

$$X_R = X_S + X_H + X_A \quad \text{(slowly biodegradable COD fraction and biomass)}$$

$$\text{and} \quad COD_I = S_I + X_I \quad \text{(inert fraction of COD)}$$

Assuming that the total COD is the sum of the S_S, X_R and COD_I fractions and by partitioning the inert COD fraction in equal parts to the BODR and BODS fractions, one obtains

$$BODR = a \times (S_S + 0.5 \times COD_I) \quad \text{and} \quad BODS = a \times (X_R + 0.5 \times COD_I),$$

where a denotes the overall ratio assumed for BOD/COD for various wastewaters (taken from Table 3.3).

Thus BODR + BODS = BOD is ensured. Using the data of Table 3.4 for the COD fractions for various types of wastewater, one obtains the BODR and BODS conversion factors as shown in Table 3.5.:

Although the underlying assumptions of this derivation might leave potential for criticism, some conversion factors will have to be assumed within this work. However, since these factors are supplied to the simulation tool as input, these can easily be changed during a simulation exercise. Sensitivity studies by Wan (1997), varying some of the conversion factors singly and jointly, showed that results obtained for DO and BOD concentrations in the river are mostly affected by the total BOD/COD ratio rather than by the individual conversion factors defined for the readily and biodegradable BOD fractions.

Table 3.4. Percentages of the various COD fractions for different types of wastewaters

Location	S_I	X_I	S_S	X_S	X_H	Source
Raw sewage						
Various European countries	20-25		20	35–45	15–20	Rauch and Harremoës (1996a), summarising data given by Henze (1992)
Simulation study	7	20	18	37	18	Härtel et al. (1995), Kollatsch (1995)
Husum/Germany	4	18	13	65		Scheer (1995)
	20		20	60		used in this work
Settled domestic sewage						
Denmark	8	19	24	49		Adapted from Henze et al. (1986)
Switzerland	12	11	32	45		Adapted from Henze et al. (1986)
Hungary	8	20	29	43		Adapted from Henze et al. (1986)
Simulation study	8	19	17	34	22	Härtel et al. (1995), Kollatsch (1995)
Zürich/Switzerland	9	8	14	47[1]	22	Sollfrank and Gujer (1991)
UK	10	18	29	43		Lessard (1989)
Switzerland	5	18	19	58		Gujer (c.1990)
		21	26	53		used in this work
Treatment plant effluent						
Dry weather	80–85		3–5	0	8–15	Rauch and Harremoës (1996a): Simulation of the effluent of a low loaded activated sludge plant)
Wet-weather	50–80		2–10	0	15–35	
Dry weather	**84**		**4**	**12**		**used in this work**
Wet-weather	**69**		**6**	**25**		**used in this work**

[1] Including rapidly and slowly hydrolysable material

Table 3.5. Derivation of conversion factors for BODR and BODS

	a (BOD/COD ratio, taken from Table 3.3)	Readily biodegrad. fraction of COD (S_S)	Slowly biodegrad. fraction of COD (X_R)	Inert fraction of COD (S_I+X_I)	BODR conv. factor (1)×[(2)+ 0.5×(4)]	BODS conv. factor (1)×[(3) +0.5×(4)]
	(1)	(2)	(3)	(4)	(5)	(6)
CSO	0.44	0.20	0.60	0.20	**0.13**	**0.31**
Storm tank	0.44	0.26	0.53	0.21	**0.16**	**0.28**
Treatment plant						
- dry weather	0.22	0.04	0.12	0.84	0.10	0.12
- wet-weather	0.31	0.06	0.25	0.69	0.13	0.18
- mean	---	---	--	---	**0.11**	**0.15**

3.3.3 Auxiliary Routines Necessary for Simulation

Besides the core simulation package, consisting of sewer system, treatment plant and river modules as outlined above, also a number of auxiliary routines has been implemented in SYNOPSIS. Among them are modules computing external inputs to the simulation model, such as river catchment runoff and the time and date dependent photosynthesis coefficient (as discussed in Section 3.2.3). Furthermore, several routines have been implemented to perform the transformation of variables at the interfaces between the simulation submodules (see the previous Section 3.3.2).

Additional pre- and post-processing modules provide various options which proved to be useful for the simulation runs, the results of which are presented in subsequent chapters.

These options include the following:

- simulation of single events, series of individual events and continuous long-term rain series;
- generation of optional graphical time series output (relevant flows and concentrations at various locations in sewer system, treatment plant and river) (examples are shown in Sections 4.2 and 4.3);
- evaluation of simulation runs, including computation of the various criteria suggested as objective functions (see below);

- preparation of gridding runs in the optimisation space (calculation of the objective function under variation of its arguments (strategy parameters); *cf.* Section 5.2);
- generation of optional graphical output for gridding runs in the optimisation space. A summary of simulation results is also provided in tables, which also include some elementary statistics (mean value and standard deviation of the objective function values; sensitivity with regard to variation of strategy parameters; *cf.* Section 5.6).

From the simulation results obtained from the sewer system, the treatment plant and the river submodules, a variety of values are computed by the auxiliary routines, which are available as options for the definition of the objective function for the optimisation (*i.e.* as criteria which can be used for the assessment of control strategies). These represent conventional criteria (such as CSO discharge volumes and loads) and river water quality based criteria (resembling, though in a simplified form, those criteria which are defined in the UPM Manual; *cf.* Section 2.2.4). Values available at the interface to the optimisation module include:

DO-M [mg/l]: Minimum concentration of dissolved oxygen in the river (minimum over all simulation time steps and all locations). Taking the minimum over all river locations avoids the problem of the results being dependent on the definition of chosen reference locations (as, for example, in Reda, 1996).

DO-DU [%]: Duration of the time periods when – at any one location in the river – the DO concentration is below a pre-defined threshold value (here: 4 mg/l). DO-DU is expressed as a percentage of the total time period covered by the simulation.

DO-E [mg/l]: Six-hours mininum of DO concentration (taken over all locations) (see Figure 3.7)

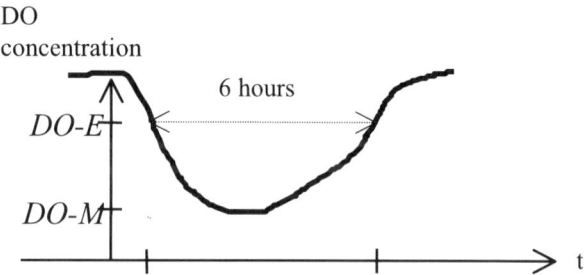

Figure 3.7. Definition of the *DO-M* and DO-E criteria

DO-E is the concentration which is the lowest concentration which has been exceeded for a <u>continuous</u> period of 6 hours at any location in the river at least once during the entire simulation period. The calculation of this value has been motivated by the Fundamental and Derived Intermittent Standards of the UPM Manual (see Section 2.2.4). The result files containing detailed simulation output also provide information about when and where in the river this exceedance value has been reached.

AMM-M [mg/l]: analogous to *DO-M*, but for ammonium

AMM-DU [%]: analogous to *DO-DU*, but for ammonium

AMM-E [mg/l]: analogous to *DO-E*, but for ammonium

BOD-M [mg/l]: analogous to *DO-M*, but for BOD

BOD-E [mg/l]: analogous to *DO-E*, but for BOD

F2 [-]: The definition of *F2* attempts to combine the values of *DO-M* and *DO-DU*, thereby expressing information about extreme DO concentrations as well as duration of low-DO periods in one single variable. This is a formalisation of an approach applied by Reda (1996) for ranking strategies.

More specifically, the value of *F2* is defined as follows:

$$F2(x_{DO-E}, x_{DO-DU}) := \begin{cases} x_{DO-M}, & \text{if } x_{DO-M} \geq DO_{thr} \\ DO_{thr} * (1 - \frac{x_{DO-DU}}{100}), & \text{otherwise} \end{cases}$$

where

x_{DO-E}: six-hour minimum of DO, as defined above (*DO-E*) [mg/l]

x_{DO-DU}: duration value *DO-DU* [%]

x_{DO-M}: DO-minimum concentration *DO-M* [mg/l]

DO_{thr}: a prespecified threshold value (here: 4 mg/l)

This definition implies that *F2* is equal to the minimum DO concentration in an (as a first, rough assumption) uncritical situation (DO above DO_{thr}). If the DO concentration drops below this critical threshold, then the duration of breaching the threshold value is of interest. The interval of [0; 100] (%) is mapped into the interval [0; 4] (mg/l) as shown in Figure 3.8. Large *F2* values represent uncritical states of the river with respect to the DO balance.

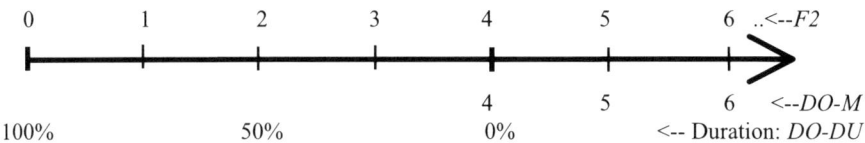

Figure 3.8. Definition of the F2 criterion

F3 [-]:   Analogous to the definition of *F2*, but for ammonium. Uncritical states of the river (with respect to the ammonium balance) are represented by small *F3* values.

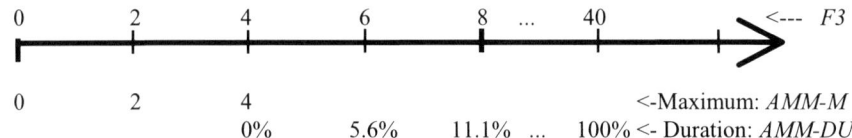

Figure 3.9. Definition of the F3 criterion

Furthermore, the following functions, based on overflow volumes and COD load, are calculated:

Q-CSO [m³]: Total overflow volume of CSOs in the sewer system

Q-ST [m³]:   Total overflow volume of storm tanks at the treatment plant

Qoverfl. [m³]: Sum of *Q-CSO* an*d Q-ST*

CODtot [kg]: Total COD load discharged into the river (from CSO, storm tank, treatment plant effluent)

CSODur [%]: Duration of CSO discharges as a percentage of the total time period covered by the simulation. Note that the smallest time step considered here is 1 hour.

Evaluations in this book will focus mainly on *F2* and *F3*, as these describe the DO and the ammonium balances in the river.

Figure 3.14 in Section 3.6 provides an outline overview of SYNOPSIS. This figure also shows how the optimisation routines, described in Section 3.5, are embedded into the tool.

3.4 Implementation of Control in SYNOPSIS

Among the requirements of a simulation tool to be used within this work is its ability to perform control actions. Since provisions are to be made for consideration of information available from all subsystems of the urban wastewater system for taking the control decision, the control module is kept as a unit which is separate from the simulation modules. System state information from all subsystems is received as input to the control module, which then takes a control decision at every control time step. The control decision taken is communicated to the simulation modules, affecting the settings of the control devices (which are listed in Table 3.7).

Table 3.6 provides an overview of system state (sensor) information which is communicated from the simulation modules to the control module. An attempt has been made to include those state variables which can be obtained by commonly applied sensors (*cf.* Section 2.4.4) and which are also available from the simulation programs implemented.

The table also contains the acronyms for the various kinds of sensor information used throughout later sections of this book. The sensor information listed in this table refers to the case study implemented, which will be described in Section 4.1.

Obviously, this list could be extended. In particular past and predicted future states could be included. Although implementation of these would be easily possible once an appropriate predictive model is included, this is not done here, since Table 3.6 already provides a fairly large number of sensor values, which can be used in control strategies to be defined and tested (*cf.* Chapter 5).

Table 3.6. Sensor information available to the control module of SYNOPSIS

Description	Acronym	Unit
Water level (degree of filling) in Basin 2 of the sewer system	B2	%
Water level (degree of filling) in Basin 4 of the sewer system	B4	%
Water level (degree of filling) in Basin 6 of the sewer system	B6	%
Water level (degree of filling) in Basin 7 of the sewer system	B7	%
Water level (degree of filling) in storm tank	BST	%
CSO discharge	QCSO	l/s
Influent flow rate to the treatment plant	QINTP	l/s
COD concentration of treatment plant influent	CODINF	mg/l
COD load of treatment plant influent	CODINL	t/d
NH_4 concentration of treatment plant influent	NH4INC	mg/l
NH_4 load of treatment plant influent	NH4INL	t/d
VSS concentration of return activated sludge	VSSRAS	mg/l
Mixed liquor volatile suspended solids (last CSTR	MLVSS4	mg/l
Sludge blanket height	SBHT	m
SS concentration of secondary clarifier effluent	SSCEFF	mg/l
COD concentration of secondary clarifier effluent	CODEFF	mg/l
NH_4 concentration of secondary clarifier effluent	NH4EFF	mg/l
Upstream river flow (multiple of plant effluent during DWF)	RIVDIL	-
Net rainfall in previous hour	RAIN00	mm

In a similar way, Table 3.7 shows those control devices within the urban wastewater system which can be set by the control module of SYNOPSIS. Thus these devices are set at every control time step according to the state of the urban wastewater system as described by the sensor information according to Table 3.6.

Table 3.7. Devices which can be set by the control module of SYNOPSIS

Description	Acronym	Unit
Pump P2 (maximum outflow from B2)	P2	× DWF
Pump P4 (maximum outflow from B4)	P4	× DWF
Pump P6 (maximum outflow from B6)	P6	× DWF
Pump P7 (maximum outflow from B7)	P7	× DWF
Maximum inflow rate into primary clarifier	PCINMX	× DWF
Emptying rate of storm tank	STTPQ	m^3/h
Threshold TP influent flow rate initiating emptying of storm tanks	STTPTH	m^3/h
Return activated sludge rate (absolute value)	RASABS	m^3/h
Return activated sludge rate (as a fraction of TP inflow)	RASPCT	-
Waste activated sludge rate	WASABS	m^3/h

n.b.: Pump rates and maximum inflow rate to the primary clarifier are expressed in multiples of average dry-weather flow

It should be noted that SYNOPSIS allows the return activated sludge rate to be defined either in absolute terms (RASABS: m^3/h) or as a fraction of the inflow rate (RASPCT). Both options will be studied in Chapters 4 and 5.

It can be seen that aeration and step-feed control options are not included in Table 3.7. Step-feed and aeration control are not considered here, since investigation of step-feed control would require major modification of the treatment plant simulation module. Since Lessard and Beck's treatment plant model does not model DO in the aeration tank in detail (see Section 3.2.2), aeration control is not investigated here. However, it is assumed that an oxygen level appropriate to the processes in the aeration tank is maintained, e.g. by local controllers (as this is successfully done in many treatment plants).

In order to synchronise the information gathered from the various submodels, which are run with different time steps (see Section 3.3.1), all sensor information used for control is averaged over one hour. Shorter time periods, which would indeed be necessary for sophisticated control of the sewer system, for example, could easily be implemented. However, since the treatment plant model is run with a time step of one hour and since the focus of this study is not the detailed analysis of sewer system operation, this simplification of using values which are averaged over one hour is believed to be justified. Following this argumentation, a control time step of one hour is assumed throughout this book.

For the demonstration of the simulation tool in Chapter 4, a very simple mode of control (employing constant set-points) will be applied. How the control module actually takes a control decision as a function of the system state will be detailed in Chapter 5 of this book, where more sophisticated control strategies will be described and studied in detail.

The next section describes how the optimisation routines selected in Section 2.5 are implemented and linked to the simulation tool. Section 3.6 provides a summary of the SYNOPSIS package developed.

3.5 Optimisation Algorithms in SYNOPSIS

As outlined in Section 2.5 (*cf.* also Figure 2.16), the optimisation procedure to be applied within the present work calls the simulation module as a means of evaluating the objective function of the optimisation problem. Therefore, the optimisation routines and the simulation tool have to be linked to each other. In the discussion of various optimisation methods (Section 2.5), the Controlled Random Search (CRS), a genetic algorithm and Powell's method for local optimisation have been identified for implementation in the simulation and optimisation tool for this work. These algorithms are presented here in some detail (Sections 3.5.1 to 3.5.3). Section 3.5.4 finally discusses how the simulation tool discussed above is embedded into the optimisation procedure.

3.5.1 Controlled Random Search

For minimisation of a $\Re^n \to \Re$ function f, subject to $x \in X \subseteq \Re^n$, where X is the feasible region, defined for example by lower and upper boundaries for the components of x, the CRS algorithm works as follows (*cf.* also Figure 3.10, which sketches the CRS algorithm described by Price (1979)).

In the initialisation stage of the algorithm, m points are randomly sampled from the feasible region. Price (1983) suggests $25n$ as a value of m, unless there are reasons to choose otherwise. These selected points constitute the initial set of trial points $A \in \Re^n$, the set $B \in \Re$ contains the function values of the elements of A. In the main loop, a new trial point y is computed by reflection of a randomly selected element of A on the centroid of n randomly selected elements of A. If the trial point lies within the feasible domain, the objective function is evaluated for this point. If this objective function value is smaller than the largest objective function value in B, y becomes a member of A, replacing the element of A with the previously largest objective function value. Similarly, $f(y)$ replaces the largest element of B. This procedure is repeated until a stop criterion is met. The algorithm terminates when either a maximum number of function evaluations is exceeded or when the ratio of the largest to the smallest objective function value in B is lower than a specified value (thus indicating that no substantial further minimisation of the function value can be expected). The element of A corresponding to the lowest value of constitutes the solution suggested by this algorithm as the global minimum of the function f.

Various modifications suggested by Price (1983; 1987) are related with the computation of the trial point in the inner loop of the algorithm. These modifications are reported to possibly lead to faster convergence of the algorithm. However, also premature convergence (to a non-global minimum) was observed. Therefore, these modifications are not implemented here.

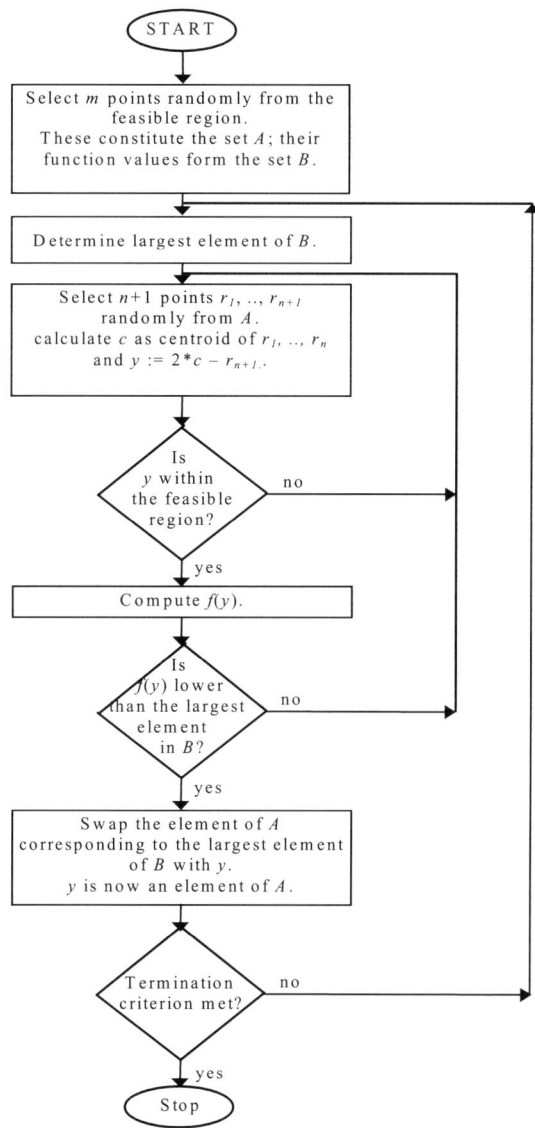

Figure 3.10. Flow chart of Controlled Random Search algorithm

3.5.2 A Genetic Algorithm

Another global optimisation procedure implemented here is a genetic algorithm procedure according to Carroll (1996). Figure 3.11 outlines this algorithm as implemented here.

The present implementation of a genetic algorithm aims at maximisation of a $\Re^n \to \Re$ function f, subject to $x \in X$, where the feasible domain X is defined by lower and upper boundaries of the components of x. Since the arguments of the objective functions are coded in genetic algorithms by strings, the arguments of the objective function are required to assume discrete values only. Most genetic algorithms, including the present implementation, use binary encoding. Use of other bases for encoding of the numerical values of the individuals is also reported to be successful (Janikow and Michalewicz, 1991).

The discretisation necessary for application of the genetic algorithms is uniform over the range of each component. Since binary representation of the individuals is applied, the number of discrete values of each argument is preferred to be a power of 2. Discretisation of the argument domain does not necessarily constitute a disadvantage, since in most applications (such as in the definition of the parameters of a control strategy), feasible values from a continuous range of values can be discretised as long as a sufficiently fine discretisation is chosen. For some parameters of a control strategy the use of discrete values may be prescribed anyway (for example, this is the case for a pump which is operated at several discrete pump rates).

The algorithm starts with a random selection of m elements of the (discretised) parameter space. These constitute the initial generation. Each time the main loop of the algorithm is iterated, a new generation is assembled and evaluated. The loop starts with evaluation of the objective function for all individuals of the current population. The objective function value determines the fitness value of each individual. The simplest, but not the most efficient (Wong, 1994) way to define the fitness function would be to set it equal to the objective function. In the present implementation, a special procedure ("niching"; see Beasley et al., 1993b) is applied, which modifies the fitness values in such a way that the algorithm is prevented from locating only one maximum, when several maxima exist with the same function value.

In the next step, pairs of individuals from the current population are selected randomly, taking into account the fitness values of the individuals, thus giving "fitter" individuals a higher chance of reproduction. With a given probability (p_x), the bit-strings of these "parents" are crossed-over, generating a new bit-string, which will become a member of the next generation. If no crossover is applied, one of the parents is "reproduced" into the next generation. According to the survey by Beasley *et al.* (1993b), none of the different options described in the literature for the crossover process (*e.g.* 1-point, 2-point, multi-point, uniform) appears to be significantly superior than the others. Uniform crossover, *i.e.* crossover by randomly swapping individual bits of the parents' bit strings, is applied in the present implementation.

In the next stage of the algorithm, "mutation" takes place for each "child" after crossover. It randomly alters each bit with a small probability (p_m). Additionally, also creep mutation (Beasley *et al.*, 1993b) is applied in this implementation, changing (with a small probability, p_c) not the bit representation of the individual directly, but its numerical value.

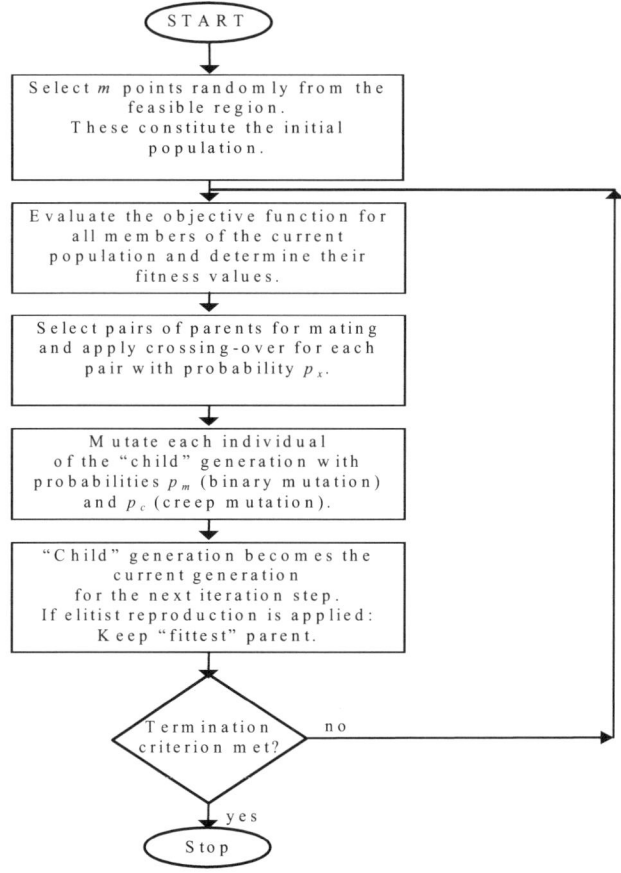

Figure 3.11. Flow chart of the genetic algorithm implemented in the present work

If elitist reproduction is chosen, the "fittest" parent is now reproduced into the "child" generation (see also Davis, 1991). Elitist reproduction ensures that a suggested solution is never worse than the best individual of the initial population. As long as the termination criterion is not met, the main loop is repeated, using the "child" generation as the new "parent" generation. Similarly to many other genetic algorithms, this algorithm terminates when a maximum number of generations (*i.e.* a maximum number of function evaluations) is reached. An alternative termination criterion could consist, for example, of a measure for non-improvement of the results as compared to the last generation.

Both key processes, crossover and mutation, are considered to be essential for this evolutionary procedure. According to Beasley *et al.* (1993a), crossover is traditionally seen as the more important process for rapid exploration of the search space, whereas mutation is seen as a means to include a random search element into the procedure, thus ensuring that no part of the feasible domain is prematurely left out in the search process. Thus, neither of these two processes should be left out in the overall process. It seems to be surprising that such a simulation of the crossover and mutation processes found in nature is quite successful in many application areas (see above). The main theoretical result supporting this observation is the schema theorem by Holland (1975). From this theorem it follows that the optimum way to explore the search space is to allocate reproductive trials to individuals in proportion to their fitness relative to the rest of the population. This resembles Darwin's "survival of the fittest" principle. The main idea of a genetic algorithm is that reproduction of "good" (in terms of the fitness function) patterns of binary strings (schemata) may result in even better patterns.

A modification of conventional genetic algorithms, named "micro GA", is proposed by Krishnakumar (1989) and is implemented as an additional option for the present work. The micro GA is characterised by a very small population size (in the order of 5, as opposed to sizes of around 50 or 100 as used in most conventional applications of genetic algorithms). Mutation is not performed in the micro GA. In all other aspects, the micro GA follows closely the procedure shown in Figure 3.11. Whenever the bit representations of the individuals of the population are very similar to each other, the algorithm is restarted, whilst the fittest population member is carried over to the new initial populations, the other individuals of which are selected randomly from the feasible domain. The micro GA requires a significantly lower number of function evaluations for optimisation of the functions tested by Krishnakumar (1989). Therefore, it appears to be an interesting alternative to conventional genetic algorithms, in particular in a context where the evaluation of the objective function is computationally demanding. Higher performance of a genetic algorithm with a lower number of individual per generation is also reported by Kuczera (1997).

Among the most common problems encountered in the application of genetic algorithms are the definition of the various parameters of the algorithm itself (such as population size and the probability values p_x, p_m, p_c, values of which are given in Table 3.8) and the definition of the fitness function. By default the objective function of the optimisation problem to be solved can be used as the fitness function. However, the performance of a genetic algorithm can be significantly improved if the fitness function, which controls the selection of individual during the evolutionary process, is defined in a prudent way. Various options are discussed by Beasley *et al.* (1993a). In the present implementation, the objective function is used as fitness function, in combination with the niching procedure discussed above.

Table 3.8. Algorithmic parameters used in various studies on genetic algorithms

	A	B	C	D	E	F	G	H	I	J	K
Crossover probability p_x	1.0	0.6-1.0	0.5 [2]	0.8	0.75	0.65	0.6	0.8	0.7	0.5	0.5
Mutation probability p_m	0.01	[1]	[3]	0.01	0.001	0.008	0.001	0.025-0.05	.002	0.02	0.02
Creeping probability p_c	n/a	n/a	[4]	n/a	n/a	n/a	n/a	n/a	n/a	0.07	0.07
Population size m	50-100	50-200	50-100	100	30	40-100	*	*	80-100	50	5 and 20
Number of generations	500-10^4	*	25-50	<1000	100	20-30	*	*	100	[5]	[5]

*: The references do not give details of these values.
[1]: 1/length of bit-string; [2]: uniform crossover; [3]: 1/population size
[4]: length of bit string/(dimension of search space × population size); [5]: various

A: Savic (1997)
B: Savic and Walters (1997)
C: Carroll (1996)
D: Stanic and Avakumovic (1996)
E: Milutin and Bogardi (1996)
F: Rauch and Harremoës (1999a)
G: Liong *et al.* (1995)
H: Cieniawski *et al.* (1995)
I: Wong (1994). Wong used various combinations of values. Those reported here are said to represent those typically used in genetic algorithms.
J: parameters used in this work
K: parameters used in this work for the micro-GA

An advantage of genetic algorithms is that they seem to be ideally suited to implementation on parallel computers, since evaluation of the objective function for the individuals of the current generation can be done in parallel, without interaction between the evaluation processes being necessary. Due to the evaluation of a population size greater than 1, genetic algorithms are particularly well suited to optimisation problems with several optima. A disadvantage of genetic algorithms is the potentially high number of function evaluations required for the determination of an optimum solution. However, the number of function evaluations can be influenced by prudent definition of the parameters of the algorithm.

Since the definition of the probability values used in genetic algorithms is of significant importance to the performance of a genetic algorithm, values chosen in various studies are given in Table 3.8.

3.5.3 Powell's Local Optimisation Method

As discussed in Section 2.5.2.1, Powell's optimisation method can be classified as a derivative-free local minimisation procedure.

The essence of Powell's method is to perform, as the first sub-step of each iteration, a sequence of line searches (minimisations of functions of one variable) from a given point in order to find the search direction of the next iteration step. Then as the second sub-step, a line search is performed along this search direction in order to determine a new iterated point. Thus, the minimisation of a $\Re^n \rightarrow \Re$ function is being substituted (in each iteration step) by a series of minimisations of $\Re \rightarrow \Re$ functions. Figure 3.12 provides an outline of the algorithm, the implementation of which follows Press *et al.* (1994).

When applying Powell's method, it has to be ensured that the search directions applied in the first sub-step do not become linearly dependent, since, if they did, the feasible space would no longer be searched in its entirety. One option to ensure linear independence would be to use the unit vectors as the set of search directions. In order to be able to use information about search directions found in previous iterations, Brent (1973) suggests a different method to guarantee orthogonal search directions. However, this method is computationally more expensive, since it is based on a singular value decomposition of the matrix of search directions. On the other hand, compared to the computational cost of the evaluation of the objective function within this work, the computational increase due to the matrix decomposition would be marginal. However, in this work, a simpler method as discussed by Acton (1970) and Press *et al.* (1994), which, for certain functions, gives up the property of quadratic convergence of Powell's original algorithm, is implemented. For the line searches within this algorithm, a procedure proposed by Brent (see also Press *et al.*, 1994) is used. It consists of a combination of parabolic interpolation and golden section search.

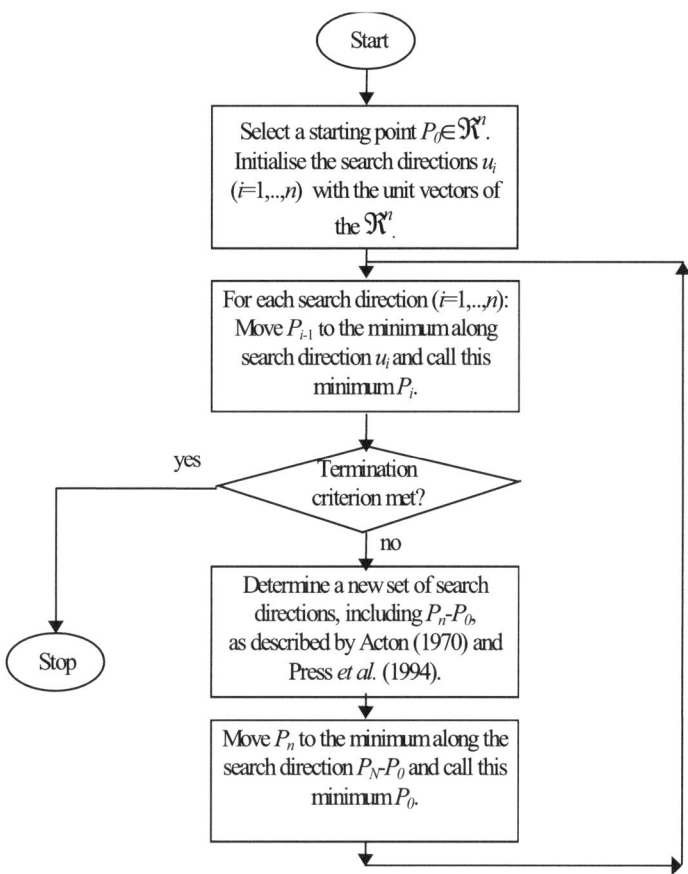

Figure 3.12. Flow chart of the local optimisation procedure implemented

The algorithm is terminated when either a maximum number of iterations is exceeded or no significant further minimisation is to be expected. Definition of the termination criteria of the overall algorithm as well as of the line-search subroutine is of crucial importance to a successful application of the algorithm. With the termination criteria inappropriately set, the search may either converge prematurely or be unnecessarily extended. The definition of the termination criteria has also to take the desired accuracy of the result into account.

It should be noted that Powell's algorithm is designed for functions without constraints on its variables. If constraints are to be considered in the optimisation problem (as it will be the case within this work), a constrained optimisation problem has to be transformed into an unconstrained one. For such a transformation, either penalty functions (adding a term to the objective function, "penalising" leaving the feasible domain) or barrier-functions (resulting in infinite values of the objective function on the boundary of the feasible domain) are commonly chosen (Fiacco and McCormick, 1968; Ryan, 1974; Swann, 1974; Gill *et al.*, 1981; Grosche *et al.*, 1984. Alternatively to modifying the objective function, also a modification of the search procedures within the optimisation algorithm would be perceivable. However, these usually increase the complexity of the algorithm significantly (Swann, 1974; Gill *et al.*, 1981).

In the present implementation, however, constraints (which are variable bounds) are optionally ignored (often the minimum determined by the procedure is found to lie within the feasible region) or a transformation is performed on the arguments of the objective function: All arguments x_i ($i=1,...,n$), which are constrained to lie within an interval $[a_i;b_i]$, are projected by the transformation (3.6) onto the entire real axis, before the optimisation procedure (now being unconstrained in all arguments) is called.

$$g(x_i) = \tan(-\frac{\pi}{2} + \pi \frac{x_i - a_i}{b_i - a_i}) \tag{3.6}$$

This procedure is simple; however, in some of the optimisation runs carried out in Chapter 5 it led to problems when the search was conducted near the boundaries of the feasible region.

3.5.4 Interfacing the Simulation Tool with the Optimisation Routines

The three optimisation routines described above have been implemented in the optimisation module of SYNOPSIS. On start of the optimisation process it has to be specified which of the three optimisation routines is to be used. Chapter 5 discusses in detail which optimisation routine was most beneficial within this work. Here, the interfaces of the simulation and optimisation modules are described. The general concept of linking these two main modules is depicted in Figure 2.16 which, for convenience, is shown here again.

Development of SYNOPSIS   173

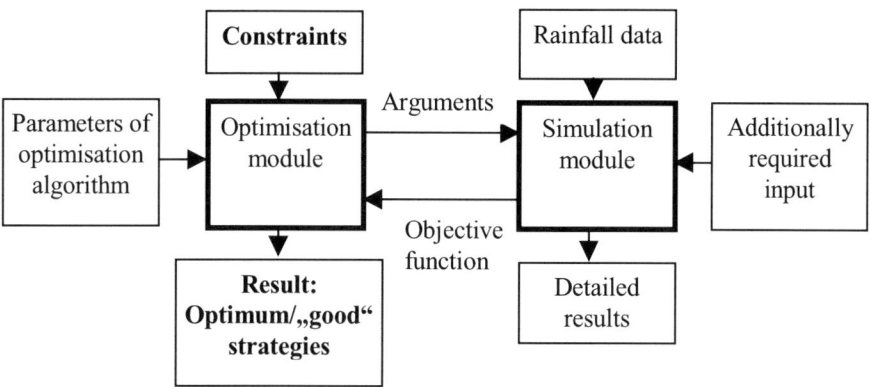

Figure 3.13 (2.16). The role of the simulation tool within the optimisation procedure

The optimisation and the simulation programs are implemented as two separate programs. The alternative approach of merging the two programs and calling the simulation as a subroutine did not appear to be feasible for two reasons. Merging these two programs would make the use of any other simulation tool within the optimisation procedure considerably more difficult (*cf.* list of requirements in Section 3.1). Furthermore, merging the two programs would exceed memory limitations, since both optimisation as well as simulation tool are programs with substantial demand on memory. Therefore, it was decided to implement them as entirely separate programs, which communicate by files. The overall control of the simulation–optimisation process rests at operating system level. It may be argued that communication by files is not an efficient means of information exchange between the programs. However, compared to the computation time required for evaluation of the objective function (*i.e.* performing a run of the simulation module), the amount of time required for reading and writing the communication files is negligible.

The procedure starts by invoking one of the three optimisation routines described above as specified at start of the procedure. When the optimisation routine requires the evaluation of the objective function, its arguments are written to a file. Then, the optimisation routine saves its internal status and terminates, thus freeing the memory. Control is now handed over to the simulation module, which now has all internal memory available. The arguments of the objective function are read from the communication file, the simulation is performed and, finally, the result (value of the objective function) is returned to the optimisation program by file. The simulation module is terminated and the optimisation routine started again. On restart, it is set to the same internal state in which it was previously terminated.

When the optimisation routine has determined a new argument for which an evaluation of the objective function is required, this procedure is repeated. The simulation–optimisation loop terminates when the termination criterion of the optimisation algorithm chosen has been met.

When calling the optimisation module, the objective function also has to be specified. Within SYNOPSIS, a variety of different options is available (see Section 3.3.3) for the specification of the objective function. The interface programs also ensure that the function value returned by the simulation module is multiplied by -1 in appropriate cases before it is supplied to the optimisation module. At this stage, the program takes into account that the implementation of the CRS and of Powell's algorithm aim at minimisation of the objective function, whilst the genetic algorithm strives for maximisation.

Besides the input data necessary for the simulation module (*e.g.* catchment characteristics, parameters of the biochemical processes) and the specification of the objective function, certain input data are required by the optimisation module. Tables 3.9 and 3.10 provide an overview of input and output data of the optimisation module.

For the implementation of simulation and optimisation programs as two independent modules interacting in the way just described, significant modifications to the optimisation routines were carried out. Test runs performed for various optimisation problems confirmed the identity of the results obtained with the unmodified and the modified version of the optimisation routines. No changes were necessary within this respect for the simulation module. As stated above, the simulation module can be substituted easily with another simulation module if so required.

The application of these optimisation procedures is demonstrated in Appendix A for a test function. In Chapter 5, these procedures will be applied for optimisation of control strategies for the urban wastewater system.

Now, a simulation and optimisation tool is available, which can be used either in a stand-alone simulation mode, modelling the urban wastewater system for control strategy or in an optimisation mode which attempts to optimise the parameters of a given strategy framework. To summarise the discussion of this and the previous sections, the following section will give an overall overview of SYNOPSIS.

Development of SYNOPSIS 175

Table 3.9. Input data of the optimisation module

Controlled Random Search (cf. Section 3.5.1)	Genetic Algorithm (cf. Section 3.5.2)	Powell's algorithm (cf. Section 3.5.3)
Name of objective function	Name of objective function	Name of objective function
Names and ranges of control strategy parameters to be optimised	Names and ranges of control strategy parameters to be optimised	Names and ranges[1] of control strategy parameters to be optimised
---	Discretisation density of control strategy parameters	---
---	---	Start value for local search
Population size; maximum number of function evaluations; Termination criterion	Population size; maximum number of function evaluations; Termination criterion	Maximum number of function evaluations; Termination criteria
---	Crossover, mutation, creeping probabilities	---
---	Flag: Elitist reproduction (Yes/No)?	Flag: Consideration of constraints in optimisation (Yes/No)?
---	Flag: Application of Micro-GA (Yes/No)?	---
Initial seed for random number generator	Initial seed for random number generator	---

[1] Ranges are required only if the algorithm is run if constraints are considered for the optimisation.

Table 3.10. Output data of the optimisation module

Controlled Random Search (cf. Section 3.5.1)	Genetic Algorithm (cf. Section 3.5.2)	Powell's algorithm (cf. Section 3.5.3)
Table (and graphs) with arguments and values of the objective function and of the other criteria listed in Section 3.3.3 for all function evaluations carried out		
Plots of population (set A; see Section 3.5.1) after 50, 100, 250, 500, 750, 1000 function evaluations	Plots of population after 50, 100, 250, 500, 750, 1000 function evaluations	---

3.6 Summary: Overview of the Integrated Simulation and Optimisation Tool SYNOPSIS

This section provides a summary of SYNOPSIS; the development and the constituent modules of which have been described in the previous sections.

SYNOPSIS can be run either as a stand-alone simulation tool or it can be used for the optimisation of parameters describing a control strategy. In both cases, SYNOPSIS requires the following input data:

- characteristics of the urban wastewater system under study. These include details of the sewer system, the treatment plant and the receiving water body. Furthermore, the relevant biokinetic coefficients used in the treatment plant and river models need to be defined;
- definition of dry-weather flow and concentration patterns of the flows entering the sewer system;
- concentrations of rainfall runoff;
- biokinetic model describing the river water quality processes;
- parameters describing upstream river catchment runoff processes (*cf.Equation* (3.5b));
- locations of CSO, storm tank and treatment plant effluent discharges into the river;
- conversion factors describing conversion between COD and BOD based state variables;
- framework of the control strategy to be applied (this will be detailed in Section 5.1);
- parameters of the control strategy (when SYNOPSIS is run in the optimisation mode, this input will be provided by the optimisation module);
- rainfall data (either continuous or as a series of individual events).

SYNOPSIS will provide various forms of output, including tables and time series plots of relevant variables in sewer system, treatment plant and river. Furthermore, various files are generated containing summary information, including the criteria as defined in Section 3.3.3. A pre-processing program allows preparation of batch-mode simulations of series of simulation runs. These prove useful for sensitivity studies or for exploratory gridding of combinations of control parameters. In these cases, additional output will include sensitivity coefficients,

coefficients of variance as well as a 3D-plot showing any of the assessment criteria (see Section 3.3.3) *vs.* any combination of two strategy parameters.

When SYNOPSIS is run in the Optimisation mode, as additional input the following information is required:

- names and ranges of those parameters of the control strategy which are to be optimised;
- name of the objective function to be used for the optimisation (*cf.* Section 3.3.3);
- selection which optimisation procedure (CRS; Genetic Algorithm; local optimisation) is to be used;
- parameters specific to the selected optimisation algorithm (*cf.* Table 3.9);
- in addition to the output for each single simulation run, which is optionally generated, output will include tabulated and graphical information on arguments, objective function values and values of the criteria described in Section 3.3.3 for all evaluations of the objective function. Additional output is available from the CRS and genetic algorithm procedures, which describes the population at various stages of the algorithm.

Constituent submodules of the simulation part are a modified version of the sewer system model KOSIM (see Section 3.2.1), Lessard and Beck's treatment plant model with modifications and a river model by Lijklema *et al.* (1996) implemented in the DUFLOW shell program (see Section 3.2.3). Additional routines simulate river catchment runoff and computation of a time-variant coefficient required for the description of photosynthetic processes (see Section 3.2.3). Various conversions between state variables are necessary and are carried out as has been detailed in Section 3.3.

The control module performs control actions in sewer system and treatment plant (see the list of control devices shown in Table 3.7). This module processes state information available from the sewer system, the treatment plant and from locations in the river which are upstream of the discharge locations (*cf.* Table 3.6). Details on how the control decision is taken depending on the parameters of a control strategy framework will be discussed in Chapter 5.

Figure 3.14 provides an overview of the integrated simulation and optimisation tool SYNOPSIS. In the next chapter, a semi-hypothetical case study site is defined which will serve as a base for the studies described in Chapter 5 of this book.

178   Modelling, Simulation and Control of Urban Wastewater Systems

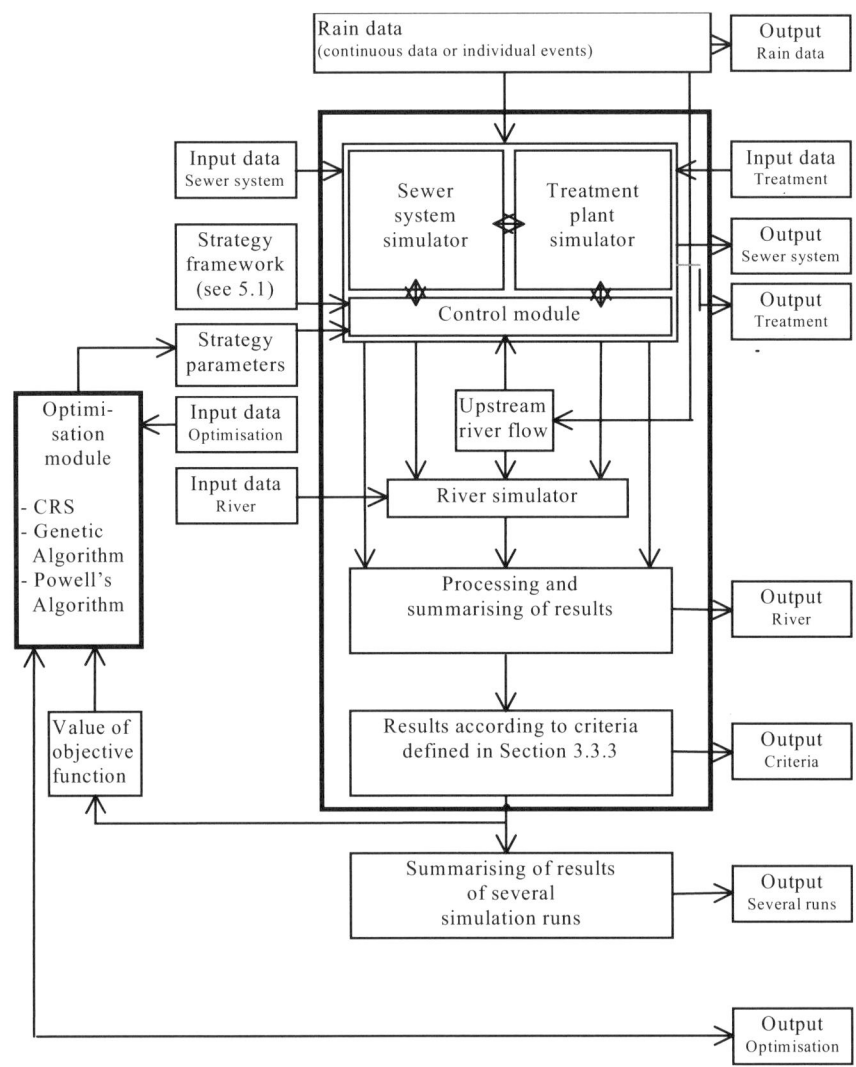

Figure 3.14. Overview of SYNOPSIS

Chapter 4
Simulation of the Urban Wastewater System Using SYNOPSIS

Following the description of the integrated simulation and optimisation tool SYNOPSIS in the previous chapter, this chapter prepares the studies to be carried out in Chapter 5 in various respects. Section 4.1 discusses the availability of data and defines the case study site used throughout this work as well as the relevant input data. Sections 4.2 and 4.3 present results of simulations of dry-weather flow and of a rainfall series, respectively, in order to illustrate the capabilities of the simulation package. Another example of the application of the simulation part of SYNOPSIS is provided in Section 4.4. Here, various settings of some of the control devices available in the urban wastewater system and their impact on receiving water quality are assessed. This prepares the analyses of control strategies in the subsequent chapters. Since the application of optimisation procedures (in Chapter 5) will be potentially demanding in terms of computing time, Section 4.5 analyses to what extent continuous long-term simulations can be substituted by simulations of series of individual events, in order to potentially reduce the time required for simulations.

4.1 Definition of a Case Study Site

Development of the simulation and optimisation package SYNOPSIS has not been done for its own sake but in order to assist in the analysis of the potential for integrated control of urban wastewater systems. In order to carry out such a study, it is necessary to have a case study site as a base for the intended simulation and optimisation runs. Ideally, such a case study site would exist in reality with a sufficient amount of data available for calibration and verification of the simulation module prior to its application. Therefore, the first subsection will review the current situation with regard to the availability of appropriate data sets. Since no data could be found which were available for use within this work and since no data collection exercise appeared to be feasible, given the time constraints of this research, hypothetical data have been used here. Although the sewer system and the river simulated in this study are not based on real data, at least the treatment plant is. Therefore, the term "semi-hypothetical" case study site will be used when referring to the origins of the system analysed in the present work.

4.1.1 Existing Data Sets

In order for a data set to be appropriate to the calibration and verification of any model simulating the complete urban wastewater system, it would ideally have been collected from all subsystems (from appropriate locations in the sewer system, the treatment plant and the receiving river) of one and the same case study site at the same time (including dry-weather and rain periods). Such a data set of flow and relevant pollution parameters would be useful for the description of the behaviour of the urban wastewater system as a whole. A survey by questionnaire, which was sent out to researchers in several European countries, showed that there did not seem to be any data set available which appeared to meet these requirements. Also other potential sources of appropriate data do not contain the required information: consultation of the description of the "Urban pollution database" (Gill and Bryan, 1994) revealed that, although provisions were made to include data from the main constituent parts of the urban wastewater system, no data from treatment plants were actually included in that database.

Since the early stages of the work described in this book when these surveys were carried out, the situation concerning data has changed (see, for example, the data collection exercises conducted within the UPM framework (DHI, 1998b)). Furthermore, various measuring campaigns are reported from Continental Europe (see Vanrolleghem *et al.* (1998), from where Table 4.1 is taken). However, since these data were not available in the early stages of this study, no detailed analysis as to whether these are appropriate for use in this study could be conducted. Of the integrated modelling studies reported in Section 2.3, a large proportion uses hypothetical data for parts of the urban wastewater system.

For these reasons, this study has to be based on a (semi-)hypothetical case study site. It is obvious that such a site should resemble the behaviour of a real system as closely as possible. Therefore the definition of the case study site (described in the subsequent sections) is based on the dimensions of the treatment plant at Norwich (it is this plant, for which the treatment plant model used in this study was originally developed). For the sewer system, an example from the literature has been chosen. In order to be compatible in size with the treatment plant, some modifications of its definition were necessary. Finally, a purely hypothetical river is defined. It will be assumed that CSO and treatment plant discharges will be led into the same river, but at different locations.

Table 4.1. List of integrated urban wastewater management studies in which important integrated measuring campaigns were conducted

Location/ River	Objective	Measurements performed upon				Ref[1]
		Waste water production	Sewer system	Waste water treatment	Rec. water	
Trondheim (N)	Min. tot. pollution	X	X	X		2 A
Lambro (I)	Risk assessment			X	X	1 TR
Bordeaux (F)	Research	X	X		X	2 A
Boran-s-Oise (F)	Research	X	X	X		2 P
Loenen (NL)	Research		X		X	n P
Vecht (NL)	Scenario analysis			X	X	n P
Hildesheim (D)	Research	X	X	X	X	2 P
Innsbruck (A)	Effect infiltration	X	X		X	1 A
Glatt (CH)	Research		X	X	X	1 TR
Tielt (B)	Min. tot. pollution	X	X	X	X	1 P
Aalborg (DK)	Real-time control		X	X	X	n P
Avedore (DK)	WWTP operation	X	X	X		1 P
Rhine-Rhône (F)	Channel design	X			X	n P
Paris (F)	Research	X	X		X	n P

[1] References given by Vanrolleghem *et al.* (1998): A: abstract, P: paper, TR: technical report. *n.b.:* X indicates monitored systems

4.1.2 Definition of the Sewer System

The sewer system used in this study is based on the example given in ATV A128 (1992). This example is also discussed by ATV 1.9.3 (1992) and has been used in a number of other studies involving the KOSIM sewer system model (Pracejus, 1994; Härtel *et al.*, 1995; Kollatsch, 1995). Therefore, a system description in the appropriate format was at hand. Figure 4.1 and Table 4.2 present the geometric characteristics of this system as reported in ATV A128 (1992).

182 Modelling, Simulation and Control of Urban Wastewater Systems

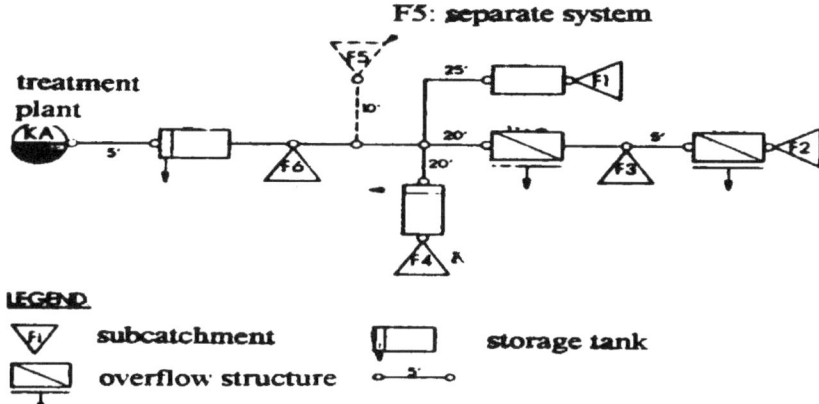

Figure 4.1. Literature sewer system (Härtel et al., 1995)

Table 4.2. Characteristics of the literature example of the sewer system

SC	Area ha	Population	DWF l/s	Extraneous inflow l/s	Parameters of reservoir cascade		Structure		
					n	K	Type	Volume m³	Max. outflow l/s
1	14	2240	4.7	1.4	3	3.592	OPT	2000	100.0
2	3	549	1.1	0.3	3	1.267	OVF	---	50.0
3	4	420	0.9	0.4	3	1.500	OVF	---	108.0
4	10	1350	2.8	1.0	3	2.083	OBT	185	12.3
5	10	1100	2.3	3.3	5	1.000	---	---	---
6	35	5600	11.7	3.5	3	6.042	OPT	1241	98.0
All	76	11259	23.5	9.9	-	---	---	3426	98.0
			Sum:	33.4					

Structure types:
OPT: On-line pass through tank (see Section 2.1.1.5); OBT: Off-line bypass tank (see Section 2.1.1.5); OVF: Simple overflow without storage

According to its original definition, subcatchment 5 represents a separate sewer system. In the present study, it will be considered as a combined sewer system (in the same way as all other subcatchments). Since this system is designed for a total dry-weather flow of 33.4 l/s, it is not compatible with the treatment plant to be simulated within this study, which processes an average dry-weather flow of 318.3 l/s. Therefore, the original definition of the sewer system (as shown above) is adapted as shown in Table 4.3.

Table 4.3. Characteristics of the sewer system simulated in this study

SC	Total area	Population	DWF	Extraneous inflow	Parameters of reservoir cascade		Structure		
							Type	Volume	Max. outflow
	ha		l/s	l/s	n	K		m³	l/s
1	66.9	13843	28.8	0	3	11.100	---	---	∞
2	66.9	13844	28.8	0	3	11.100	OPT	2800	1)
3	28.7	6647	13.9	0	3	3.915	---	---	∞
4	38.2	5867	12.2	0	3	4.635	OPT	1400	1)
5	95.5	17465	36.4	0	3	6.437	---	---	∞
6	95.5	25624	53.4	0	5	3.090	OPT	2000	1)
7	334.3	69493	144.8	0	3	18.672	OPT	7000	1)
All	725.8	152784	318.3	0	n/a	n/a	---	13200	1)

SC: Subcatchment; OPT: On-line pass through tank (see Section 2.1.1.5)
[1] The maximum outflow rates of subcatchments 2, 4, 6, 7 are set by the control module (simulating pumps at the outlet of these subcatchments/basins).

The required adaptation of the dry-weather flow rates is achieved by enlarging the catchment areas and the population numbers and by joint consideration of dry-weather flow and inflow from extraneous sources. In order to consider the increased size of the subcatchments, the storage constant K is multiplied by the square root of the factor by which the areas are multiplied. This is done since K can be related to the flow time within the subcatchment (see Section 2.1.1.2), which is assumed here to increase with the square root of the catchment area.

Subcatchment 1 of the original example is split into two subcatchments in order to consider longer flow times between catchments. These are set as shown in Figure 4.2. Thus the largest flow time within the system is 65 minutes, as opposed to 30 minutes in the original example.

Four of the seven subcatchments of the sewer system are grouped together in pairs of two subcatchments each (SC1-SC2; SC3-SC4). Each of these pairs has a storage basin at its downstream end, whilst dry-weather flows and rainfall runoff are generated in both subcatchments.

The specific storage volume for each group of subcatchments is set to 21 m³/ha. This value results from application of the A128 guidelines for the design of storage tanks (ATV A128, 1992) to this system, assuming a maximum discharge of all pipes of three times average DWF (using a value of, for example, seven times average DWF would result in a specific storage of 12 m³/ha). The assumption of uniform distribution of storage, which is made here, appears to be justified, given the fact that no spatial distribution of rainfall is assumed (due to the non-availability of spatially distributed rainfall data); therefore, no benefits from control utilising storage available in parts of the system with lower loading than in other parts of the system can be expected.

The storm tank at the treatment plant represents an additional storage volume, resulting in a total storage of 28 m^3/ha. This value is in the lower range of specific storage volumes reported from other case study sites, *e.g.*, Fehraltorf: 30 m^3/ha (Almeida, 1994); Bremen (left and right bank side): 120 m^3/ha (Schilling, 1990; Rohlfing, 1993a); Oslo: 200 m^3/ha (Weinreich and Schilling, 1995).

The maximum flow from the upper towards the lower subcatchment within each pair is set to a (practically) infinite value. This ensures that no overflow will occur in the upstream subcatchment due to potentially inappropriate definition of the connecting pipe. Therefore the study can focus on the basins and related CSO discharges at the downstream catchment of each pair. Doing otherwise could lead to impaired performance of the system solely caused by its design and not by the application of any control strategy.

Figure 4.2 provides an overview of the sewer system case study site defined for this study.

The outflow of each of the four storage tanks (located in subcatchments 2, 4, 6 and 7) can be controlled by a pump. Thus part of the inflow to subcatchment 7 (besides the runoff generated within subcatchment 7) and the flow towards the treatment plant can be regulated. Each pump is assumed to be operated in discrete pump stages. These are set here to 2, 3, 4 and 5 times average dry-weather flow, respectively. Thus a setting of, for example, 2 DWF for P7 means here that all flows up to 2 DWF can pass, whereas any excess will be kept in the storage tank upstream of the pipe (and discharged if the capacity of the tank is exceeded). Settings of the pumps in the simulation are effected by the control module of SYNOPSIS (see Section 3.4). For the "base case" scenario of control, these pumps are set to a pump rate of five times average dry-weather flow. This value may appear to be rather high. However, this setting allows a larger portion of the wastewater to be led to the treatment plant, which may provide a larger control potential at the plant for this study. Setting the maximum pump rates to considerably lower values would result in more water being discharged at the CSO structures – which would be hardly affected by any control actions at all.

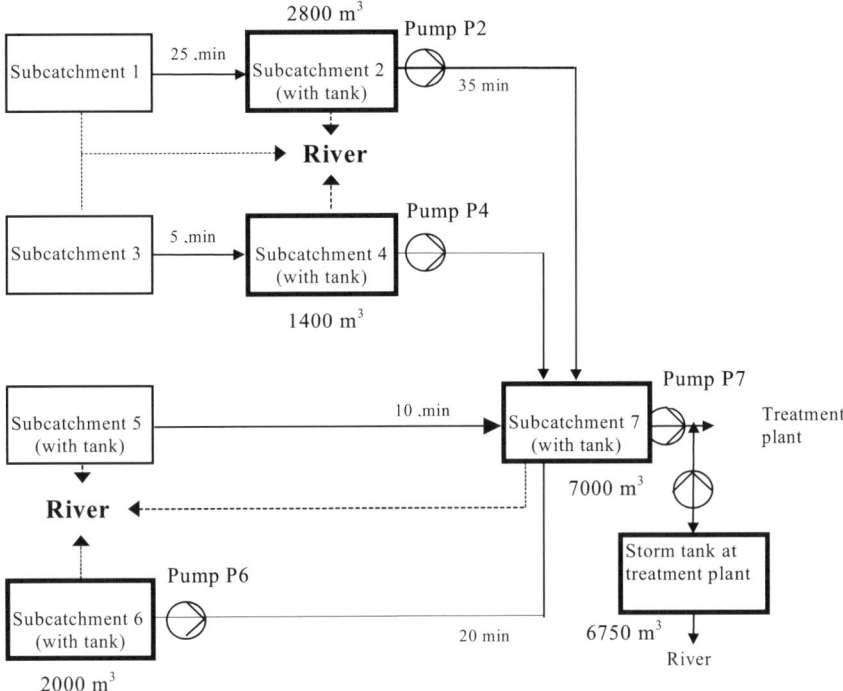

n.b.: The volumes shown in this figure indicate the storage volume of the basins; the time values indicate the flow time assumed between the various subcatchments of the sewer network.

Figure 4.2. Overview of the sewer system defined for this study

Input data: Dry-weather and rainfall runoff concentrations

Information about pollutant concentrations in dry-weather flow and in rain-runoff is required as input for their simulation in the sewer system. Table 4.4 summarises values reported in the literature and states those which will be used throughout this study. The values chosen for domestic wastewater are obtained from the data set described by Lessard (1989), which was sampled from the inlet of the treatment plant in Norwich.

The values chosen as rainfall runoff concentrations are based on values from various references (*cf.* Table 4.5). For the determination of the values for VSS and soluble COD, an assumption of equal ratios for dry-weather flow and for rainfall runoff of VSS/SS and CODs/COD is made, since no data on concentrations of these parameters in rain-runoff were found in the literature. It can be seen from the table that the selected values fall well within the ranges reported in other studies.

Table 4.4. Concentrations of municipal dry-weather flows as reported in various studies

Dry-weather flow in combined sewers (concentrations in mg/l)						
	COD	COD$_{soluble}$	SS	VSS	NH$_4$+NH$_3$	NO$_3$
Aalderink, 1990	755		392			
Butler, 1994	750		300		30	
Dauber and Novak, 1982	498				24.6	
Degrémont, 1991	300–1000		150–500		20–80	<1
Gujer, c.1990	350	180	150		20	1
Hammer, 1996			240	180		
Härtel *et al.*, 1995	572				41	
Lessard, 1989	606	281	335	245	27.7	0
Tebbutt, 1992	700		400		40	<1
Vanrolleghem *et al.*,1996a	622		222			
Used in this work	**606**	**281**	**335**	**245**	**27.7**	**0**

Table 4.5. Concentrations of rainfall runoff as reported in various studies

Rainfall runoff (concentrations in mg/l)						
	COD	COD soluble	SS	VSS	NH$_4$+NH$_3$	NO$_3$
Butler, 1994	100		190		2	
Chebbo and Saget, 1995	83–339		182-456			
Dauber and Novak, 1982	60–220					
Ellis, 1989a	85		190		1.8	
Härtel *et al.*, 1995	80				2	
House *et al.*, 1993	20–365		21–2582		0.2–4.6	
Mance and Harman, 1978			112			1.7
Vanrolleghem *et al.*, 1996	130		500			
Xanthopoulos and Hahn, 1993	47–120				0.2–0.8	0.02–1.8
Used in this work	**100**	**46**	**190**	**139**	**2**	**0**

The variety of values reported in these tables reflect the many factors which affect the composition of urban wastewaters. Besides type (combined; separate), configuration and extent of the sewer system, these factors also include the location at which the pollutant data were sampled, temperature, catchment characteristics (*e.g.,* size and usages), time of the day, presence or absence of industrial wastewater and sociological characteristics of the connected population (Thornton and Saul, 1986; Nielsen *et al.*, 1992; Almeida, 1999a).

Concentration values derived from those reported in Table 4.4, most of which were measured at downstream locations of a sewer system, will be used for the simulations as concentration values for inflows upstream in the sewer system. This simplifying assumption appears to be justified, since no in-sewer degradation processes are modelled within this study (*cf.* description of the model KOSIM; Section 3.2.1). Similarly, although the sewer model provides the option of simulating sedimentation and resuspension processes in the sewers and in the tanks, no use is made of these options, since appropriate parameter values would have to be defined for the sedimentation and wash-off properties. As discussed in Section 2.2.1, the issue of "first-flush" effects is the subject of intensive research and discussion on which a great variety of results and opinions can be found in the literature.

Input data: Diurnal patterns of flows and concentrations

In order to consider variations of flow and concentrations of dry-weather flow during the day in the simulations, diurnal patterns for dry-weather flow and its pollutant concentrations are derived from the three-hourly measurement data over four days from the influent of the treatment plant in Norwich (Lessard, 1989). These patterns, which will be used throughout the present study, are shown in Figure 4.3 together with patterns derived from samples taken from the downstream end of the sewer system of Great Harwood, Lancashire (Thornton and Saul, 1986) and with patterns implemented as default patterns in the commercial software packages MOSQITO (Gent, 1992) and KOSIM (itwh, 1995). In MOSQITO, a distinction is made between weekday and weekend patterns, of which only the former are shown here. The default patterns of the KOSIM and MOSQITO programs represent no specific pollutant. This may explain their relatively poor agreement with the measurement data from Great Harwood and Norwich.

188 Modelling, Simulation and Control of Urban Wastewater Systems

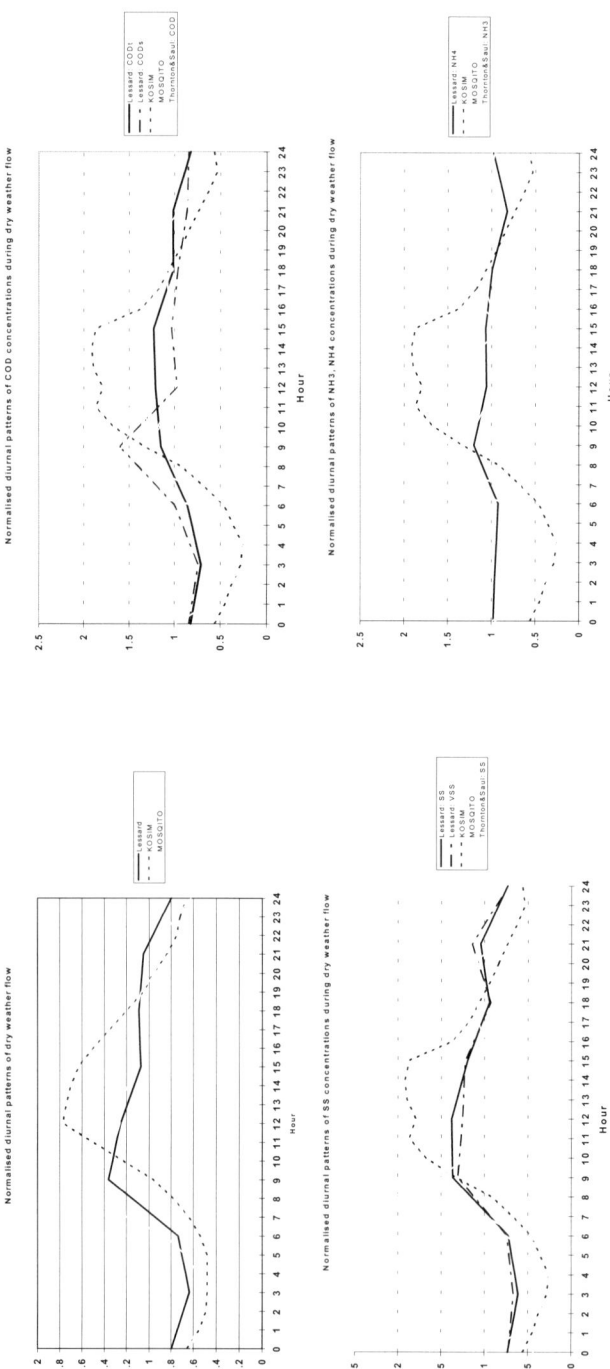

Figure 4.3. Concentrations of dry-weather flow and runoff concentrations defined for use within this work

4.1.3 Definition of the Wastewater Treatment Plant

As discussed earlier, the nitrifying activated sludge plant in Norwich is used as the basis of the treatment plant simulations. This appears to be an obvious choice since the treatment plant model used here has been calibrated and validated on data from this plant (Lessard, 1989). Thus only a short description of this plant is given here, further details can be found in Tong *et al.* (1980), Beck (1984), Lessard (1989), Lessard and Beck (1993).

The Norwich plant is designed for a dry-weather flow of about 55000 m^3/d and serves a population equivalent of about 200000. Part of the influent is treated by trickling filters, the other part by a conventional activated sludge process. In the present work, as in the study by Lessard and Beck referred to above, the treatment of only half of the influent in the activated sludge stream of the plant is considered. Thus the average dry-weather flow amounts to 27500 m^3/d (*i.e.* 318.3 l/s).

The layout of the plant corresponds to the structure of the model, which is shown in Figure 3.1 in Section 3.2.2.1. Preliminary treatment in Norwich consists of screening and degritting; for primary clarification three tanks are in use. Aeration is carried out in six aeration lanes, whilst secondary clarification is performed in four settlers. For simplicity of modelling, these tanks are substituted by one tank with the overall volume for modelling purposes. Table 4.6 provides the dimensions of the plant.

The arrangement of the storm tank simulated here corresponds to an off-line pass-through tank (see Section 2.1.1.5 and Figure 2.4) without an emergency overflow. Thus wastewater led into the storm tank is allowed to settle in the tank and settled wastewater is discharged to the receiving river whenever storm tank overflow occurs. Filling of the storm tank (again, for modelling purposes the four storm tanks actually present at Norwich are jointly considered as one tank) is controlled by the maximum inflow rate to the primary clarifier. The tank is emptied (back to the plant inlet) at a certain pumprate, as soon as the inflow rate to the plant drops below a prespecified threshold value. These two rates (named STTPQ, STTPTH throughout this book) are among the control parameters which are set by the control module (*cf.* Section 3.4).

Table 4.6. Dimensions of the treatment plant simulated in this work

Dimension					
Average dry-weather	flow: 27500 m³/d (*i.e.* 318.3 l/s)				
Storm tanks			**Aerators**		
Number	4		Number	6	
Length	26.8	m	Length	60	m
Width	18.0	m	Width	2 at 6	m
Depth	3.5	m		4 at 12	m
Total capacity	6750	m³	Depth	2.8	m
			Total capacity	10400	m³
			Retention time at DWF	9	h
			Air flow rate	30000	m³/h
Primary clarifiers			**Secondary clarifiers**		
Number	3		Number	4	
Diameter	30	m	Diameter	25.5	m
Total area	2118	m²	Total area	2040	m²
Total capacity	6785	m³	Total capacity	6600	m³
Retention time at DWF	6	h	Retention time at DWF	6	h
Overflow rate at DWF	0.5	m/h	Overflow rate at DWF	0.56	m/h

Also the maximum inflow to the primary clarifiers of the treatment plant (subsequently called PCINMX), which is commonly set to three times average dry-weather flow, is among the control parameters used here. For most of the simulation studies described in this book, the maximum inflow to the aeration tank is restricted to 3 DWF, thus implying that the aerator bypass is active whenever the inflow to the plant exceeds this threshold. However, a set of simulations in which the inflow to the aerator is not limited has also been carried out (*cf.* Section 5.6.4).

The aeration rate is set in the same way as was done by Lessard (1989), *i.e.* so that the DO concentrations in the aeration basins should be higher than 1.5 mg/l during periods of dry-weather. Therefore, DO concentrations should not be a factor limiting growth rate during normal operation. The aeration scheme applied is tapered aeration, *i.e.*, different aeration rates are applied in different parts of the aeration tank.

Return activated sludge is taken from the secondary clarifier. The waste sludge is taken also from this line. In Norwich, the waste sludge is pumped into consolidation tanks, the overflows of which are pumped back to the primary clarifier. This recycle line is not included in the model. However, information about these waste sludge returns is read from the input file and considered in Module 4 of the treatment plant program (*cf.* Section 3.2.2). Since no data on these returns are available, the assumption is made that they (150 m³/h) occur once every six hours.

This resembles in its frequency the pattern used by Lessard in his studies (see also Section 3.2.2).

As in Lessard's base case of control, the pump rate for return activated sludge, which is pumped back to the head of the aerator, is set to a constant value of 600 m^3/h (corresponding to about 50% of the dry-weather flow rate), unless stated otherwise. However, variation of this setting as well as the use of a RAS rate proportional to the inflow rate as means of treatment plant control will be the subject of the analyses described in subsequent sections.

In Lessard's studies, the waste sludge rate was set to a constant value of 15 m^3/h. However, using this value, no steady-state which would correspond to dry-weather behaviour of the plant could be reached for long simulations of dry-weather flow. Figure 4.4 shows results of a simulation using a setting of DWF for the WAS rate for the treatment plant effluent (with the diurnal variation as described in Section 4.1.2). It should be noted here that similar observations were also made using the program without the extensions by Vazquez-Sanchez (1996). Since this problem could not be resolved by setting the return sludge rate to a different value and despite the fact that a change of the waste sludge rate affects the sludge age, the waste sludge rate was set to 27.5 m^3/h, giving an (almost) steady-state effluent, as will be shown and discussed in detail in Section 4.2.

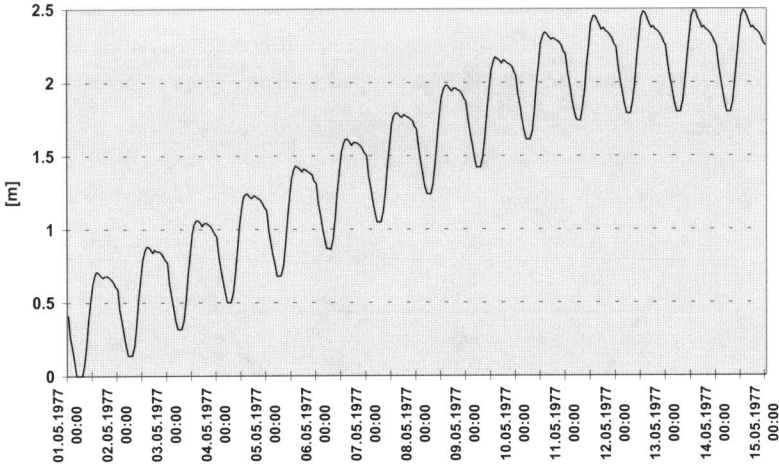

Figure 4.4. Dry-weather flow simulation with the original setting of WAS rate (15 m^3/h) – Sludge blanket height

Table 4.7 gives the settings of the control devices which defined the so-called "base case" control. Control strategies investigated in the following chapters will be compared against this scenario.

Table 4.7. Values of control parameters in the treatment plant defining the base case of control

Description	Acronym	Value
Maximum inflow rate into primary clarifier	PCINMX	3 × DWF
Emptying rate of storm tank	STTPQ	500 m^3/h
Threshold TP influent flow rate initiating emptying of storm tanks	STTPTH	1000 m^3/h
Return activated sludge rate (absolute value)	RASABS	600 m^3/h
Waste activated sludge rate	WASABS	27.5 m^3/h

Variations of these settings as well as implementation of (time-variant) settings of some of these parameters within control strategies will be analysed in Sections 4.4 and Chapter 5.

4.1.4 Definition of the River

As opposed to the sewer system and the treatment plant simulated in this work, which are based at least to some extent on real systems, the river defined for this study is of purely hypothetical nature.

The hypothetical river consists of 40 sections of identical geometry as defined below. The cross-section of each section is assumed to be trapezoidal as shown in Figure 4.5. The river is assumed not to be equipped with any weirs or other ancillary structures.

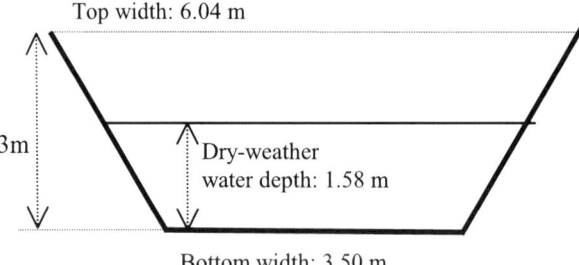

Figure 4.5. Cross-section of the river simulated in this study

As impacts of the treatment plant effluent and any overflow discharges on the water quality in the river are not to be mitigated to too large an extent simply by a large dilution ratio of river base flow and discharges into the river, the river base flow was set to 1.5 m³/s. This results in a dilution ratio of dry-weather treatment plant discharges of about 5. This is significantly smaller than the dry-weather dilution ratios of the Innerste (Lammersen, 1997a) and Cam (Reda, 1996) rivers of about 8 and 12, respectively.

Dry-weather flow velocity and water levels (shown in Figure 4.5) are derived from the Manning-Strickler formula for flow in open channels flow (Imhoff and Imhoff, 1990) (assuming a slope of 0.0001 and a Manning number of 0.035 $m^{-1/3}$ s^{-1}, which is the default value in the DUFLOW model (IHE, 1992)). Hence, the initial and boundary conditions of river flow under dry-weather conditions are defined. Additional boundary conditions are given by the discharges from CSOs, storm tank and treatment plant, which are calculated by the corresponding simulation submodules of SYNOPSIS. The location of these discharges into the hypothetical river are shown in Figure 4.6. Upstream catchment runoff caused by rainfall and calculated as described in Section 3.2.3 is modelled as additional inflow into the river at Section 02. No additional catchment runoff from catchment areas discharging into river sections further downstream is assumed. Each of the river sections shown in Figure 4.6 is assumed to have a length of 1 km.

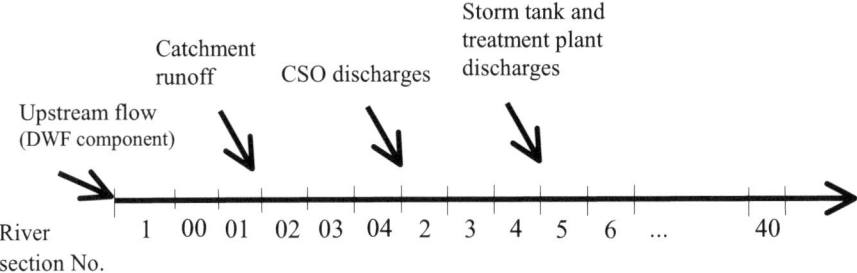

Figure 4.6. Layout of the river with discharge locations used in this study

The total length of the simulated river stretch amounts to 45 km. It is assumed that any pollution which is observed further downstream of this stretch is not relevant to the evaluation of assessment of river water quality. This seems to be a realistic assumption if, for example, the river flows into another water body (*e.g.,* an ocean) for which the assumption is made that pollution from this river is of no major significance.

Similarly, for the water quality variables, initial and boundary conditions have to be defined. Table 4.8 provides an overview of the values of some of the water quality variables simulated by Reda (1996). These describe three different upstream scenarios.

Table 4.8. Definition of water quality parameters for upstream river scenarios

Parameter	"Dry" scenario	"Extra dry" scenario	"High algae" scenario
Base flow	2.0 m^3/s	1.6 m^3/s	2.0 m^3/s
BOD	1.8 mg/l	2.5 mg/l	3.5 mg/l
Amm-N	0.09 mg/l	0.15 mg/l	0.3 mg/l
DO	9.0 mg/l	8.0 mg/l	9.0 mg/l

For the present study, upstream concentrations are defined as shown in Table 4.9. These settings of the water quality variables resemble Reda's values for his "dry" scenario. Upstream river flow is set to 1.5 m^3/s as described above. All upstream BOD is assumed to be slowly biodegradable, since readily biodegradable might well have been biodegraded in sections upstream of the urban discharges.

Table 4.9. Initial and upstream boundary conditions of river water quality variables

Variable	Initial condition (all sections) and upstream boundary conditions
Flow	1.50 m^3/s
AMM	0.09 mg/l
BODR	0.00 mg/l
BODS	1.80 mg/l
DO	9.00 mg/l

Concentrations of discharges into the river are calculated by the appropriate submodules of SYNOPSIS. Since DO concentrations are not modelled by the sewer system and the treatment plant models and also for simplicity, constant values are assumed for the DO concentrations of the various discharges (coefficients c_7, c_8 and c_9 in Figure 3.6). Each of these coefficients is set to a value of 5 mg/l. This corresponds well with the assumptions made by Reda (1996) about storm tank overflows, the value of DO concentration during dry-weather flow chosen by Lessard (1989) and also roughly with the assumption by Rauch (1996b) of the DO concentration being equal to 60% of the saturation concentration.

As discussed in Section 3.2.3, the river water temperature can be defined either as a fixed pattern or as an external time series which is to be supplied as an input file. In order for the effects of control actions and the impacts of temperature variations not to be amalgamated into the assessment of controls strategies, river water temperature is assumed to be constant throughout this study. The significance of the influence of temperature on the oxygen balance in the river is obvious since not only do the biokinetic rates depend on the temperature (see Equation (3.3)), but also the DO saturation concentration itself depends on the temperature (see Section 2.1.3.3). A constant value of 17 °C is chosen here. This value corresponds to the temperature of the Innerste river during the summer months (reported to be around 15°C to 20°C (Lammersen, 1997a)). Also, the temperature values assumed by Reda (1996) for his "dry" scenario are in the range of 17°C to 19°C.

In order to consider increased river flow due to (upstream) catchment runoff, the coefficients γ_k and δ_k (see (3.5b)) have to be defined. Lacking catchment rainfall runoff data, these are set for the present study as follows:

$$\gamma_k := 0 \quad \forall k = 1,..,576$$

$$\delta_k := \begin{cases} 1, & for 37 \leq k \leq 324 \\ 0, & otherwise \end{cases}$$

Considering that rainfall information is available in time steps of five minutes, the additional river base flow due to catchment runoff is thus defined as the sum of the rainfall of the last 24 hours, taken with a 3-hour delay, mulitplied by coefficients for catchment area and runoff. These coefficients are chosen in such a way that a relatively intense rainfall event (21.7 mm on 7/8 February – see Figure 4.9) causes an upstream flow in the river which amounts to about twice the base flow. This increase in flow is comparable to that of one of the rainfall events (of higher intensity: 15 mm in 4 hours) simulated by Reda in his studies. Averaging the rainfall values of earlier time steps is done with a 3-hour delay (*cf.* definition of δ_k), thereby ensuring that increased river flow due to catchment runoff occurs later than the CSO discharges. Reda's research suggests a time-lag for this effect of about (at least) 2 hours.

Finally, the latitude of the river location, which is required for calculation of the photosynthesis factor β (see (3.4)), is set to $\varphi = 52°$. This corresponds to large parts of the south and east of England and Northern Germany, where the catchments of the Cam and Innerste rivers are located, both of which assisted in the definition of the hypothetical river for this work.

4.1.5 Overview of the Case Study Site Defined

The previous sections provided the definition of the case study site simulated in the remaining part of this book. To assist the reader in the study of the subsequent sections and chapters, Figure 4.7 presents an overview of the simulated site. Table 4.10 summarises the definition of the base case scenario for control. Under this scenario, all control devices listed in Table 3.7 are operated with constant settings. The base case will serve in the subsequent parts of the book as a reference case for the evaluation of control strategies.

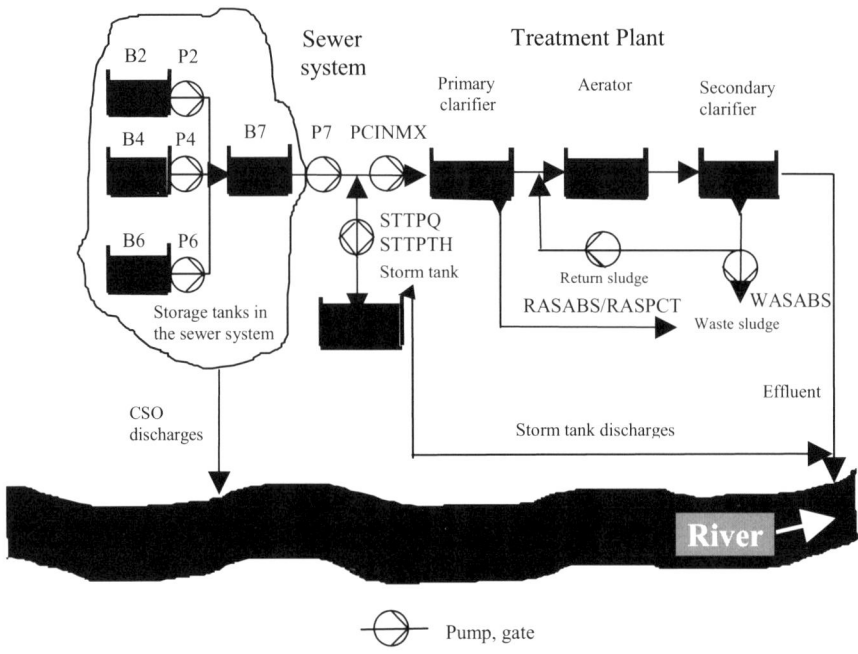

Figure 4.7. The (semi-hypothetical) case study analysed in this work

It should be stressed that the hypothetical case study defined here represents merely an example of an urban wastewater system used for the development and analysis of the tools and methodology presented in this work. As soon as an appropriate description of a real case study site is available, this could serve as a base for further studies.

Table 4.10. Definition of the base case control scenario

Description	Acronym	Value
Pump P2 (Maximum flow capacity)	P2	5 DWF
Pump P4 (Maximum flow capacity)	P4	5 DWF
Pump P6 (Maximum flow capacity)	P6	5 DWF
Pump P7 (Maximum flow capacity)	P7	5 DWF
Maximum inflow rate into primary clarifier	PCINMX	3 DWF
Emptying rate of storm tank	STTPQ	500 m^3/h
Threshold TP influent flow rate initiating emptying of storm tanks	STTPTH	1000 m^3/h
Return activated sludge rate (absolute value)	RASABS	600 m^3/h
Waste activated sludge rate	WASABS	27.5 m^3/h

Finally, a statement about computation time can be made at this stage. Simulation of the proposed operating over a period of one year takes about 75 minutes on a Pentium 133 computer. Therefore, the simulation module seems to be sufficiently fast to be used in the optimisation procedure briefly outlined in Section 2.5.1 and detailed in Chapter 5. Also its use within some on-line control applications would appear to be feasible.

4.2 Simulation of Dry-weather Flow

Now the simulation tool has been established and a case study site defined, simulations of the system under study can be carried out. This and the next section provide detailed results of simulation runs with dry-weather flow (this section) and a rainfall time series (next section) as input. These results not only illustrate the capabilities of the simulation developed, but also help to assess the plausibility of the results obtained with this tool.

A zero rainfall time series is provided as the input time series to the simulations described in this section. Thus the flow and concentration pattern in the sewer system follow the patterns defined in the previous section. Similar diurnal patterns are observed for the influent to the treatment plant (Figures 4.8a to d). Figure 4.8e shows the effluent concentrations of the primary clarifier. The peaks with regular periodicity which can be seen in Figure 4.8f (treatment plant effluent flow) are caused by the returns from sludge treatment into the primary clarifier, which are assumed to take place once every six hours (*cf.* Section 4.1.3). Furthermore, a sharp decrease of flow in the treatment plant effluent can be observed every 24 hours. This is due to sludge wastage from the primary clarifier, which is assumed to be effected at this frequency. Figures 4.8g and h show that the effluent concentrations during dry-weather flow are around 72 mg/l for COD, around 20 mg/l for suspended solids, around 28 mg/l for NO_3 and, around 0.2 mg/l for NH_4. Thus, nitrification is achieved. Assuming a BOD/COD ratio of 0.22 (*cf.* Section 3.3.2), it can be said that the "20:30 standard" is met. Figures 4.8i to l show the values of key parameters in the activated sludge process. These include the mixed liquo volatile suspended solids (MLVSS) concentrations of the first and the last tank The concentration of volatile suspended solids in the return sludge and the dynamics of the height of the sludge blanket are shown. Finally, Figures 4.8m to p show time series plots of some of the state variables in the river. Although the ammonium concentration (slightly) increases during the first days, it then assumes a steady-state (at diurnal maximum values of 0.17 to 0.25 mg/l, depending on the location within the river) at the end of the simulation period shown. Thus, overall the results correspond to those reported by Lessard for the dry-weather case of his study and can be considered to be plausible.

Simulation of the Urban Wastewater System 199

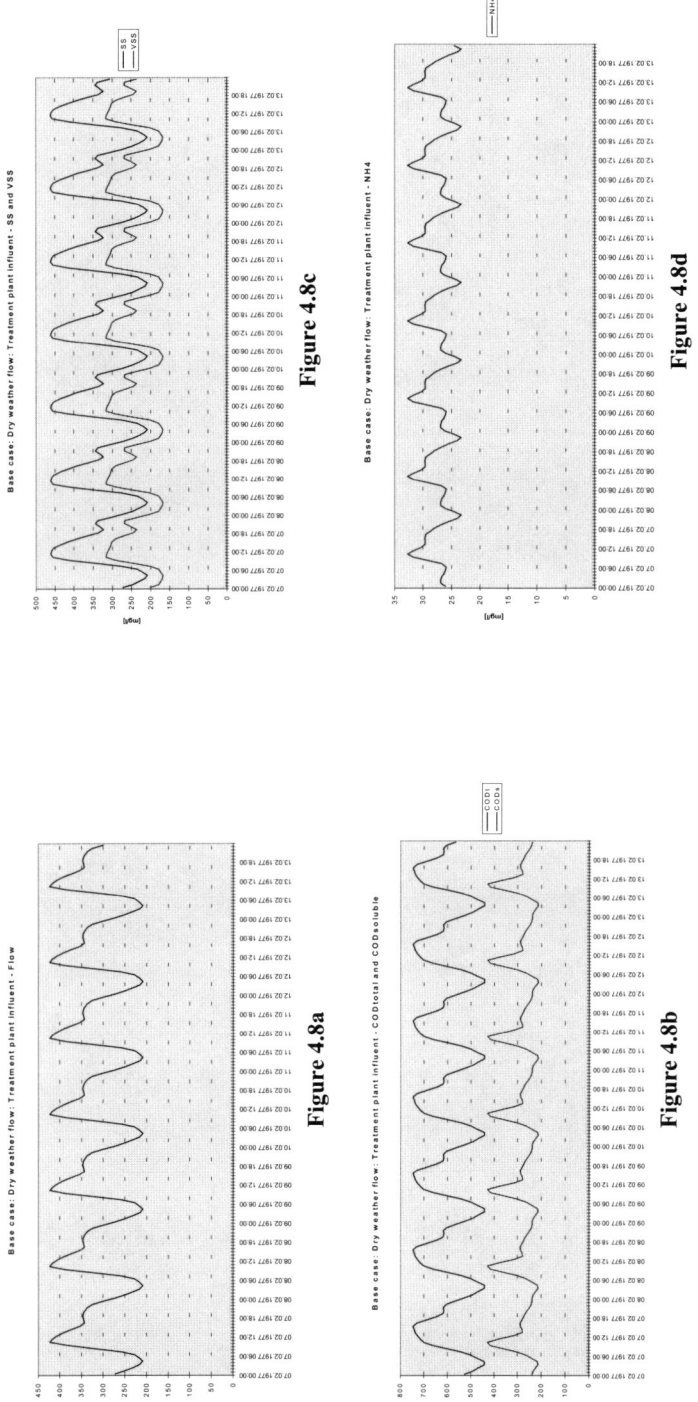

Figure 4.8. Simulation results for the base case under dry-weather flow

200 Modelling, Simulation and Control of Urban Wastewater Systems

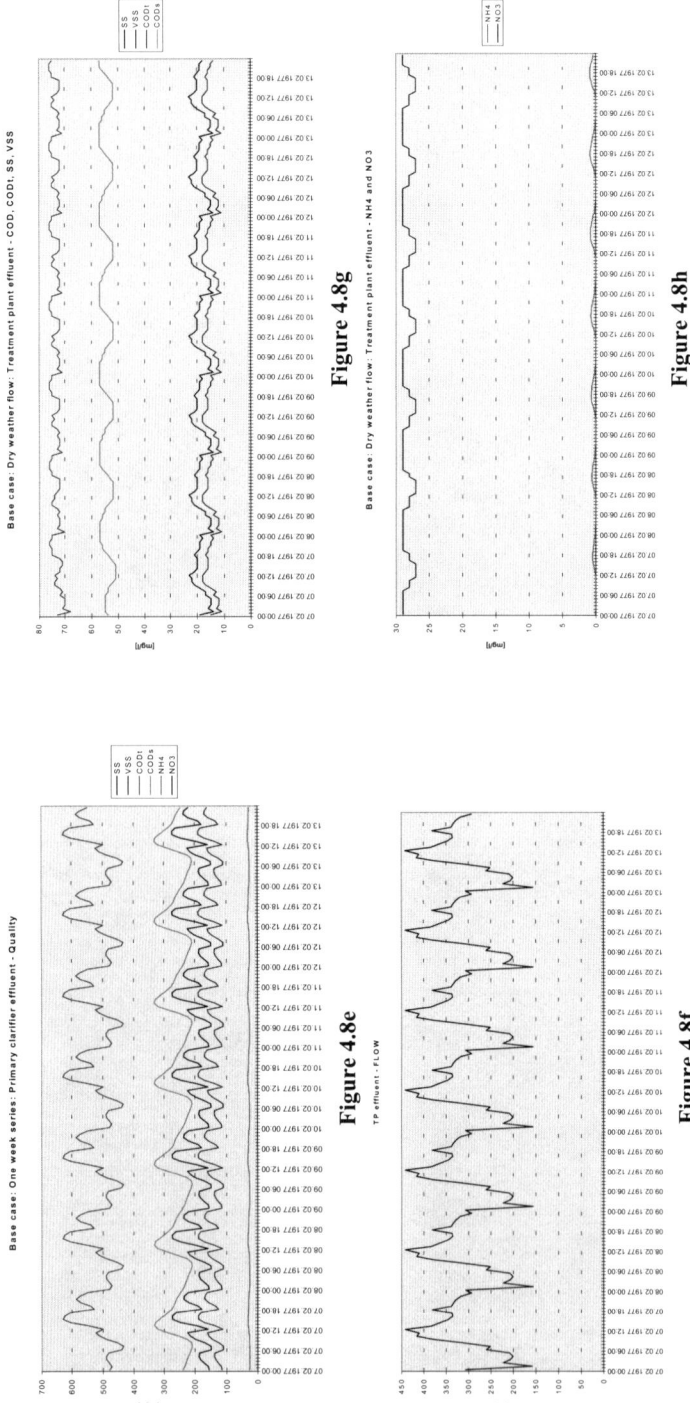

Figure 4.8. Simulation results for the base case under dry-weather flow (continued)

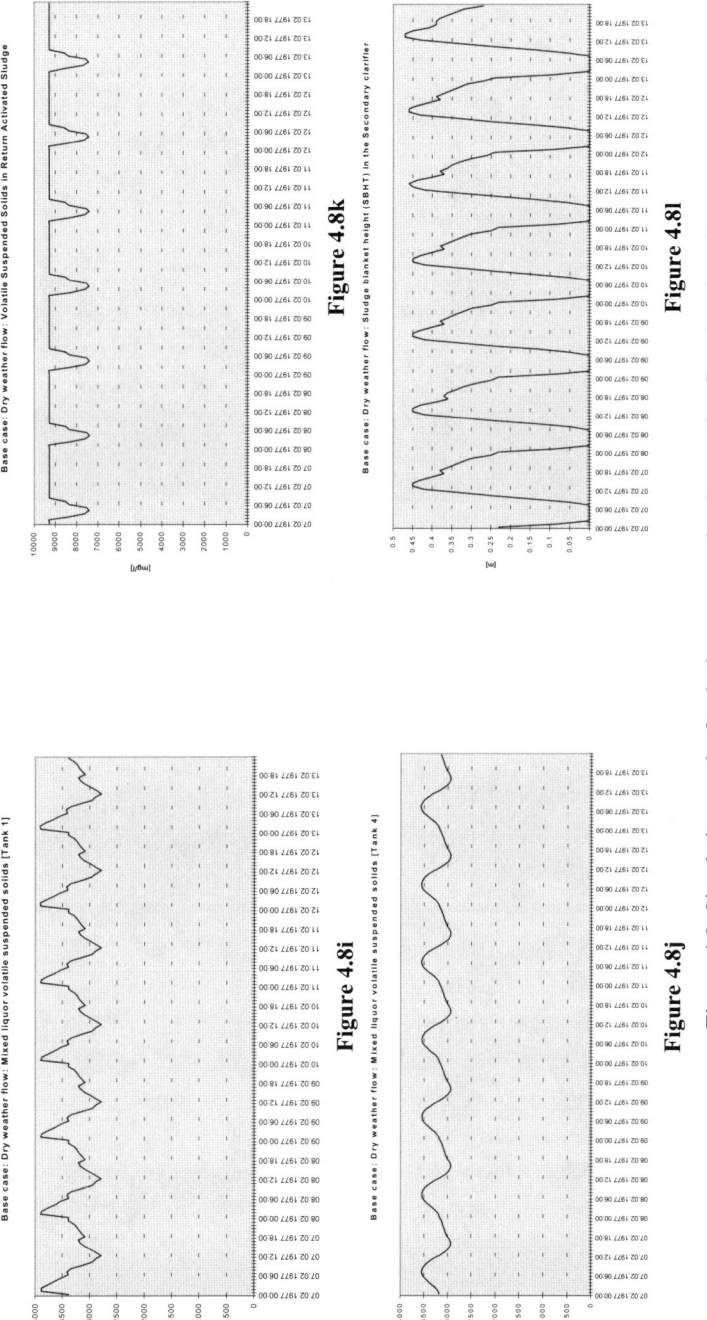

Figure 4.8. Simulation results for the base case under dry-weather flow (continued)

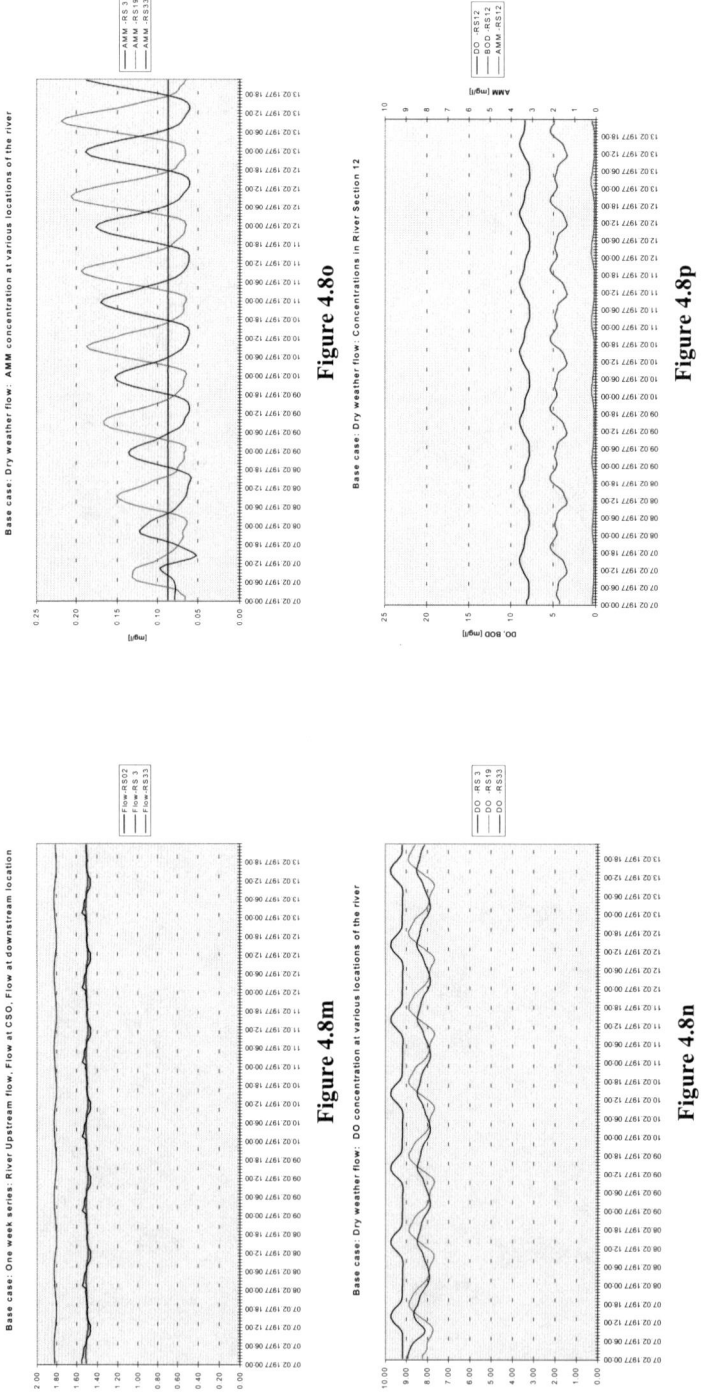

Figure 4.8. Simulation results for the base case under dry-weather flow (continued)

4.3 Simulation of a Rainfall Time Series

This section shows results obtained from the simulation using a time series of rainfall data as input. For this demonstration a simulation period of one week of the year 1977 has been selected from the rainfall data available from Fuhrberg/Northern Germany (see Appendix C for the rain data of the entire year and Figure 4.9 for the selected subseries). The annual precipitation measured at Fuhrberg in 1977 amounts to 621.2 mm, which is close to the annual mean of 634.2 mm (taken over the total time period of 11 years for which rainfall data were available). The total rainfall depth during the first four days of the selected week (7 – 10 February) amounts to 27.1 mm. The rainfall series has two main peaks (one on 7/8 February, with a total of 21.7 mm and the second on 10 February with 5.4 mm depth). Care was taken to select a significant, though not extreme event, for this simulation. Not only the simulation run detailed in this section, but also most of the optimisation runs described in Chapter 5 use this rainfall series as input. The data series contains rain data in 5-minute time steps (note that Figure 4.9 displays the rain data in hourly values).

After the end of the rain period itself, the simulation is continued over three additional days in order to allow any acute effects that could be a direct consequence of that rainfall to be considered. For these three days the rainfall was set to zero (the original data show 8.8 mm additional rainfall during these days). As the simulation results will show, this post-simulation period of three days is sufficient for this rainfall series. The simulation is preceded by simulation of one day of dry-weather flow (not shown in the figures) to allow the system to reach an initial steady-state.

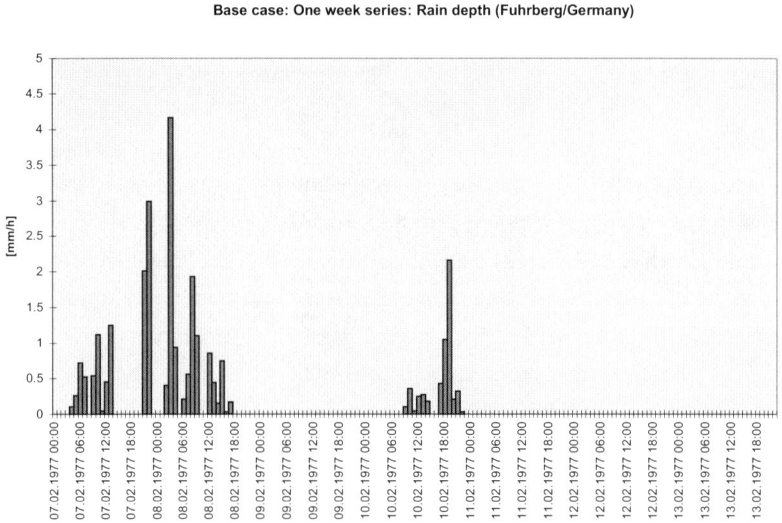

204 Modelling, Simulation and Control of Urban Wastewater Systems

Base case: One week series: Rain depth (Fuhrberg/Germany)

Figure 4.9. Rainfall series used in this study (adapted from the Fuhrberg data) (7 – 13 February 1977)

Simulation with this input rainfall series is carried out for the semi-hypothetical case study site described in Section 4.1. Control is performed as defined for the base case scenario (*cf.* Section 4.1.5).

Figure 4.10 provides some of the results obtained for this input.

As Figure 4.10a shows, CSO discharges can be observed twice in this period (the figure shows the sum of the discharges over the four overflows in the sewer system). It should be recalled here that in the base case scenario which is considered in this section, the maximum flows in the sewer system are set to 5 DWF, thereby resulting in relatively few overflows. As can be seen from Figure 4.10b, the storm tank at the treatment plant inlet is used twice during this time period. Its full capacity is used for a period of 22 hours, during which the tank overflows twice (Figure 4.10c). Viewing the treatment plant influent data (Figures 4.10d and e depict flow and COD concentrations, respectively), the dilution effect caused by less polluted rainwater can clearly be recognised (it has to be recalled that sedimentation processes in sewers are not simulated here; *cf.* Section 4.1.2). The treatment plant effluent flow (Figure 4.10f, not including storm tank discharges) is restricted to essentially 3 DWF due to the aerator bypass being active for flows exceeding this threshold. Effluent concentrations for solids and organic matter are shown in Figure 4.10g, and those for ammonium and nitrate can be found in Figure 4.10h. The loss of nitrification over parts of the simulation period can be seen clearly. The solids concentration in the secondary clarifier effluent appears to be limited to 39 mg/l. This upper limit results from the way the effluent SS concentration is calculated in the model (see the detailed discussion in Section 3.2.2.1). Since the sludge blanket height does not exceed a value of 1.58 m, the modifications to the secondary clarifier model performed by Vazquez-Sanchez do not come into effect. Finally, Figure 4.10i shows the river flow at three locations (upstream; between CSO and treatment plant discharge locations; downstream), which is increased due to the rainfall. The impact of the rainfall on the oxygen and ammonium balances can be seen in Figures 4.10j and k. Note that the lowest DO concentration in the river is observed at Section 33 at a time considerably later than the rainfall. Figure 4.10l shows the concentrations of BOD and its readily and slowly biodegradable fractions for a selected river section. The former fraction can be seen to be increased during periods of wet-weather discharges into the river compared to dry-weather periods.

Again, the simulation results seem to show plausible behaviour and fall into reasonable ranges.

An evaluation of this simulation run applying the various criteria defined in Section 3.3.3 results in the values shown in Table 4.11.

Table 4.11. Evaluation of the base case scenario applying the criteria defined in Section 3.3.3

DO-M	Minimum DO	3.5 mg/l	F2	combined from DO-M and DO-DU	3.86
DO-DU	Duration of DO below 4 mg/l	3.6%	F3	combined from AMM-M and AMM-DU	6.60
DO-E	6-hour minimum	4.4 mg/l	QCSO	CSO volume	3214 m^3
AMM-M	Maximum ammonium	5.8 mg/l	QST	Storm tank overflows	23567 m^3
AMM-DU	Duration of ammonium above 4 mg/l	7.2%	Qoverf	Total overflows	26781 m^3
AMM-E	6-hour maximum	1.8 mg/l	CODtot	Total COD load	22556 kg
BOD-M	Maximum BOD	21.1 mg/l	CSODur	CSO duration	3.01%
BOD-E	6-hour maximum	16.8 mg/l			

206 Modelling, Simulation and Control of Urban Wastewater Systems

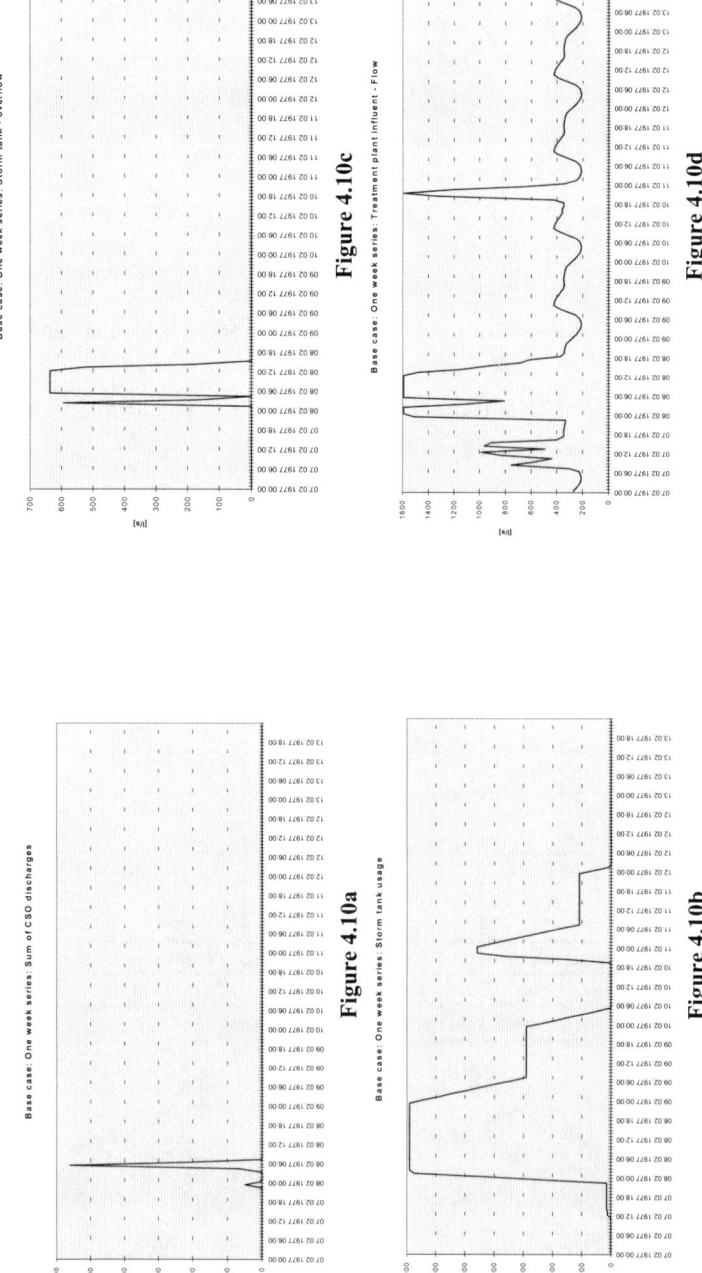

Figure 4.10. Simulation results for the base case for the input rainfall series 7 – 13 February 1977

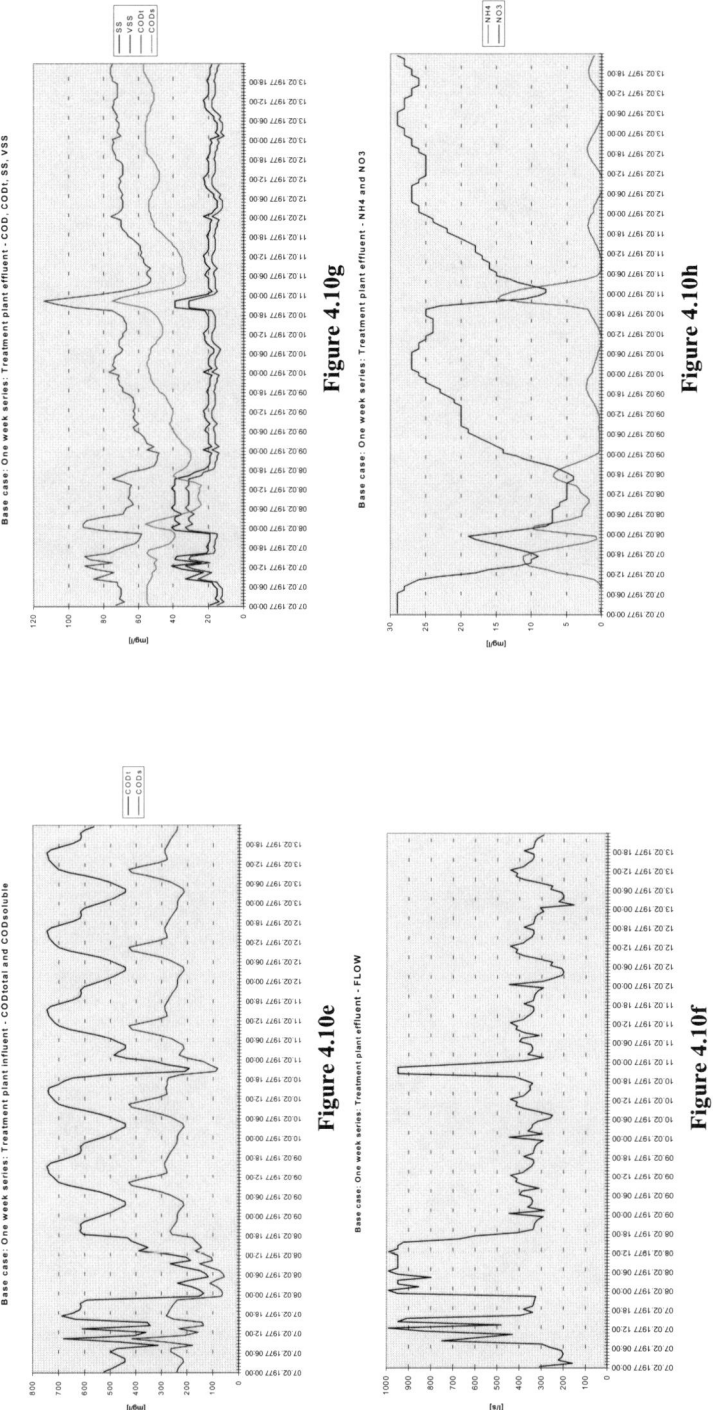

Figure 4.10. Simulation results for the base case for the input rainfall series 7 – 13 February 1977 (continued)

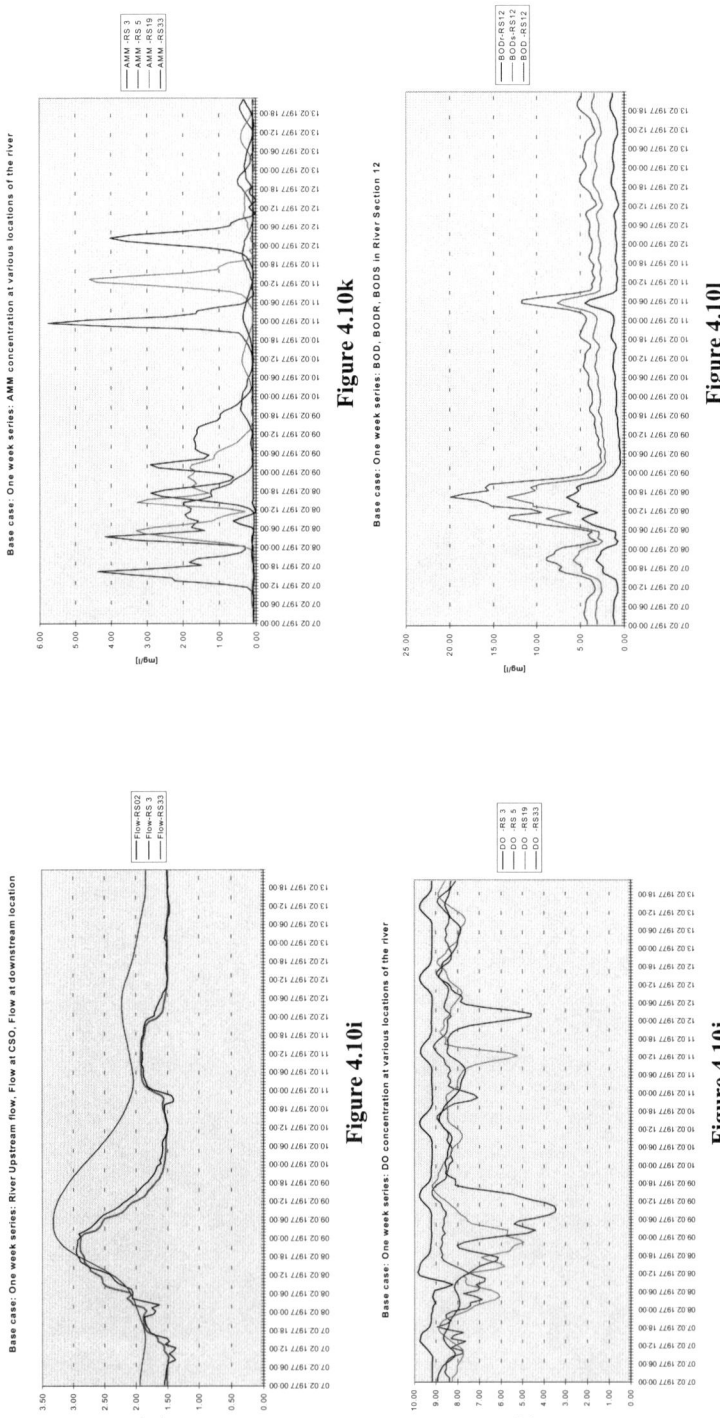

Figure 4.10. Simulation results for the base case for the input rainfall series 7 – 13 February 1977 (continued)

4.4 Analysis of the Control Devices of the Urban Wastewater System

In the previous sections simulation results were shown for two different input rain series (one of them representing dry-weather), applying the control devices of the urban wastewater system as defined in the base case scenario (*cf.* Section 4.1.5). This section analyses the effects of variations of constant settings of these devices. This should not only extend the set of simulations serving to demonstrate the capabilities of the simulation tool, but also give insights into the behaviour of the system under study and provide information about the feasible ranges of settings of the control devices. This will prove to be useful for the definition of the optimisation problems discussed in the subsequent chapters.

A series of simulation runs is performed for each of the control devices of the urban wastewater system listed in Table 3.7, in which the effect of various settings of the control devices on DO and ammonia concentrations in the river is analysed. Each of the simulation runs covers the first six months of the 1977 rainfall series (see Appendix A.3). Starting from the base case, a one-variable-at-a-time approach is chosen, varying the constant setting of each individual control device over a wide range of values. All other control variables are kept at their base case value. Figures 4.11 to 4.20 illustrate the results obtained in terms of the criteria describing DO and ammonium levels in the receiving river (see Section 3.3.3 for a definition of these criteria).

As outlined above, pumps in the sewer system are controlled in discrete pump rates. A pump rate of x l/s means that up to x l/s is pumped (discharged) to the downstream part of the system. For the sake of simplicity and comparability, the pump rates within the sewer system are expressed as multiples of average dry-weather flow of the part of the catchment which is upstream of the pump. The settings for each pump are varied here within a range from two to seven times average DWF.

Figure 4.11 shows the results for various settings of pump P7. This pump is furthest downstream in the sewer system. Therefore, it completely controls the inflow into the treatment plant and storm tank. How much water will flow into the primary clarifier and how much is diverted into the storm tanks is determined by the setting of the maximum inflow rate into the treatment plant (*PCINMX*). The setting of this parameter is discussed further below in this section.

With respect to the total overflow volume and to the DO concentration in the river, the results suggest that the optimum setting for pump P7 is at around four times average DWF. A setting lower than this causes more CSO discharges from within the sewer system, higher settings cause more storm tank discharges and increased inflow loads to the treatment plant.

The ammonium based criteria perform better for low pump rates towards the treatment plant, since for these rates increased loads deteriorating the performance of the plant are avoided. However, observations like this also strongly depend on the concentrations assumed for ammonium in the sewage and on the dilution in the river (*cf.* Sections 4.1.2 and 4.1.4).

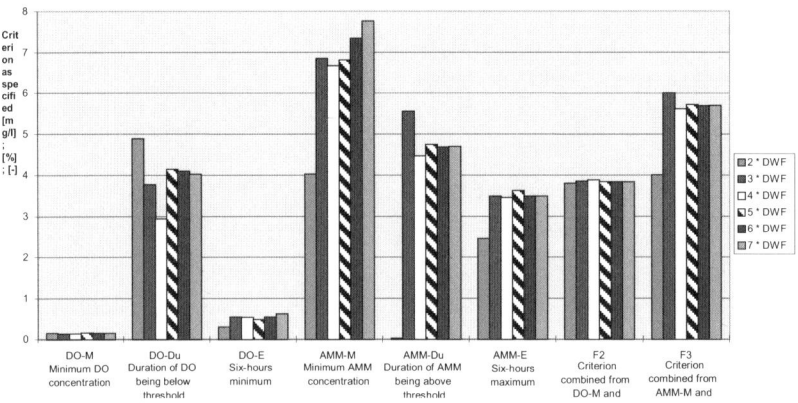

Figure 4.11a. Effect of settings of pump P7 – DO and ammonium criteria

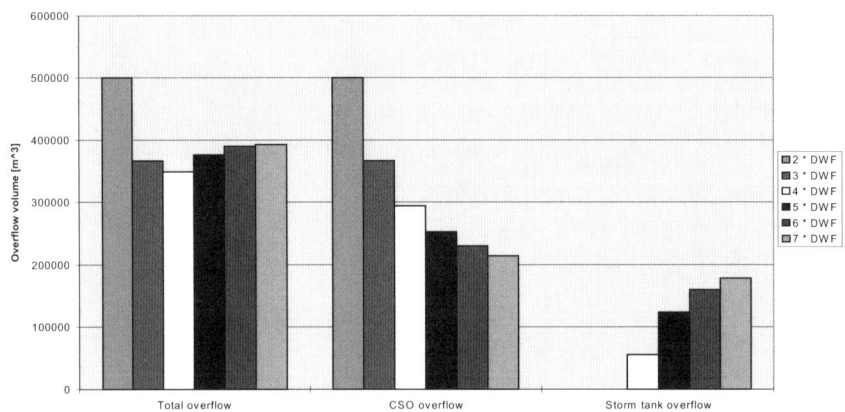

Figure 4.11b. Effect of settings of pump P7 – Overflow volumes

The figures also show the values of the F2 criterion, which will be used throughout this thesis. As defined in Section 3.3.3, the F2 value condenses the minimum concentration of DO and the duration of DO being below a threshold value of 4 mg/l into one single variable. The F3 criterion, based on the ammonium levels, is defined in a similar way.

Figures 4.12 to 4.14 provide similar information for the other pumps in the sewer system. As these pumps (P2, P4, P6) are located further upstream in the system, higher settings of these, resulting in increased outflows from the related subcatchments, also result in higher inflows further downstream. This explains why higher settings can lead to increased overall CSO discharge volume.

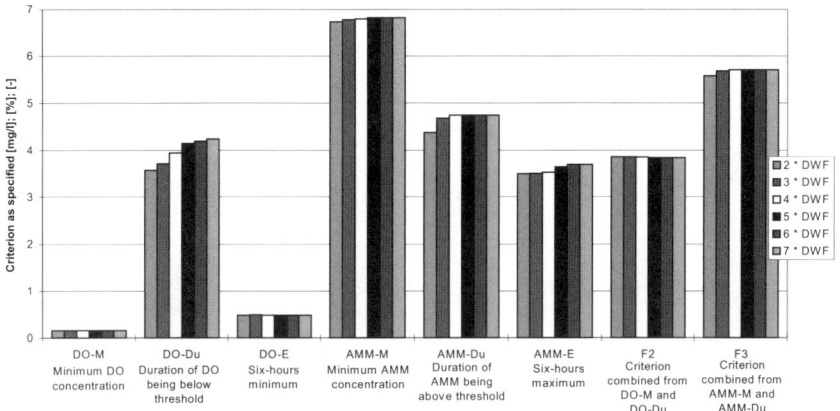

Figure 4.12a. Effect of settings of pump P6 – DO and ammonium criteria

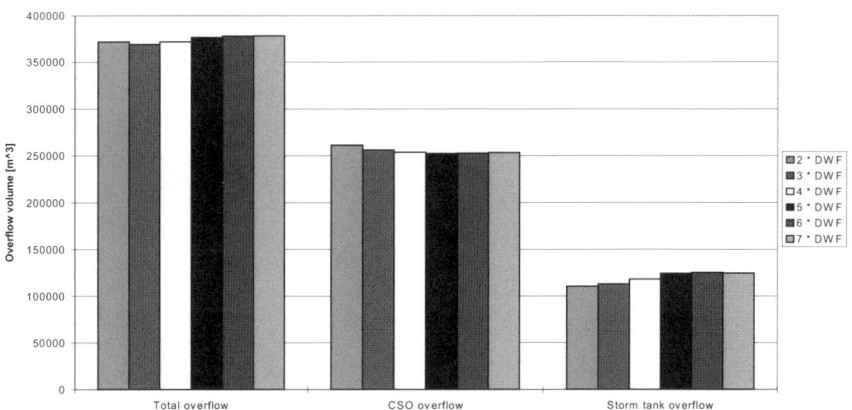

Figure 4.12b. Effect of settings of pump P6 – Overflow volumes

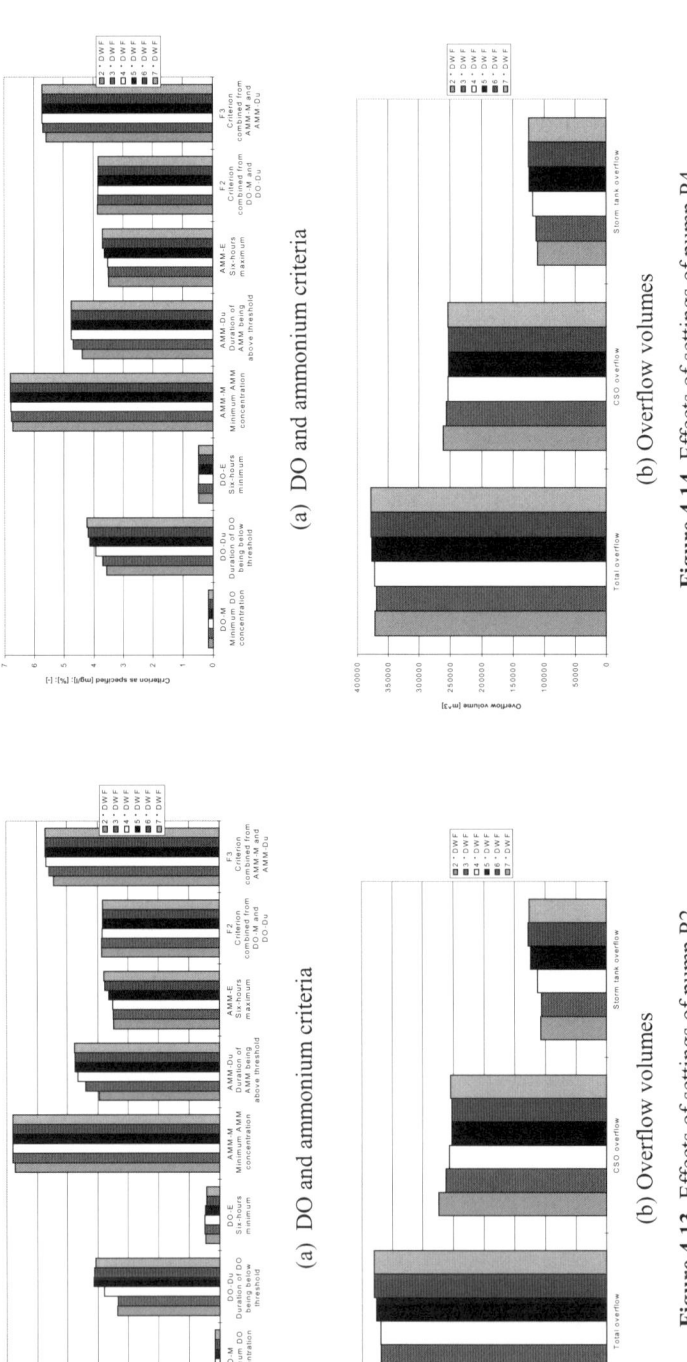

Figure 4.13. Effects of settings of pump P2

Figure 4.14. Effects of settings of pump P4

Figures 4.15 show results obtained in a similar manner – now for the control devices at the treatment plant.

Figure 4.15 shows the impact of various settings of the maximum inflow rate to the plant (set to 3 DWF in traditional treatment plant design). As can be seen, the optimum setting with regard to the DO level for the plant simulated here seems to be at a value slightly higher than the traditional design value. Similar observations were also made in other studies (Lessard, 1989; Guderian *et al.*, 1998).

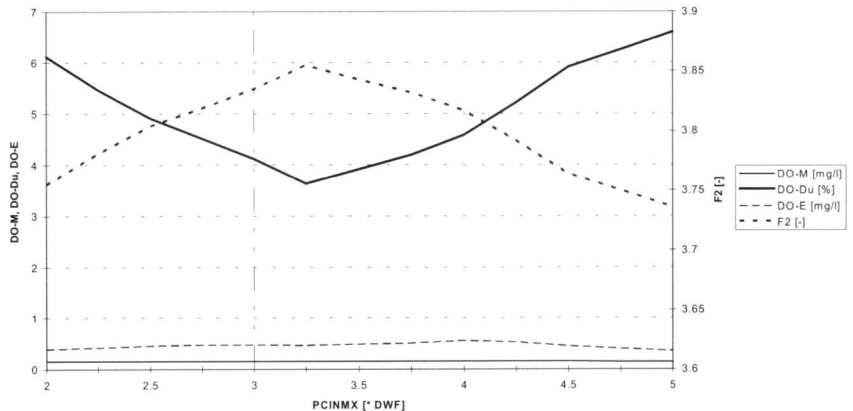

Figure 4.15a. Variation of the maximum permissible inflow to the treatment plant
– Effects on DO

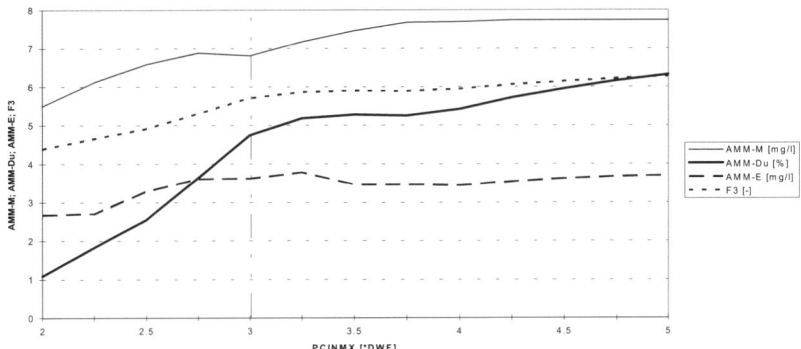

Figure 4.15b. Variation of the maximum permissible inflow to the treatment plant
– Effects on ammonium

Since the minimum DO concentration is lower than 4 mg/l, the value of F2 directly mirrors the duration of DO being below 4 mg/l (*cf.* the definition of F2).

An increase in the maximum permissible inflow rate to the treatment plant leads to a deterioration in terms of the ammonium criteria. This behaviour meets expectations. For maximum inflows to the treatment plant greater than 3 DWF, however, only the maximum ammonium concentration in the river and the duration of its concentration greater than 4 mg/l increase. As opposed to these two criteria, the maximum of the ammonium concentration taken over a period of six hours (marked as AMM-E in Figure 4.15b) does not show substantial variation. Overall, it can be said that for the plant simulated here, the traditional approach of restricting the maximum inflow to the plant to 3 DWF seems indeed to be a reasonable setting.

Figures 4.16 and 4.17 show the effects of different settings concerning the operation of the storm tank located at the treatment plant inlet.

From Figure 4.16 it becomes obvious that the rate with which the storm tank is emptied back towards the treatment plant does not have much influence on DO. With regard to ammonium levels, hardly any effects of the storm tank emptying rate can be distinguished. The duration of high ammonium concentrations in the river is slightly larger for low emptying rates, which cause prolonged higher influent loads to the plant.

As Figure 4.17 shows, the threshold value of the inflow to the treatment plant which triggers emptying of the storm tank seems to be more significant than the emptying rate itself. However, effects of variations of this threshold can be observed only in a fairly narrow interval (around 700 to 1300 m^3/h). This is not surprising, given that the lowest flow within the sewer system (dry-weather flow at its daily minimum) amounts to 734 m^3/h. Flows towards the plant exceeding 1300 m^3/h are observed only for relatively short time periods (compared to the total period covered by the simulation).

Simulation of the Urban Wastewater System 215

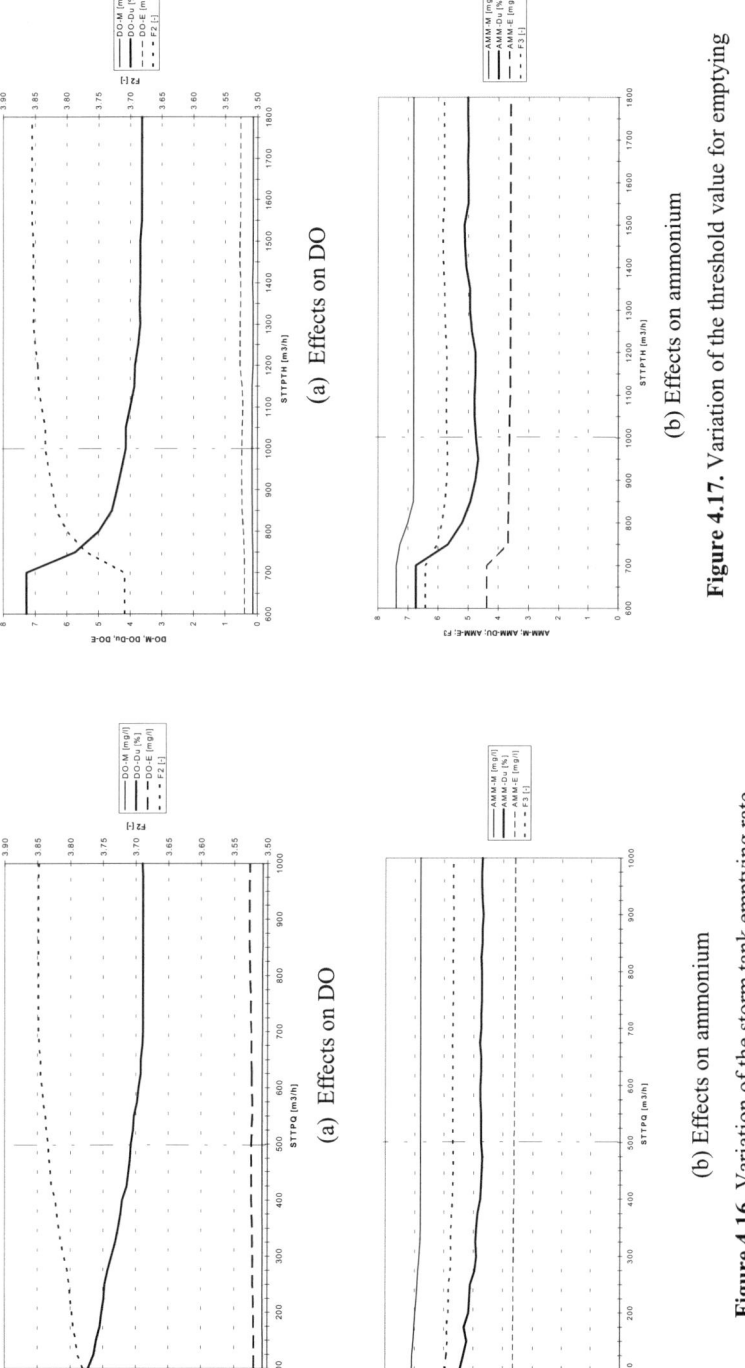

Figure 4.16. Variation of the storm tank emptying rate

Figure 4.17. Variation of the threshold value for emptying

The impact of variations of the settings of return and waste activated sludge rates is illustrated in Figures 4.18 to 4.20. Figure 4.18 shows results obtained from simulations in which the RAS rate is kept constant over the entire simulation period (here: 6 months). As can be seen, the impact (for settings of the RAS rate not too far from its base case setting) is more pronounced on the DO levels rather than on the ammonium levels. The base case setting of 600 m^3/h corresponds to about 50% of the dry-weather inflow rate. Also this parameter of the base case scenario seems to be close to its optimum setting.

Figure 4.19 shows results of simulations in which the RAS rate is set to a fixed fraction of the inflow rate to the treatment plant, which represents a frequently applied type of RAS rate control. Under dry-weather flow the range of the RAS fraction analysed here (0.25 to 0.9) corresponds to the range of the fixed RAS rates shown in Figure 4.18. Therefore, these figures allow a direct comparison of the effects of the use of fixed or proportional setting of the RAS rate. According to the results shown here, use of a proportional RAS rate leads to slightly better results in terms of ammonium. However, the differences between the effects of the two types of RAS rate control do not seem to be very significant.

Figure 4.20 illustrates the effects of different constant settings of the waste activated sludge rate. The impacts on minimum and 6-hours minimum concentration of DO appear to be almost negligible. However, low WAS rates result in long periods of the DO concentration being below the critical threshold of 4 mg/l. An explanation for this might be the build-up of the sludge blanket in the secondary clarifier at low WAS rate conditions (see also the discussion in Section 4.1.3), which results in increased solids concentrations in the effluent and possibly loss of sludge from the clarifier.

With regard to ammonium concentrations in the river it can be said that these seem to be affected by the WAS rate only at relatively high settings.

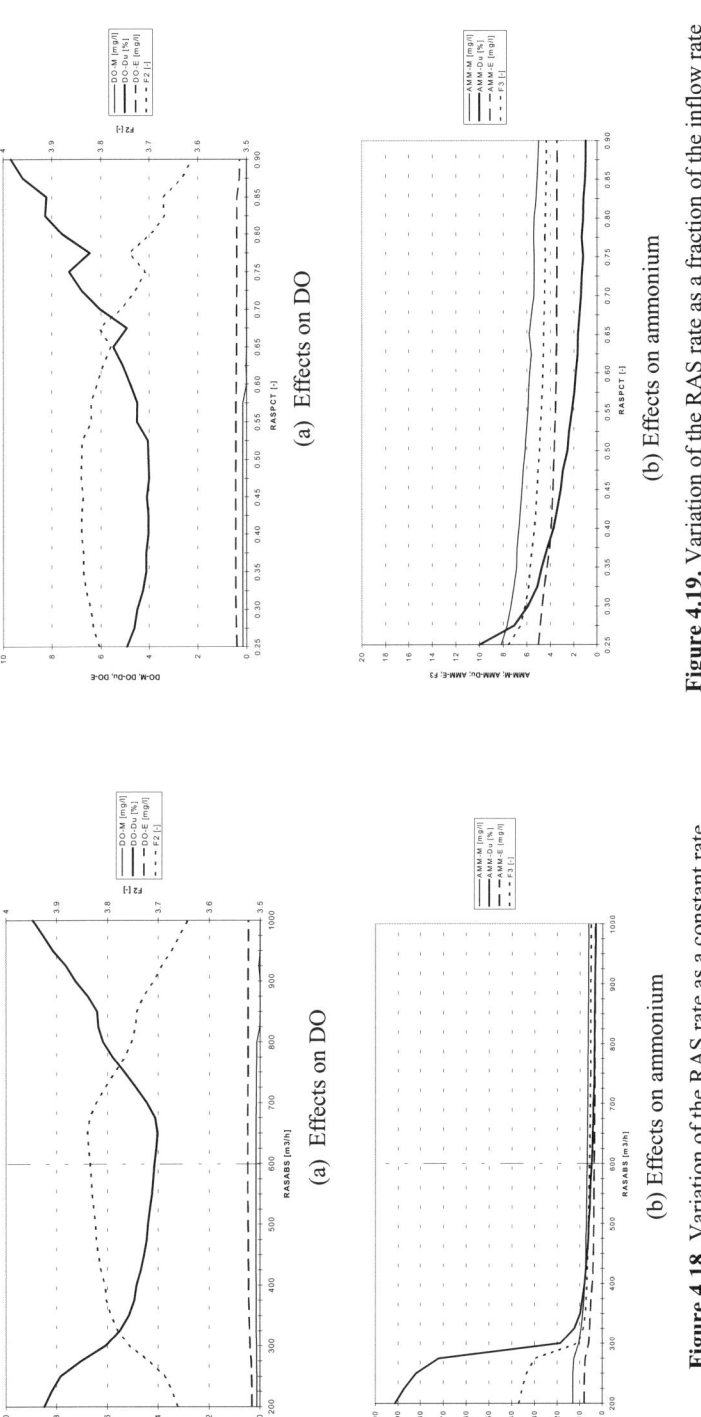

Figure 4.18. Variation of the RAS rate as a constant rate

Figure 4.19. Variation of the RAS rate as a fraction of the inflow rate

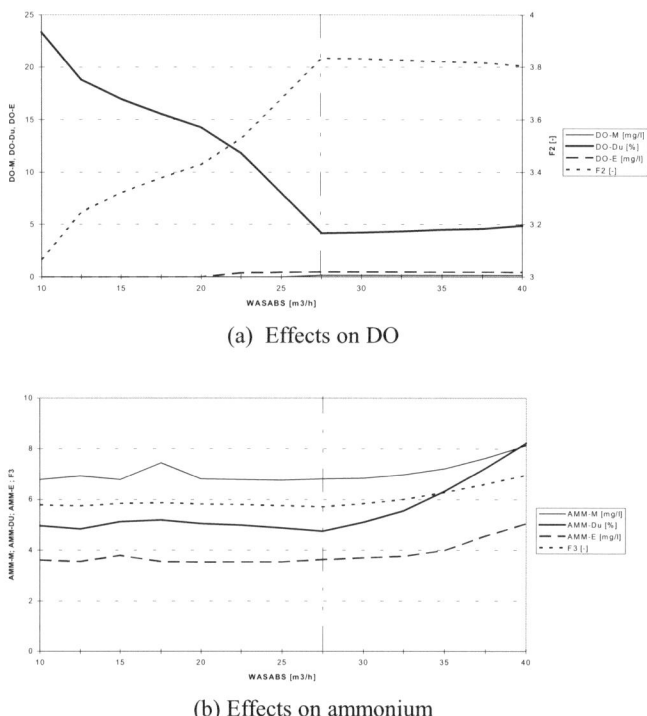

(a) Effects on DO

(b) Effects on ammonium

Figure 4.20. Variation of the waste activated sludge rate

4.5 Potential of Reduction in Simulation Time by Selective Simulation

4.5.1 Separation of Rainfall Events

Among the most obvious issues to consider when a fairly large number of simulation runs is planned (as necessary for the simulation and optimisation approach to the off-line development of strategies, as outlined in earlier chapters) is the question of ow the computational demand can be kept to a minimum. One solution to this problem would consist of the use of a less detailed model for the description of the process behaviour. The potential of the option to perform the optimisation on a series of substitute models (rather than on the "full" model), such as pursued in the approach of the Response Surface Methodology (see Section 2.5.2.1), will be briefly discussed in the comparative evaluation of optimisation methods of Section 5.3

Besides embarking on this option or on alternative approaches to reduce the computation time, which include the use of faster hardware or less demanding optimisation procedures, one also could attempt to keep the input rain data series as short as possible. This section analyses whether and to what extent gains in computation speed can be expected by the simulation of individual events rather than a continuous time series of rain data.

The approach to simulate a number of selected rainfall events is widespread, in particular in simulation exercises involving detailed sewer system models. A related approach is to simulate a series of hypothetical events, assuming that this selected series represents the "typical" or most relevant scenarios. Common to both approaches is the inherent need (and possibly difficulty) to define what is meant by "event", or by "typical". The overall aim when defining (or separating) individual events is to ensure their independence from each other: two events are considered to be independent if the system state of interest is influenced only by the input (*e.g.*, rain) within this period and not dependent on the system state prior to the event. Common simulation practice in sewer system simulation uses various criteria, such as inter-event period and minimum rainfall depth per event, for the separation of individual events. When considering the urban wastewater system as a whole, any definition of the term "event" has to incorporate potential effects of transient pollution events not only within the sewer system, but also at the treatment plant and the receiving water body.

It should be noted that the definition of an "event" (as the period during and after an incident of interest) depends on the particular processes of interest. For example, in an analysis of increased river flows caused by CSO discharges on the one hand and in an analysis on long-term effects of heavy metals on the aquatic ecosystem within a river on the other hand, quite different definitions of the term "event" will have to be used. A typical duration of one event would be from several hours up to a couple of days in the former case, whereas in the latter example consideration of periods of months or years, if not decades, would be more appropriate (*cf.* House *et al.*, 1993).

As the investigations within this project focus on short-term impacts, which are centred around the oxygen balance in the river (under simplification of in-river sedimentation processes), the time-span to be considered per event becomes significantly shorter than in the above mentioned more general cases. Therefore – and only because of this focus of this project – the approach of considering individual events becomes potentially feasible and motivates a further study aiming at the identification of the potential savings in computing time achieved by simulating individual events rather than a continuous series.

Essentially, there are two options to define a series of events. One of them consists in the definition of hypothetical events, which would cover a range of "interesting" (relevant and/or typical) scenarios for the system under investigation. The task remains to find a good and representative set of events meeting these requirements. As the example of design storms in sewer design practice shows, this is not always an easy task. The questions "What is a typical event?" and "How can it be ensured that the constructed events really do cover the processes and effects of interest?" would still have to be replied to.

The alternative approach is to define a set of criteria which could be applied to select a subseries, *i.e.,* the really relevant (with respect to the objectives of the investigation) parts of the available historical data. One advantage of using historical (as opposed to hypothetical) data is that statistical analyses can be performed without having to assign statistical properties (such as return periods) to constructed events. This argument is obviously only valid if the available series of data covers a sufficiently long period of time. Given this, statistical evaluations can be carried out on the pollution data (resulting from simulations using historical rainfall data as input), and would not have to be restricted to a statistical analysis of rainfall data. As discussed in Section 2.2.4, some of the river water quality criteria described in the literature (Spildevandskomiteen, 1985; FWR, 1994) are based on such a statistical evaluation.

This approach of selecting events from a historical data series is analysed in this section. As a starting point, rainfall events are selected from a continuous historic rainfall series, using various settings of conventionally applied event separation criteria. These criteria aim at defining the end of a rainfall event as that moment when the last rain-drop has left the sewer system. Such criteria include inter-event time (minimum dry-weather periods necessary to separate two events from each other) and minimum total rain depth per event. A criterion such as the latter one filters out all events which are considered to be irrelevant to the study, since their total rainfall is insignificant, potentially not even generating rainfall runoff. A typical value for this criterion (within the context of sewer system simulation) is 3 mm (*cf.* the Danish rainfall series ODE1571 series, which covers all events with a total depth of at least 3 mm out of a period of 30 years (Spildevandskomiteen, 1984)).

When defining event separation criteria for a specific study, the characteristics of the case study site (*e.g.,* the maximum flow time in the sewer system; time required to empty all storage tanks in the system) are to be taken into account. For the case study site analysed here, the maximum flow time in the sewer system is 65 minutes, and the time required to empty the storm tanks at the treatment plant (emptying the tanks in the sewer system takes less time) at the lowest rate (of 300 m^3/h, which is in the feasible range defined for some of the optimisation runs in Chapter 5) amounts to 22.5 hours. When taking into consideration the treatment plant (and, finally, the receiving water body as well), the time period to be considered for the definition of the term "event" will be longer since rainfall events generally have impacts on the processes in these parts of the urban wastewater system even (possibly significantly later) after the actual rainfall. Such impacts include, for example, the potential deterioration of the treatment plant performance due to increased hydraulic loads, potentially sludge losses in the secondary clarifier, and the shape of the oxygen-sag curves in the river)

It is also interesting to note in the context of defining inter-event times that a recent CIRIA report (Ainger *et al.*, 1996) recommends a delay period of at least seven consecutive days of dry-weather after rainfall has stopped before taking samples of dry-weather flow sewage. This suggests that quality impacts of storm events influence water quality parameters within the sewer system for up to seven days.

To summarise the discussion of this subsection, it can be stated that a selection of a series of relevant events is influenced not only by the rainfall data, but also by the case study site under investigation, by the control strategy applied to this system and also by the criteria which are used for a particular study (*e.g.,* concentration of DO in the river). The definition of inter-event times has to consider the processes within all parts of the urban wastewater system.

Rather than investigating, in detail, all these processes at this stage, a preliminary investigation of the potential savings in computation time has been carried out, based on some relatively crude, but not over-cautious, assumptions. After this, a more detailed investigation, including simulation studies, is carried out for various event separation criteria.

4.5.2 Potential Savings in Simulation Time

This section briefly evaluates the potential savings in computing time which can be achieved by simulation of selected events only ("long-term series simulation").

All analyses described here were performed on a subseries covering six months (January to June 1977) of the Fuhrberg rainfall data shown in Appendix A.3.

As criteria for the first investigation, a minimum rainfall depth of 0.3 mm and initial intensity of 0.06 mm/5 min (exceeding evaporation losses) have been chosen. As a (minimum) inter-event time, four days is considered to be appropriate, following the discussion above. This value is larger than the maximum time period required for emptying storage tanks, it also covers the extent of the DO-sag in the river as observed in the simulation runs described in Section 4.3. A borderline of 96 hours was also given by Hvitved-Jacobsen (1986) to discriminate between short and long-term impacts of CSOs. However, analyses of the rain data were performed also for a wide range of other values for the inter-event time (*cf.* Figure 4.21).

Figure 4.21 provides an analysis of the rain data with regard to the potential of decreasing simulation time by selective simulation. For different settings of the inter-event time in the event selection process, the resulting number of events and the percentage of total rainfall covered by the selected series is shown. Other information included in the figure are the percentage of the event time (this includes also dry-weather periods, which are shorter than the inter-event time defined; furthermore, it contains the inter-event time which, by definition, is characterised by (almost) dry-weather) and of the rain time of the total time period considered. The rain time (per event) is defined here as the event duration minus the inter-event time. In a simulation of a series of rainfall events, an allowance has also to be made for initialisation procedures and to achieve a steady-state of the simulated system prior to the simulation of the event itself. The computational demand for pre-rain simulation increases linearly with the number of simulated events. It therefore outweighs some gains in computation time one gets from the definition of very short inter-event times.

However, some savings in computation time can be achieved if – after the last rain-drop of an event has fallen – only the relevant portion of the inter-event period is simulated, not the complete period. In order to determine the actual minimum DO concentration in the river during this event or the duration during which the DO concentration in the river is lower than 4 mg/l, one does not necessarily have to model the complete DO-sag curve. Following these considerations, one can make a distinction between the inter-event time (required for proper separation of events) and a shorter post-rainfall simulation time. More specifically, the crude assumption is made here that the part of the DO-sag which is relevant to the determination of the values of the DO criteria defined above can be observed within 48 hours after the end of any rainfall. The potential savings in computation time achieved by this procedure can be seen in Figures 4.21 and 4.22 when comparing the bars marked "% of total time to be simulated" with the bar "% Event of total time". The total time period to be simulated, including all considerations of this and the previous paragraphs, is indicated by the bar "% of total time to be simulated" in the figures.

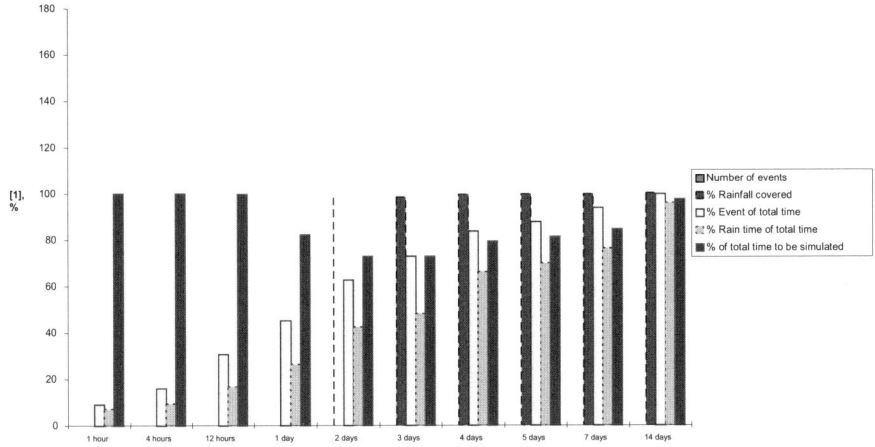

Figure 4.21. Event and simulation times for various inter-event times

Figure 4.21 is based on a minimum rain-depth per event of 0.3 mm. This is a cautious assumption (compare this to the value of 3 mm used for setting up the Danish rain series). Subsequent investigations (*cf.* Table 4.11) show that savings gained by definition of a larger minimum rain depth (*e.g.,* 3 mm) are not too substantial.

Figure 4.21 shows, for example, that when setting the inter-event time to a value of 4 days (as the above discussion shows, this is not overly cautious), one would have to simulate approximately 80% of the total time period. This, consequently results in savings of slightly more than 20%. The use of shorter inter-event times (such as 2 or 3 days) would result in a slight, though not impressive, reduction of simulation time, with the minimum simulation time achieved for an inter-event time of 2 days. Even in this case the savings do not exceed 30%.

Applying less stringent criteria for the separation of events (neglecting periods of very low rainfall during an event, which will avoid the duration of an event too long just because a negligible amount of rain has fallen), one would achieve savings of nearly 40% (Figure 4.22) when defining an inter-event time of four days.

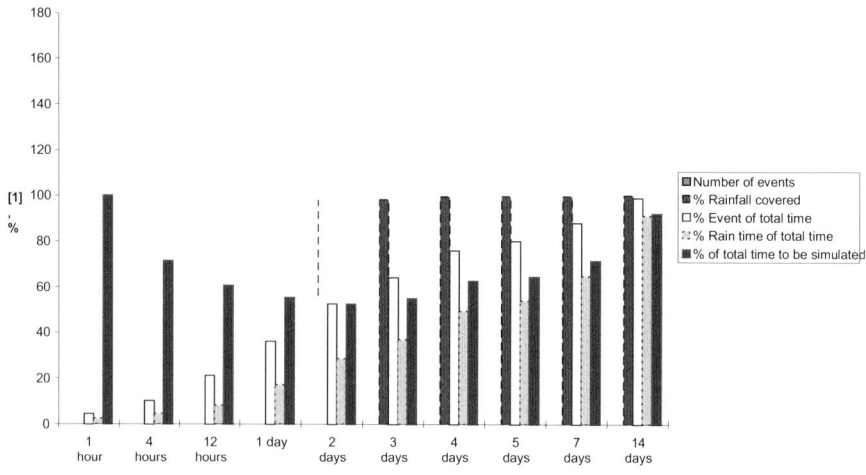

Figure 4.22. Event and simulation times for various inter-event times (less strict event separation criteria)

Complementing the data shown in Figures 4.21 and 4.22, Table 4.12 lists the savings in computation time which can be expected for various settings of the event separation criteria. Simulation runs of series of events defined according to these criteria (selected from continuous rain data of 6 months) are discussed below.

Table 4.12. Various event separation criteria and resulting savings in total simulation time

Rain series	Pre-/Post event time [h]	Inter-event time [days]	Minimum total depth [mm]	Insignificance limit [mm]	Number of events	Potential savings in time [%]
R1	24 + 48	4	3.0	0.00	8	25.9
R2	24 + 48	4	0.3	0.00	8	25.9
R3	24 + 48	4	3.0	0.06	11	43.6
R4	24 + 48	4	0.3	0.06	13	41.1
R5	24 + 48	2	3.0	0.00	16	46.7
R6	24 + 48	2	0.3	0.00	19	43.9
R7	24 + 48	2	3.0	0.06	13	61.6
R8	24 + 48	2	0.3	0.06	22	53.4

n.b.: The column "Insignificance limit" specifies up to which depth (within 5 mins) rain data are ignored during an event.

4.5.3 Selective Versus Continuous Simulation

Several of the various rainfall series defined in Table 4.12 are investigated further. Whenever the additional savings to be expected resulting from applying less strict (*i.e.,* less cautious) event separation criteria to each of the four pairs in this table are not significant, then these stricter criteria are chosen. This motivates the use of the series R2, R4 and R6 for a more detailed analysis. Since out of the pair R7-R8, simulation of R7 promises considerably larger savings in computing time, R7 also is investigated in more detail.

A set of simulation runs is carried out for the rainfall series resulting from the application of the event separation criteria as listed under R2, R4, R6 and R7 in Table 4.12. Different scenarios for the settings of the control devices in the system are simulated for each of these series of events, resulting in 16 simulation runs each. These settings are to account (in a very simplified manner) for potentially applied control in pump rates and return and waste sludge rates. The detailed investigation of control strategies is the topic of later chapters; here just a very simple attempt is made to consider various settings of control variables for the investigations with respect to separation of rain events.

When performing simulations of selected events only (rather than the complete series), one would aim to get results which are related (for example yielding similar ranking of options or control strategies) to the results one would obtain in a simulation of the complete data series. In order to check whether the simulation of a subset of the complete data gives comparable results, Table 4.13 gives, as a simple measure to assess the comparability between the results obtained using the complete continuous rainfall series and the series containing the selected events, their correlation coefficients (indicating whether a linear relationship exists). Correlation coefficients are calculated for several criteria describing DO and ammonium levels in the river (see Section 3.3.3 for their definition).

Table 4.13. Correlation of results of simulation of full and partial rain series

Rain series	DO-M	DO-DU	DO-E	AMM-M	AMM-E	AMM-DU	F2
R2	0.927	0.866	0.999	0.989	0.993	0.998	0.865
R4	0.927	0.727	0.988	0.989	0.992	0.996	0.727
R6	0.755	0.582	0.999	0.990	0.987	0.996	0.584
R7	0.804	0.328	0.966	0.952	0.807	0.974	0.325

Out of these four rainfall series, only R2 appears to give results fairly close to those of the original series. Simulations carried out for R4 (for which similar, but stricter event separation criteria were used than for R2) lead to fairly poor results with regard to the DO-DU criterion. Therefore, it can be concluded that from the series shown in Table 4.13, only R2 seems to give representative results. Table 4.14 lists the events of series R2.

Since simulation of the series R2 would not result in significant savings of computation time (see Table 4.12), an attempt at systematic selection of events is made. Such an approach has been applied (within a different context) successfully by Schütze (1990). The idea of this approach is to sort the list of events obtained by application of some event separation criteria according to maximum depth and then to pick, say, every third event (*i.e.,* the first, the fourth, the seventh, ...). This ensures to some degree the "representativeness" of the selected rain events at least with respect to the sort criterion applied. Picking the second, fifth *etc.,* event of this list results in a similar set which then can be used for validation purposes.

Table 4.14. Event series R2 in chronological order

Event start date	Event start time	Event end date	Event end time	Event duration [days]	Total depth [mm]	Sub-series No.
01.01.1977	05:05	01.02.1977	05:55	31.08	37.9	R2X1
03.02.1977	21:35	24.02.1977	15:00	20.73	53.6	
25.02.1977	03:55	22.03.1977	07:15	25.14	18.5	
27.03.1977	08:55	05.05.1977	06:25	38.90	77.4	R2X2
08.05.1977	08:55	18.05.1977	18:10	10.39	14.9	R2X1
20.05.1977	07:00	25.05.1977	07:30	5.02	26.8	R2X2
30.05.1977	09:50	21.06.1977	06:00	21.84	88.3	R2X1
25.06.1977	14:10	30.06.1977	00:55	4.46	10.7	R2X2

Two subseries of events of R2 found by application of this method are marked in Table 4.14, identified by "R2X1" and "R2X2". In a similar way, subseries of events (R4X1, R6X1, R7X1) are identified from the other series defined in Table 4.12. The results of the corresponding correlation analyses are shown in Table 4.15.

Table 4.15. Correlation of results of simulation of full rainfall series and selected events. *n.b.*: The column "Time saving" indicates the savings in computation timecompared to simulation of the complete (continuous) rain series

Rain series	DO-M	DO-DU	DO-E	AMM-M	AMM-E	AMM-DU	F2	Time saving [%]
R2X1	0.927	0.862	0.999	0.945	0.868	0.952	0.850	65.2
R4X1	0.927	0.727	0.988	0.989	0.992	0.996	0.727	69.6
R6X1	0.739	0.467	0.966	0.997	0.967	0.988	0.468	65.7
R7X1	0.601	0.395	0.966	0.968	0.869	0.964	0.398	80.6
R2X2	0.349	0.810	0.758	0.962	0.963	0.960	0.810	72.9%

As the results shown in this table suggest, selection of events according to the method described above does not lead to satisfactory results. When comparing the results shown for the individual assessment criteria one can observe that the ammonium based criteria appear to be less sensitive to event selection than the DO based criteria. The series R2X1 gives results similar to those of R2; however, this does not necessarily support the chosen approach for the selection of events, since the series R2X2 (which has been obtained in a similar way to R2X1) does not show high correlation (except for the ammonium based criteria) with the results of the continuum. The effects observed might be due to an extreme event which is contained in the series R2 and R2X1, but not in the R2X2 series.

4.5.4 Conclusions

As the discussion of the previous subsections suggests, selection of individual events from a historic rainseries can result in some savings in simulation time. However, these savings are not of significant magnitude. Futhermore, great care has to be taken when defining and separating individual events. This also involves additional uncertainty about the validity of the results obtained from using the selected subseries for simulations.

From the rain series investigated in this substudy, only R2 (defined by cautious assumptions for the event separation criteria) provides results comparable to those gained by simulation of the full series. A significant reduction of simulation time can (only) be achieved by selection of a subset of the individual events of the full series. However, application of the proposed method does not generally lead to satisfactory results (with respect to the water quality criteria applied in this study). An exception is the series R2X1, which might be due to an extreme event covered by this selection.

Since the results of the analysis described in this section are not very encouraging, the use of a complete continuous time series (rather than the selection of individual events) seems to be necessary for the simulation and optimisation of urban wastewater systems. Therefore, such a series will be used throughout the simulations described in the remainder of this book. Since a large number of simulations and optimisations is to be carried out, the results of which are to be compared with against each other, the relatively short data series introduced in Section 4.3 (covering one week in February 1977) will be used throughout the remaining sections of this text, unless stated otherwise. However, Section 5.6.3 will briefly discuss the issue of required length of the input rain series with respect to the optimisation of control strategies.

Chapter 5
Analysis of Control Scenarios by Simulation and Optimisation

This chapter applies the simulation and optimisation tool SYNOPSIS to the evaluation of various control scenarios for the semi-hypothetical case study defined in the previous chapter. Some fundamental definitions are given in Section 5.1. Section 5.2 provides an example of the application of the simulation and optimisation procedure to the determination of the parameters of a control strategy.

The topic of the subsequent sections is the definition and analysis of various frameworks defining conventional and integrated control scenarios, thereby approaching the initial question "Is integrated control superior to conventional control?". Since the definition of appropriate strategy frameworks is of importance for the success or failure of a control scenario, two different main routes for the definition of frameworks (called "top-down" and "bottom-up" approaches) are chosen here. Subsequently, strategy parameters are determined by the optimisation techniques discussed earlier on. Section 5.3 describes a top-down approach (starting from the point of view of the urban wastewater system, *i.e.,* from the conceptual "top") for the definition of frameworks. Section 5.3 also compares various optimisation procedures, including those discussed in Section 3.5, and draws conclusions about their performance within the given context of optimising strategy parameters. Section 5.4 presents an alternative approach to the definition of frameworks, starting from the point of view of individual controllers ("bottom-up"). A comparison of the results obtained in Sections 5.3 and 5.4 as well as related conclusions are presented in Section 5.5. Finally, Section 5.6 briefly discusses various matters related to the simulation and optimisation procedure applied in this study (e.g. sensitivity of solutions found, consideration of multiple objectives).

5.1 Definitions and Methodology

Prior to the application of SYNOPSIS to the development and refinement of control strategies, a few definitions are provided here, which are fundamental to the subsequent discussion.

A *strategy framework* is defined by a list of devices to be controlled, a list of sensor information to be used for control of these devices and a procedure describing how the settings of the control devices are determined from the available sensor information. A *control strategy* is then defined by the (finite dimensional) vector of parameters describing such a procedure. Note that this definition of the term "control strategy" is in accordance with the definition provided by Schilling (1989) (see also Section 2.4.1). The components of the vector defining a strategy are called *strategy parameters*. These can, for example, include the following:

- fixed settings of control devices;
- set-points of controllers;
- controller parameters;
- threshold values triggering control actions or changes of controller set-points;
- numerical values in condition and action parts of if-then rules.

Hence, a strategy framework describes the general concept of a strategy, without, however, assigning specific values to its parameters. A control strategy represents an instantiation of a framework, with defined values of its parameters. Determination of optimum settings of these is achieved in this work by the application of optimisation methods (*cf.* Figure 3.13). The term *control scenario* denotes a general concept of control (e.g. local, integrated), which then can be specified by one or several frameworks. Overall, the process of finding optimum (or good) strategies consists of two phases: finding an adequate framework and subsequent determination of optimum (or, at least, good) settings of its parameters.

For the evaluation of different control scenarios using SYNOPSIS, corresponding strategy frameworks have to be defined and supplied as input to the software (*cf.* Figure 3.14), which then optimises the settings of the strategy parameters. This procedure is demonstrated in Section 5.2 and applied in subsequent sections to the analysis of the driving question of this book "Is integrated control superior to conventional control?"

All of the above listed options for the parametrisation of control strategies are implemented, in a simple manner, in the control module of SYNOPSIS. In addition to the option of defining constant settings for the control devices listed in Table 3.7, two-point controllers defining the setting of a control device as a function of the value provided by one sensor ("single input – single output") can also be simulated by SYNOPSIS as described below. Two-point controllers represent a simple type of discrete controller (Schilling, 1989), the use of which appears to be appropriate within this work since control decisions are taken at discrete time steps (*cf.* Section 3.4). Also the use of proportionality factors for control (such as frequently applied for the RAS rate) is implemented. Furthermore, in Section 5.3, a set of threshold values describing various states of the urban wastewater system, which are used in if-then rules, is defined and analysed. This represents an example of use of numerical values in if-then rules (*cf.* the last two options for the parametrisation of control strategies listed above).

A two-point controller is defined by a set of four parameters $(x_0, y_0, \Delta x, \Delta y)$ as shown in Figure 5.1. As long as the measurement value of the sensor is lower than $x_0+\Delta x$, the control device is set to the value y_0. As soon as the sensor gives a value greater-equal than $x_0+\Delta x$, the control device is set to the value $y_0+\Delta y$. It is switched back to y_0 when the sensor value drops below x_0. The hysteresis defined by Δx prevents frequent switching of the control device whenever the sensor value is close to the threshold value x_0. Two important special cases are obtained for certain settings of the parameters: setting the parameter Δx to zero eliminates the hysteresis and allows simple threshold control. Similarly, setting Δy to zero results in the control device being operated with constant setting.

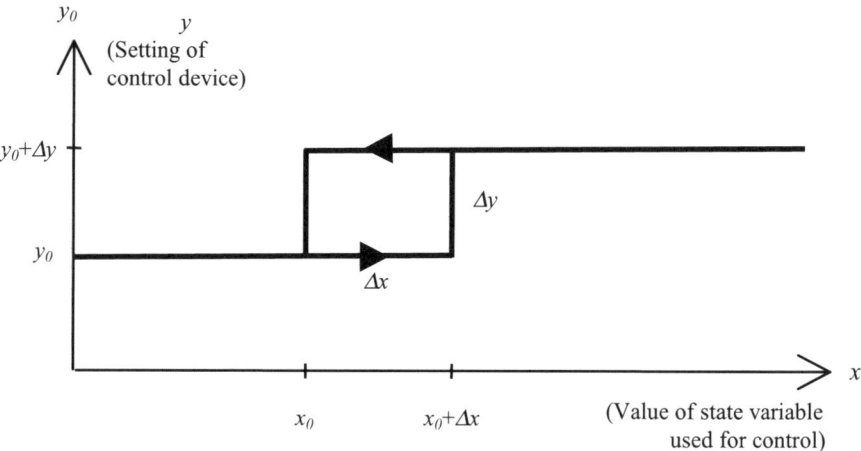

Figure 5.1. Two-point controller with four parameters $(x_0, y_0, \Delta x, \Delta y)$ as implemented in SYNOPSIS

In addition to this type of controller, control proportional to the inflow rate can be activated for the return activated sludge rate. For simulations with SYNOPSIS, any control device listed in Table 3.7 can be controlled by any sensor variable listed in Table 3.6. Which control device is influenced by which sensor is defined by the strategy framework. Various extensions to the concept of such a two-point controller are conceivable (*e.g.*, three-point controller (Schilling, 1989) or a combination of several two-point controllers as used for the definition of pump settings in the sewer system simulation program HYSTEM-EXTRAN (Fuchs *et al.*, 1994b) or the use of other controller types). Another extension would take into account the information provided by more than one sensor for control of a device. However, although implementation of these concepts within SYNOPSIS would be straightforward, these are not considered in the present study, since even the simple controller type implemented here allows for a vast number of (theoretically) possible frameworks to be investigated. Assuming that each of the nine devices listed in Table 3.7 can be controlled (at least, in theory) by each of the 19 sensor variables listed in Table 3.6 (plus operation by constant setting as an additional option), a total of 9^{20} ($\approx 1.22 \times 10^{19}$) possible frameworks would be obtained. Even if only half of the control devices is assumed to be controlled by half of the number of available sensors, this still would result in $5^{10} = 9765625$ potential combinations. Obviously, the number of practically reasonable frameworks constitutes only a fraction of these numbers. However, still only a very small number of frameworks can be considered here.

In summary, the following procedure will be applied for the evaluation of a control scenario:

1. For each of the control devices available in the system, define the sensor information which is to be used for control. Also the definition of a constant set-point for a controller is among the feasible options.

2. For each control device, define the type of controller to be used (here: two-point controller or, for RAS rate only, proportional control). For each of the controller parameters (x_0, y_0, Δx, Δy; proportionality constant) decide whether a constant value can be defined *a priori* or whether its optimum value is to be found by the optimisation procedure. Finally, define appropriate values of the fixed strategy parameters and feasible ranges of the variable controller parameters. Steps 1 and 2 constitute the definition of a strategy framework.

3. Select the criterion according to which the control strategies are to be assessed (within SYNOPSIS, any of those defined in Section 3.3.3 can be used).
4. Select an input rainfall series to be used for simulation of the urban wastewater system.
5. Perform the optimisation of parameters defined under 2. The specifications made under 1 and 2 define the dimension of the optimisation problem and the restrictions on the variables. The decision made under 3 defines the objective function.

As discussed in earlier sections, since no strong assumptions on the characteristics of the optimisation problem can be made, use of one of the global optimisation procedures implemented in SYNOPSIS is recommended. Step 5 constitutes the optimisation of the strategy parameters.

This procedure closely resembles the common procedure for off-line development of control strategies (see Figure 2.14). Here, however, modification and testing of new strategies is done automatically by the simulation and optimisation tool. Thus, this procedure is considered to represent an improvement over the conventional laborious procedure of manually defining, evaluating and modifying control strategies.

The next section provides a simple example to illustrate the application of this procedure. Subsequent sections will apply the procedure demonstrated here to more sophisticated frameworks.

5.2 Analysis of Strategy Parameters – an Example

This section presents an example of the application of the procedure for the optimisation of the parameters of a control strategy. Before the optimisation itself is demonstrated in Subsection 5.2.3, the underlying problem is defined in Section 5.2.1 and plots of the objective function are presented in Subsection 5.2.2. In order to keep the example simple, a framework with just two strategy parameters is defined here. Therefore, the objective function and the progress of the optimisation algorithm can be plotted easily. This will assist in the illustration of the procedure.

5.2.1 Definition of a Strategy Framework

Here, the optimum parameters of a strategy are determined which sets the maximum inflow to the treatment plant as a function of the ammonium load in the influent.

The idea behind the definition of this framework is to reduce the inflow to the treatment plant whenever the ammonium load in the influent is high, thus providing the plant with a more constant ammonium load. All other control devices within the system are assumed to be used with constant settings as in the base case (see Section 4.1.5). Table 5.1 summarises and Figure 5.2 illustrates the strategy framework defined. Since only two strategy parameters are kept variable (x_0 and y_0), plots can easily be generated which show a water quality criterion as a function of the strategy parameters (see next subsection).

Table 5.1. Strategy framework defined for the illustrative example

Control device	Controlled by ...	Value/Strategy parameters
Pump P2 (maximum flow capacity)	(constant value)	5 × DWF
Pump P4 (maximum flow capacity)	(constant value)	5 × DWF
Pump P6 (maximum flow capacity)	(constant value)	5 × DWF
Pump P7 (maximum flow capacity)	(constant value)	5 × DWF
Maximum inflow rate into the treatment plant (*PCINMX*)	*NH4INL* (Ammonium load in the influent)	$x_0 \in [0;2]$ $y_0 \in [2;6]$ $\Delta x = 0.5$ (fixed) $\Delta y = -0.5$ (fixed)
Emptying rate of storm tank (*STTPQ*)	(constant value)	500 m³/h
Threshold Emptying storm tanks (*STTPTH*)	(constant value)	1000 m³/h
Return activated sludge rate (*RASABS*)	(constant value)	600 m³/h
Wasteactivated sludge rate (*WASABS*)	(constant value)	27.5 m³/h

n.b.: Those parameters the settings of which are to be found by optimisation are highlighted.

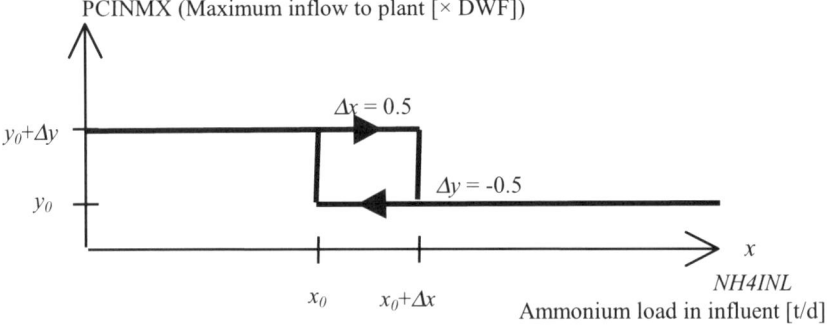

Figure 5.2. Controller for *PCINMX* as defined for the illustrative example

5.2.2 Exploring the Parameter Space by Gridding

In this subsection, the parameter space defined in the previous section is explored by a "gridding" procedure, *i.e.*, of simulations for which both parameters are varied across their ranges. This helps to establish a general idea of the shape of the function. The criteria F2 and F3 (defined in Section 3.3.3) are evaluated and shown for two different series of input data. Figures 5.3 and 5.4 show the results obtained for the series of one week's rainfall data used in Section 4.3, whereas Figures 5.5 and 5.6 present the results for the six-months series of continuous rainfall data from the period January to June 1977 (see Appendix C).

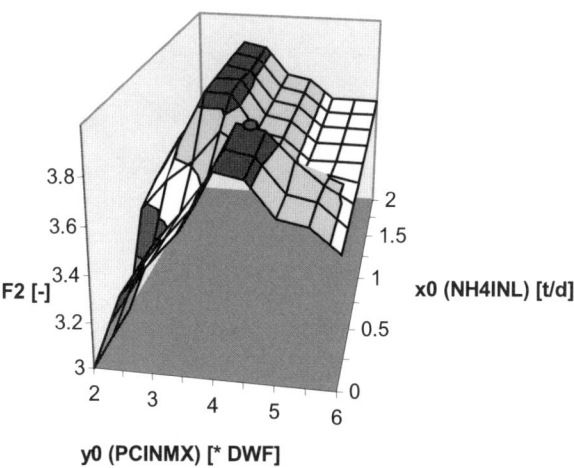

Figure 5.3. Gridding (1 week): results of control of maximum inflow rate to the treatment plant (*PCINMX*) as a function of ammonium load in the influent (*NH4INL*) – DO criterion F2. The solution (0.65; 4.1) found by the optimisation process described below is marked.

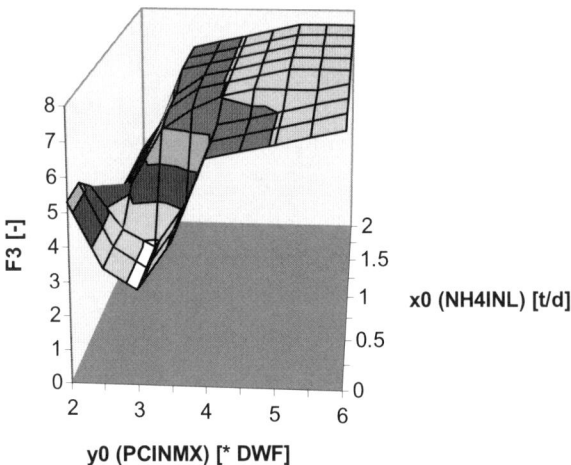

Figure 5.4. Gridding (1 week): results of control of maximum inflow rate to the treatment plant (*PCINMX*) as a function of ammonium load in the influent (*NH4INL*) – ammonium criterion F3

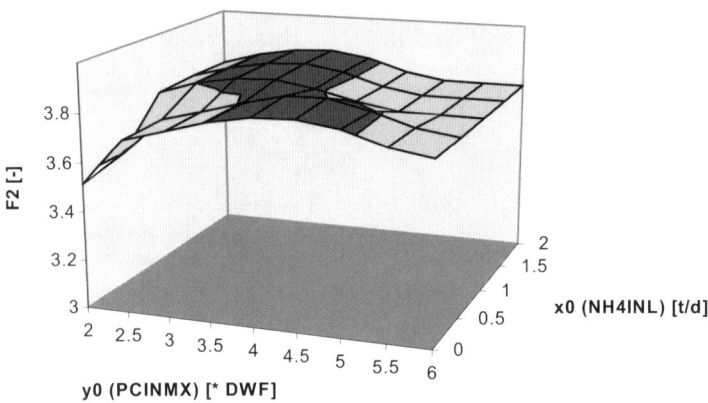

Figure 5.5. Gridding (6 months): results of control of maximum inflow rate to the treatment plant (*PCINMX*) as a function of ammonium load in the influent (*NH4INL*) – DO criterion F2

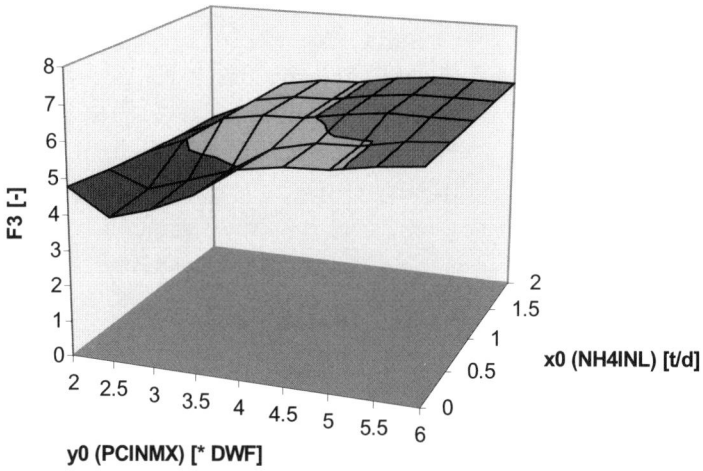

Figure 5.6. Gridding (6 months): results of control of maximum inflow rate to the treatment plant (*PCINMX*) as a function of ammonium load in the influent (*NH4INL*) – Ammonium criterion F3

From Figures 5.3 and 5.5 it becomes obvious that best control with regard to DO (F2 criterion to be maximised) can be obtained for two regions of the strategy parameter domain. A similar observation can be made with regard to ammonium (F3 to be minimised) from Figures 5.4 and 5.6. From the figures it also becomes immediately evident that optimum control with regard to DO is generally not optimum with regard to ammonium. In subsequent discussions emphasis will be made on DO (expressed by F2) as a performance criterion. Multiobjective optimisation will be briefly touched upon in Section 5.6.2.

This example also illustrates the need to use optimisation methods which do not focus their search on a single local maxima (or minima) alone.

Another interesting observation can be made from comparison of Figures 5.3 and 5.5 and Figures 5.4 and 5.6, respectively. Although the simulation period on which Figures 5.3 and 5.4 are based consists of just one week, and is therefore considerably shorter than the period of six months used for Figures 5.5 and 5.6, the location of the interesting regions of the parameter domain is similar. This encourages the use of the shorter rainfall series for optimisation of other strategy frameworks as well, despite the less encouraging results obtained in Section 4.5 with regard to selective simulation. A more detailed analysis of the feasibility of using shorter time periods for optimisation is briefly introduced in Section 5.6.3. For the purpose of demonstration of the optimisation procedure, however, use of the shorter rainfall series appears to be justified.

Furthermore, it can be observed from a comparison of Figures 5.3 and 5.5 (and Figures 5.4 and 5.6, respectively) that the variations of the values of F2 and F3, which are based on the simulations of a long time series (Figures 5.5 and 5.6) are less pronounced than those which are derived from a considerably shorter time series (Figures 5.3 and 5.4). Hence, effects of different strategies seem to be smoothed over longer time periods, since not only significant (or "interesting") events are selectively considered, but also a complete series of rainfall including dry-weather periods.

5.2.3 Optimisation of Strategy Parameters

This section demonstrates the principles of optimisation by Controlled Random Search for the example discussed in the previous subsections. For a study involving just two variables (as in this example) one would not necessarily embark on optimisation, since reasonable results may also be obtained by a gridding procedure. However, as soon as the number of strategy parameters (*i.e.*, the dimension of the optimisation problem) increases, gridding very quickly becomes inappropriate due to the exponential increase in the number of simulation runs required. Then the application of more formal optimisation algorithms becomes advantageous.

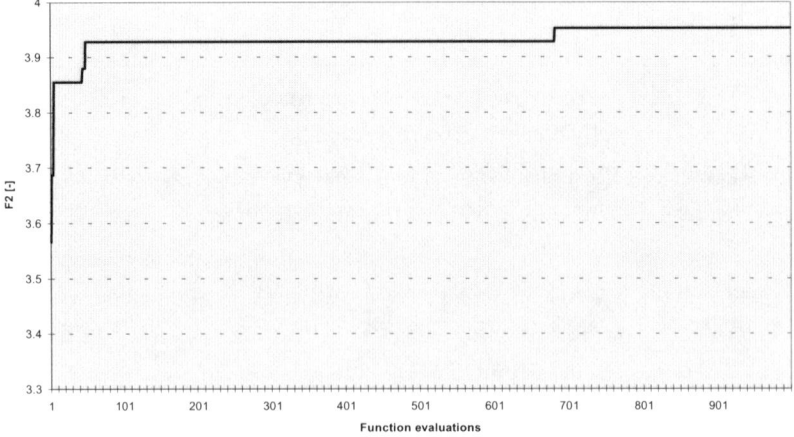

Figure 5.7. Optimisation of F2 by application of the Controlled Random Search algorithm for this example

The following figures demonstrate the optimisation of the two parameters x_0 and y_0 of the above defined controller with regard to the F2 criterion. Feasible ranges $x_0 \in [0;2]$ and $y_0 \in [2;6]$ are defined. For the optimisation, the Controlled Random Search procedure (as detailed in Section 3.5.1) is applied here. Figure 5.7 plots the best F2 value obtained during the optimisation process against the number of function evaluations (simulation runs with the simulation module of SYNOPSIS).

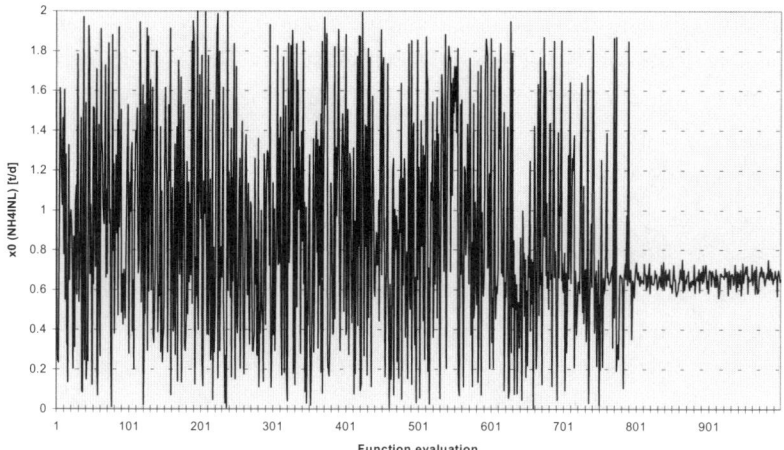

Figure 5.8. Illustrative example: optimising strategy parameter x_0

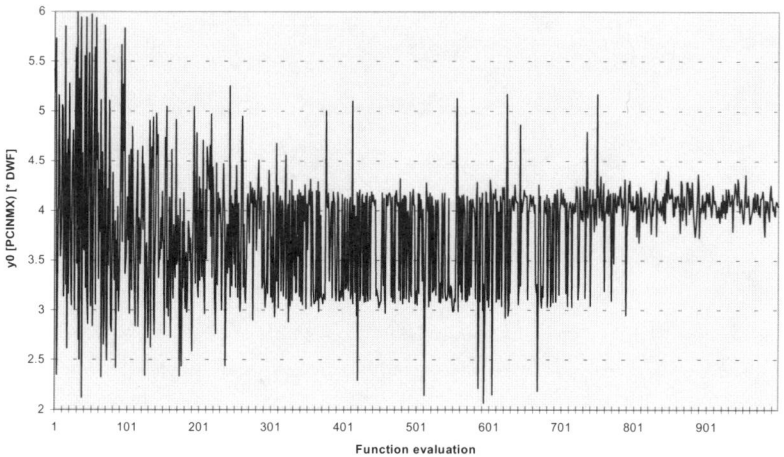

Figure 5.9. Illustrative example: optimising strategy parameter y_0

Figures 5.8 and 5.9 show the values of the arguments in each single call of the objective function. It can clearly be seen that after evaluation 800, the algorithm focuses its search solely on the global maximum of the function. However, values of x_0 and y_0 which lead to a value of the objective function close to the best value obtained at termination of the algorithm has been found already in function evaluation No. 49. Thereafter, only a small improvement is made in terms of the best F2 value found so far (see Figure 5.7).

As can be seen from these figures (in particular from Figure 5.9), the CRS algorithm soon focuses its search on the most promising region (between about three and four times DWF), before (at about evaluation No. 800) it narrows its search further down to the region around 0.65 t/d and 4 DWF, respectively.

The series of plots in Figure 5.10 show the candidate solutions which are kept in the set "A" used by the CRS algorithm (see Section 3.5.1). This set is used to generate other trial points. At the same time, it also indicates which regions of the search space the algorithm is currently focusing its search on. Figures 5.10a to g show this set at the start of the algorithm and after 100, 250, 500, 750 and 1000 function evaluations, respectively. The x-axis denotes the value of x as *NH4INL* [t/d], whereas the y-axis shows y as *PCINMX* [× DWF]. Figure 5.10a clearly shows that the initial population is randomly distributed within the feasible region. After less than 250 evaluations, the CRS algorithm focuses its search on two promising regions, until later (after about 800 evaluations), it focuses on one region (described by values of x_0 of about 0.65 t/d and y_0 of about 4 DWF).

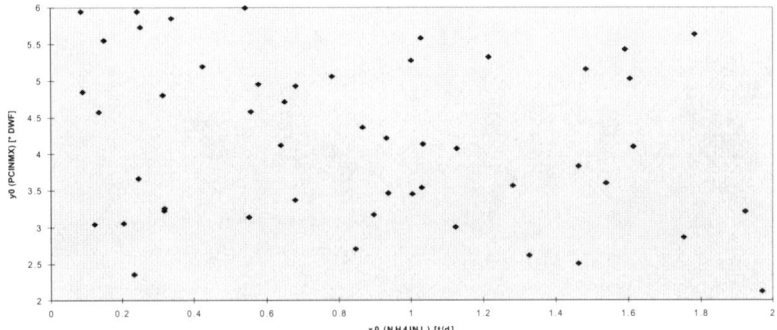

Figure 5.10a. Illustrative example: Controlled Random Search: initial population

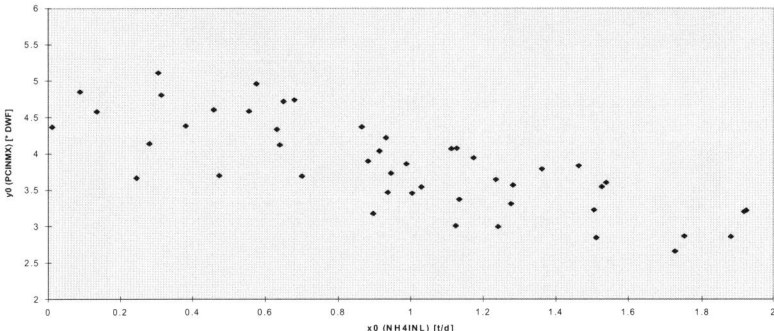

Figure 5.10b. Illustrative example: Controlled Random Search: population (set A) after 100 evaluations

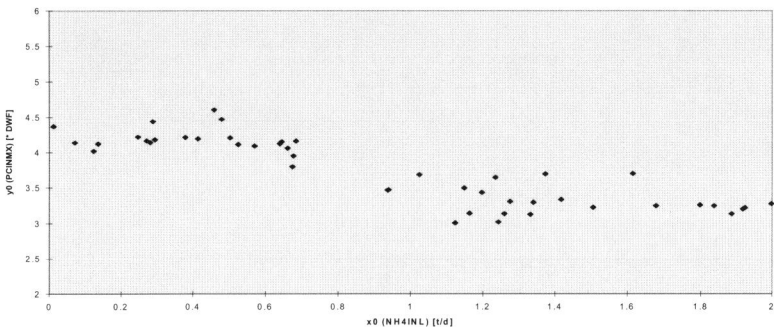

Figure 5.10c. Illustrative example: Controlled Random Search: population (set A) after 250 evaluations

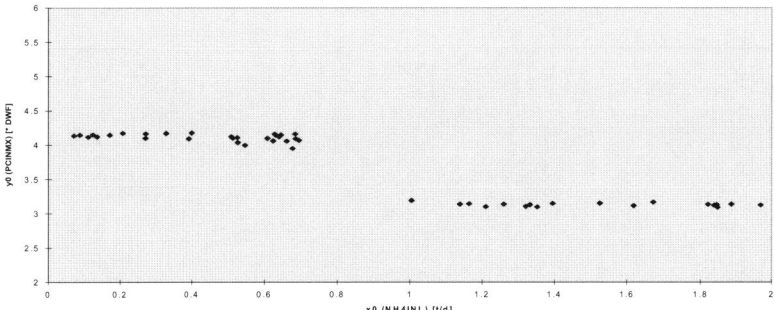

Figure 5.10d. Illustrative example: Controlled Random Search: population (set A) after 500 evaluations

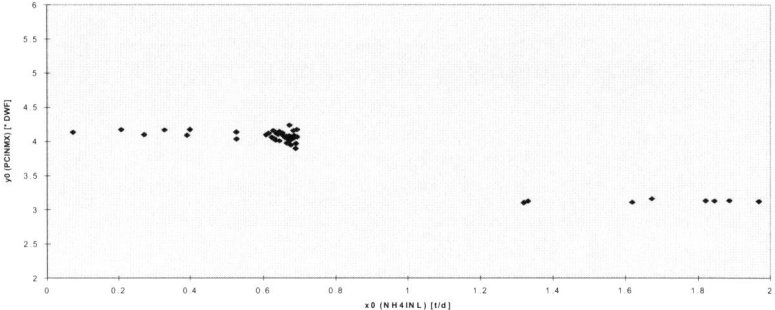

Figure 5.10e. Illustrative example: Controlled Random Search: population (set A) after 750 evaluations

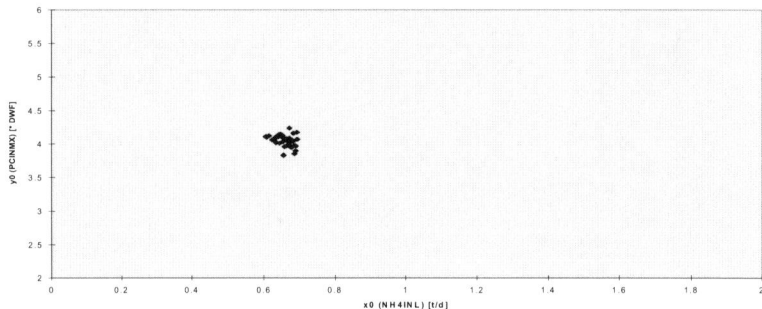

Figure 5.10f. Illustrative example: Controlled Random Search: population (set A) after 1000 evaluations

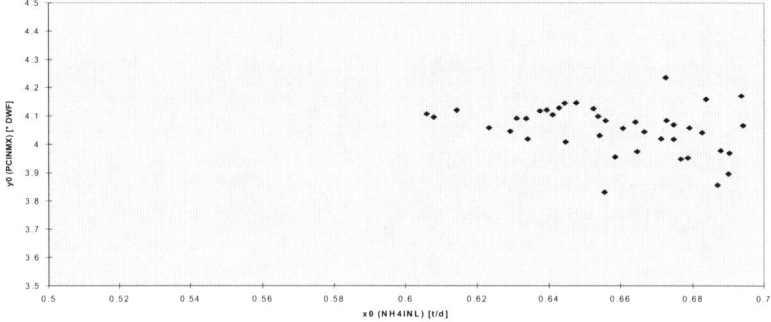

Figure 5.10g. Illustrative example: Controlled Random Search: population (set A) after 1000 evaluations (zoomed in)

The Controlled Random Search algorithm terminates after 1000 evaluations of the objective function, suggesting as solution $(x_0; y_0) = (0.65; 4.125)$, which results in a value of F2 of 3.95. After just 49 evaluations, the algorithm provided a solution (0.64; 4.120) with F2 = 3.93. These F2 values are higher than the value of F2 = 3.86 obtained for the base case (as defined in Section 4.1.5) when evaluated over the same simulation period. Therefore, this analysis suggests that this form of control studied here (setting the maximum inflow rate into the plant depending on the ammonium concentration in the influent) can lead to results superior to the base case scenario, when evaluated in terms of DO concentrations in the receiving river.

In the example discussed in this section, the framework of the strategy to be analysed was given. Subsequently, its parameters were optimised. In subsequent sections, more general approaches to the identification of successful strategies will be taken.

5.3 A Top-down Approach to the Definition of Control Strategies

As stated above, this section analyses different frameworks for strategies, which represent examples of local and integrated contro. Definition of these frameworks (in Section 5.3.1) is based on the urban wastewater system as a whole. For its subsystems, system states are defined which determine the control actions to be taken. Besides the comparison of strategy frameworks, Section 5.3.2 also compares the optimisation methods implemented in SYNOPSIS and evaluates them with regard to their ability to determine strategy parameters within a given amount of computation time. Conclusions of these studies are provided in Section 5.3.3.

The term "local control" as used in this and subsequent sections refers to control where only sensor information is used (if at all) which is available from within the same subsystem (*i.e.,* sewer system, treatment plant, river) to determine the setting of a control device. Note that what conventionally would be called "global control" in control of sewer systems (see Section 2.4.1) is termed "local control" here, since, although information is transferred across distances in the sewer system, only information from within the same subsystem is used for this type of control. As opposed to local control in this sense, integrated control involves transfer of information between different subsystems of the urban wastewater system (*cf.* the definition of this term, Section 2.4.1).

5.3.1 Definition of Various Frameworks

In this section, two different control scenarios are defined and compared with each other as well as with the base case defined in Section 4.1.5. The first scenario represents an extension of the base case and constitutes a simple example of local control involving three strategy parameters (constant settings of control devices). Furthermore, a hierarchical control scenario is defined in order to consider the state of the entire urban wastewater system when taking a control decision. Here, the settings of the control devices are overridden in certain cases (extreme system states), which in turn are defined by another set of parameters found by the optimisation procedures.

Scenario 1

A framework representing one form of local control is defined by a straightforward extension of the base case in the sense that three of its controllers are operated with constant settings for which optimum values are found by the optimisation routines. As the analysis of Section 4.4 suggests, the discharge of pump P7, the maximum inflow capacity and the threshold triggering emptying of the storm tank are among those parameters which have the most significant impact on the DO level in the river. Therefore, these three parameters are chosen here. Variation of the settings of the RAS and WAS rates are not considered, since, due to the slower process dynamics, these would require a longer simulation period for their adequate assessment. The ranges of the strategy parameters to be optimised here are defined in such a way that they include the settings applied in the base case as well as the ranges which showed to be reasonable in Section 4.4. Table 5.2 summarises the definition of this framework.

Scenario 2

This framework represents an extension of the framework of Scenario 1. In a similar way, operation of pump P7, of the maximum inflow rate into the plant and of the pump emptying the storm tank is defined by constant settings (these are subsequently called z_1, z_2, z_3). However, the settings for P7 and the maximum flow rate to the plant are overridden in certain states of the overall system, namely when some of its parts suffer from overloading whilst others still have spare capacities. Thus, optimum use of spare capacities in one part of the urban wastewater system can be made when others are overloaded. This option of coordinating the use of the parts of the urban wastewater system would not be possible if these were operated as individual units. Table 5.3 defines a total of sixteen states of the urban wastewater system with suggestions for appropriate settings of the control devices.

Table 5.2. Strategy framework defined for Scenario 1 (local control)

Control device	Control	Value/Strategy parameters
Pump P2 (maximum flow capacity)	(constant value)	5 DWF
Pump P4 (maximum flow capacity)	(constant value)	5 DWF
Pump P6 (maximum flow capacity)	(constant value)	5 DWF
Pump P7(maximum flow capacity)	(constant value)	$x_0 = 0$ (fixed)
		$z_1 \in [2;5]$
		$\Delta x = 0$ (fixed)
		$\Delta y = 0$ (fixed)
Maximum inflow rate into the treatment plant (*PCINMX*)	(constant value)	$x_0 = 0$ (fixed)
		$z_2 \in [2;5]$
		$\Delta x = 0$ (fixed)
		$\Delta y = 0$ (fixed)
Threshold: Emptying storm tank (*STTPTH*)	(constant value)	$x_0 = 0$ (fixed)
		$z_3 \in [700;1300]$
		$\Delta x = 0$ (fixed)
		$\Delta y = 0$ (fixed)
Emptying rate of storm tank (*STTPQ*)	(constant value)	500 m^3/h
Return activated sludge rate (*RASABS*)	(constant value)	600 m^3/h
Waste activated sludge rate (*WASABS*)	(constant value)	27.5 m^3/h

n.b.: Those parameters the settings of which are to be found by optimisation are highlighted.

A similar way to describe the potential states of the system to be controlled was chosen by Vitasovic and Andrews (1987). They defined rules for each of the $2^4 = 16$ states defined by the STOUR (specific total oxygen uptake rate) level in four reactors. In a similar manner, Koskinen and Viitasaari (1990) define $3^3 = 27$ different states of a treatment plant by all combinations of "high", "optimum" and "low" levels of each of the three parameters sludge age, sludge volume index and oxygen uptake rate.

Each of the four main parts of the urban wastewater system (within this section, the storm tank is considered as a separate part) is assumed to be in one of two states "heavily loaded" and "not heavily loaded", which may result in potentially different control actions. The decision whether a subsystem is heavily loaded or not is based on the value of an appropriate state variable, for which several options are listed in Table 5.4. The threshold values z_4, z_5, z_6 and z_7, marking the borderline between "heavily loaded" and "not heavily loaded" for each of these four subsystems, will also be found by the optimisation procedure.

Table 5.3. Sixteen different states of the urban wastwater system and suggested control actions – Scenario 2

(1)	(2)				(3)	(4)	(5)	(6)
No.	Subsystem heavily loaded?				Quick intuitive comments	Action to be taken	Suggested setting of P7	Suggested setting of PCINMX
	Sewer system	ST	TP	River				
0.	No	No	No	No	Be happy!	Default (local control)	z_1	z_2
1.	No	No	No	Yes	Avoid discharges to the river			
2.	No	No	Yes	No	Avoid further load to the treatment plant			
3.	No	No	Yes	Yes	Use all existing storage in the system			
4.	No	Yes	No	No	Empty storm tanks			
5.	No	Yes	No	Yes	Use sewer storage and plant capacity wisely, possibly risk CSO discharges			
6.	No	Yes	Yes	No	Use sewer storage wisely; risk CSO discharges	Use storage within system	$z_1 - 1$	z_2
7.	No	Yes	Yes	Yes	Use sewer storage wisely; avoid CSO discharges			$z_2 + 1$
8.	Yes	No	No	No	Pump towards storm tank and treatment plant	Increase flows to TP; utilise storm tanks	$z_1 + 1$	z_2
9.	Yes	No	No	Yes	Pump towards storm tank and treatment plant		$z_1 + 2$	
10.	Yes	No	Yes	No	Utilise storm tank, possibly risking overflows		$z_1 + 1$	1
11.	Yes	No	Yes	Yes	Utilise storm tank			
12.	Yes	Yes	No	No	Ease the situation in the sewer system and storm tanks; increase flows downstream	Increase flows towards TP	$z_1 + 1$	$z_2 + 1$
13.	Yes	Yes	No	Yes	Avoid overflows; increase flows to the TP	Increase flows to TP	$z_1 + 2$	
14.	Yes	Yes	Yes	No	System is heavily loaded: balance between TP and overflows; try to gain some capacity	Allow maximum flows d/s to gain some capacity	Maximum possible value	$z_2 + 1$
15.	Yes	Yes	Yes	Yes				

Table 5.4. Examples of state variables for which the definition of the term "heavily loaded" for various parts of the urban wastewater system could be based. Those which are used in this study (Scenario 2) are highlighted

Subsystem	System state "heavily loaded" described by exceedance of a given threshold value for ... (examples)
Sewer system	• Water level in a basin downstream in the sewer system (B7) • Flow rate downstream in the sewer system • Concentration downstream in the sewer system • Rainfall • CSO discharge rate • Load of CSO discharge
Treatment plant	• Influent flow rate to the treatment plant • Concentrations of treatment plant influent • Concentrations of treatment plant effluent
Storm tank	• Water level in storm tank • Concentrations in storm tank • Loads in storm tank
River	• River flow rate (Scenario 2a) • Concentrations of treatment plant influent (COD) (Scenario 2b) • Concentrations of the river

For the definition of the status "heavily loaded" for the river, two alternative definitions are tested here – one is based on the river flow (thus representing the dilution capacity of the river) and the other one uses the COD concentration of the treatment plant effluent.

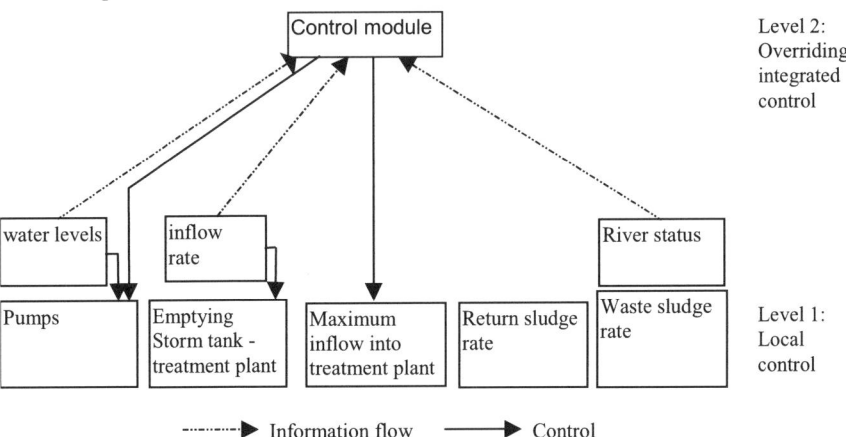

Figure 5.11. Illustration of control Scenario 2 (an example of integrated control)

Since information from the treatment plant and the river is used for control actions performed within the sewer system and the treatment plant, this framework represents an example of integrated control. At the same time, it is also an example of hierarchical control, since control is performed on two levels (cf. Figure 5.11).

Finally, Table 5.5 summarises the definition of the strategy framework for control Scenario 2.

Table 5.5. Strategy framework defined for Scenario 2 (integrated control)

Control device	Control	Value/Strategy parameters
Pump P2 in the sewer system	(constant value)	5 DWF
Pump P4 in the sewer system	(constant value)	5 DWF
Pump P6 in the sewer system	(constant value)	5 DWF
Pump P7 in the sewer system	(constant value; potentially overridden; see Table 5.3)	$x_0 = 0$ (fixed) $z_1 \in [2;5]$ $\Delta x = 0$ (fixed) $\Delta y = 0$ (fixed)
Maximum inflow rate into the treatment plant (*PCINMX*)	(constant value) potentially overridden; see Table 5.3)	$x_0 = 0$ (fixed) $z_2 \in [2;5]$ $\Delta x = 0$ (fixed) $\Delta y = 0$ (fixed)
Threshold: Emptying storm tank (*STTPTH*)	(constant value) potentially overridden; see Table 5.3)	$x_0 = 0$ (fixed) $z_3 \in [700;1300]$ $\Delta x = 0$ (fixed) $\Delta y = 0$ (fixed)
Emptying rate of storm tank (*STTPQ*)	(constant value)	500 m³/h
Return sludge rate (*RASABS*)	(constant value)	600 m³/h
Waste sludge rate (*WASABS*)	(constant value)	27.5 m³/h
Threshold values defining the state "heavily loaded" in the various subsystems (cf. Table 5.4):		
Sewer system (water level in Basin B7) [%]	(constant value)	$z_4 \in [0;100]$
Storm tank (water level in tank) [%]	(constant value)	$z_5 \in [0;100]$
Treatment plant (Inflow rate) [× DWF]	(constant value)	$z_6 \in [3;5]$
River: Scenario 2a: Flow rate [× DWF TP] Scenario 2b: COD in plant effluent [mg/l]	(constant value)	$z_7 \in [4.5;4.75]$ $z_7 \in [75;120]$

n.b.: Those parameters the settings of which are to be found by optimisation are highlighted.

It should be noted that Scenario 2 consists of two alternative definitions of the state "river heavily loaded" (represented by strategy parameter z_7): Scenario 2a uses the flow rate of the river (thus its dilution capacity) as a criterion to assess the state of the river, whereas Scenario 2b is based on the COD effluent concentration of the treatment plant. The feasible ranges for the parameters to be optimised are defined as in Scenario 1 (*cf.* Table 5.2). The feasible ranges for the newly introduced parameters z_4, z_5, z_6 and z_7 are derived from the results of the base case simulations described in Sections 4.2 and 4.3.

5.3.2 Evaluation of the Optimisation Algorithms

For each of the three strategy frameworks defined above (1, 2a, 2b), a variety of optimisation routines is applied for the determination of the strategy parameters. The aim of the analysis described in this section is the comparative evaluation of the optimisation procedures. Since only a limited amount of computation time is available for any detailed optimisation study, the following comparison aims at identifying those routines which return the best solutions for the optimisation problems defined by the frameworks 1, 2a, 2b (*i.e.*, involving three and seven parameters, respectively) within a limited number of function evaluations. Due to the large number of optimisation methods applied for each of these problems, the maximum number of function evaluations was limited to 250 for the runs reported in this section. Tests carried out with a larger maximum number of function evaluations are discussed further below. Again, the F2 criterion (see Section 3.3.3 for its definition) is used here as objective function for the optimisation of strategy parameters.

Optimisation methods evaluated in this section include the following:

1. Controlled Random Search (as presented in Section 3.5.1) with a size of set A of 50;
2. Genetic Algorithm (as presented in Section 3.5.2) with a population size of 50;
3. Micro GA (as presented in Section 3.5.2) with a population size of 20;
4. local optimisation after Powell (as presented in Section 3.5.3), applying the tangent transformation (3.6) for the consideration of constraints;
5. gridding, using four values per dimension. Therefore, no gridding results can be obtained for the seven-dimensional problems 2a and 2b within 250 evaluations;

The following options for developing simple substitute models, which are then used for optimisation. The substitute models provide the advantage of very fast evaluation of the objective function. Also their analytical optimisation is possible.

6. Development of a quadratic substitute model over the entire feasible region. The $(n+1)(n+2)/2$ coefficients of the quadratic polynomial in n variables

$$f(x_1,...,x_n) = \sum_{\substack{i,j=1 \\ i \geq j}}^{n} \alpha_{ij} x_i x_j + \sum_{i=1}^{n} \alpha_{i0} x_i + \alpha_{00} \tag{5.1}$$

are determined by least-squares regression over all 3^n combinations of the smallest, mean and largest value of the range of each of the n strategy parameters. The function (5.1) is then maximised. The solution vector obtained is considered to be the optimum set of strategy parameters. The corresponding F2 value is obtained by a simulation run using as input the strategy parameters just obtained. In total, this approach requires $3^n + 1$ simulation runs with SYNOPSIS. Therefore, only for framework 1 (for which n has a value of 3) can results be obtained with less than 250 function evaluations.

7. Development of a new quadratic substitute model. Here, the model is not built over the entire feasible region, but around the solution found in 6. The parameter ranges used for definition of this substitute model do not cover the entire feasible region, but extend over a range of 20% of the ranges of each dimension in the feasible domain. Thus, the substitute model represents a local approximation of the full simulation model around the parameter values determined in 6.

8. Development of a linear substitute model (with mixed terms) over the entire feasible region. The $n(n+1)/2 + 1$ coefficients of the polynomial in n variables

$$f(x_1,...,x_n) = \sum_{\substack{i,j=1 \\ i > j}}^{n} \alpha_{ij} x_i x_j + \sum_{i=1}^{n} \alpha_{i0} x_i + \alpha_{00} \tag{5.2}$$

are determined by least-squares regression over all 2^n combinations of the smallest and the largest value of the range of each of the n strategy parameters. Function (5.2) is then maximised. As in the procedure described under 6, the solution vector obtained is considered to be the optimum set of strategy parameters. The corresponding F2 value is obtained by a simulation run using as input the strategy parameters just obtained. Since only two values are considered per parameter, the total number of simulation runs required amounts to $2^n + 1$, which is considerably lower than the number of runs required in the approach described under 6. Thus, frameworks 2a and 2b can be considered here.

9. Since the analyses of Chapter 4 suggest that the default values for the control devices defined for the base case are not too far from the optimum settings, a substitute model is built around these default values. As described under 7, the ranges considered for each parameter cover 20% of the total range of each parameter. For framework 1, a quadratic model (as in (5.1)) is built, requiring $3^3+1 = 28$ runs of the full model, whereas a linear model with mixed terms (as in (5.2)) is used for frameworks 2a and 2b, requiring $2^7+1 = 129$ runs for each of these two.

It should be noted that the analysis described under 9 represents the first iteration of the Response Surface Methodology (*cf.* Section 2.5.2.1) applying substitute models. In a similar way, the analyses under 6 and 7 can be considered as the initial iterations of a RSM procedure based on quadratic substitute models. Full application of the RSM approach would continue at this stage by building a new substitute model around the solution found in 9 (or 7). This is not investigated here, since the other optimisation procedures evaluated in this section provide satisfactory results within the given maximum number of simulation runs without user interaction (as is required for the RSM approach).

Having outlined the scenarios defined and the optimisation procedures applied, Table 5.6 summarises the results of the various optimisation runs.

5.3.3 Conclusions

Two conclusions can be drawn from the results shown in Table 5.6.

Different optimisation methods appear to give different results. This may be due to a number of reasons, including the definition selections of the parameters of the respective procedures itself or the initial seeds for random number generation applied in the stochastic algorithms (*i.e.*, 1 to 3 in the previous subsection). However, when ranking the performance of the different strategy frameworks (which is what, in the end, constitutes the aim of the optimisation study), a general pattern can be observed. Most results suggest that Scenarios 2a and 2b (representing forms of integrated control) give better results (in terms of the DO criterion F2) than does optimised local control (Scenario 1). All of these perform better than operation of the system using the settings of the base case, which results in an F2 value of 3.86 (*cf.* Table 4.11). Detailed results as well as time series plots of the urban water system, when controlled by the strategy defined by Scenario 2a can be found in Sections A.7 and A.8.1. Figure 5.12 summarises the results obtained by application of the Controlled Random Search procedure for the various scenarios.

Table 5.6a. Evaluation of various optimisation methods – best F2 values found **after 100 function evaluations**

Scenario	1. Controlled Random Search	2. Genetic Algorithm	3. micro GA	4. Local optimisation	5. Gridding	6. 3^k-substitute model (1st iteration)	7. 3^k-substitute model (2nd iteration)	8. 2^k-model (1st iter.)	9. Local substitute model
No. 1	3.93	3.90	3.86	3.93	3.93	3.88	3.93	3.69	3.86
No. 2a	4.35	3.93	4.00	4.04	---	---	---	---	---
No. 2b	4.31	4.19	4.48	4.04	---	---	---	---	---

Table 5.6b. Evaluation of various optimisation methods – best F2 values found **after 250 function evaluations**

Scenario	1. Controlled Random Search	2. Genetic Algorithm	3. micro GA	4. Local optimisation	5. Gridding	6. 3^k-substitute model (1st iteration)	7. 3^k-substitute model (2nd iteration)	8. 2^k-model (1st iter.)	9. Local substitute model
1	3.93	3.90	3.88	3.93	3.93	3.88	3.93	3.69	3.86
2a	4.35	4.09	4.32	4.04	---	---	---	3.70	3.86
2b	4.31	4.25	4.48	4.04	---	---	---	3.72	4.04

n.b.: No results for frameworks 2a and 2b are shown for gridding and substitute modelling approaches, since no solutions can be obtained within 100 or 250 function evaluations for those. Best results obtained for each framework are highlighted.

Analysis of Control Scenarios 253

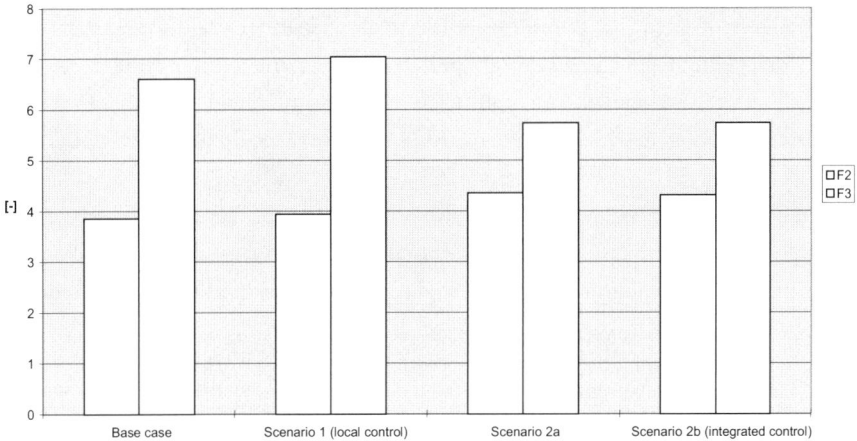

Figure 5.12. Results for the base case and for Scenarios 1, 2a and 2b obtained by optimisation (CRS) with regard to F2 and F3

From Figure 5.12 it can be seen that local control (as defined in Scenario 1) slightly improves the DO levels in the river (expressed by an increase of the F2 value). However, a deterioration in terms of ammonium (increase of F3) can be observed at the same time. Integrated control using river water flow (Scenario 2a) or COD effluent concentration (Scenario 2b) as a decision criterion describing the state of the river yields results which again represent an additional improvement in terms of F2. Here, perhaps surprisingly, a better (lower) F3 value is obtained also, despite the fact that the optimisation aims only at maximisation of F2.

Comparing the performance of the various optimisation algorithms, it can be concluded that the Controlled Random Search performs best (providing the best solutions, *i.e.*, resulting in the largest F2 value within the given maximum number of function evaluations) for most of the runs shown here. It is interesting to note that the micro GA suggests a solution of 4.48 for framework 2b (see Table 5.6b), which is far better than the solutions provided by the other routines. Further analyses of this algorithm appear to be of interest, but are not conducted here. The local optimisation procedure appears to have got stuck in a local minimum for scenarios 2a and 2b, since no improvement of the function value is observed between the 100th and the 250th function evaluation and since the best value obtained is far below the solutions provided by the other procedures. The approaches based on the development of substitute models perform less favourably than the other procedures. However, as outlined above, complete application of the RSM procedure would involve several iterations similar to those described in the previous section under 9.

It may be argued that permitting a larger number of function evaluations could result in different statements about performance of the optimisation runs being made. Figure 5.13 shows best optimisation results obtained with the Controlled Random Search, genetic algorithm, micro GA and local optimisation techniques (*cf.* 1 to 4 in the previous subsection) for a maximum number of 500 function evaluations. For CRS, the genetic algorithm and for the micro GA, also various population sizes were tested. For local optimisation, both versions, with and without consideration of constraints using the tangent transformation (3.6), were tested.

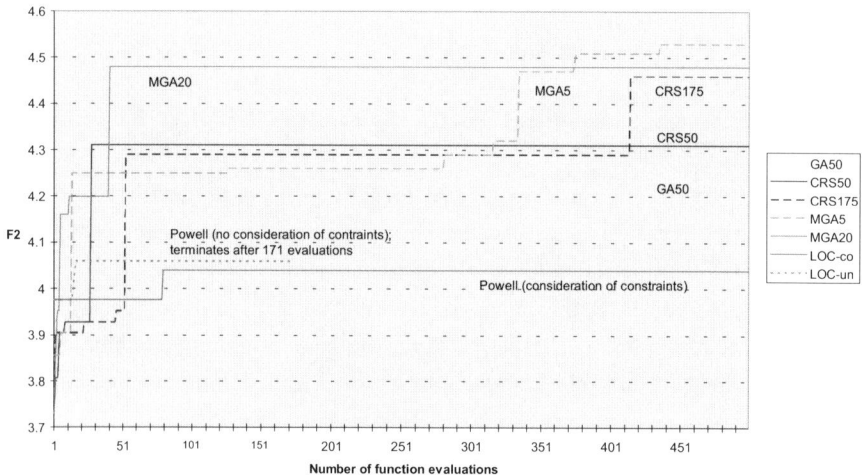

GA50:	Genetic Algorithm with population size 50
CRS50, 175:	Controlled Random Search with population sizes 50 and 175, respectively
MGA5, 20:	micro GA with population sizes 5 and 20, respectively
LOC-co, -un:	Local optimisation with/without consideration of constraints on the parameters

Figure 5.13. Evaluation of various optimisation methods for Scenario 2b – F2 values found by various optimisation procedures

It can be seen from Figure 5.13 that most of the procedures applied here do not yield an improved solution when the optimisation is extended from 250 to 500 function evaluations. However, the micro GA with a population size of 5 and the Controlled Random Search with a population of 175 (this value is motivated by the suggestion by Price (1983) to define a population size of 25 times the dimension of the optimisation problem) return better values (of around 4.5) after 330 and 410 evaluations, respectively. This may suggest the use of the micro GA for further analyses. However, due to the small population size of the micro GA (which results in a poorer ability to locate several optima), the Controlled Random Search algorithm will be applied for the optimisation runs reported in the remaining sections of this work. Since a number of function evaluations considerably larger than 250 would result in a large time demand for the multitude of optimisation runs to be carried out in the next section, this algorithm will be run subsequently with a population size of 50 and a maximum number of function evaluations of 250.

The tangent transformation (3.6), which was introduced in Section 3.5.3 as a very simple means of considering constraints in the local optimisation procedure, showed not to be very successful in several optimisation runs, including the one depicted in Figure 5.13. In those cases where the search procedure approaches a region close to the boundary of the feasible region, the tangent transformation (expanding even small differences in the parameters in such a region) prevents convergence or further successful search steps of the procedure.

As noted in the description of the various algorithms (see Section 3.5, *cf.* also Chapter 6), improvements would be possible for the algorithms evaluated here. However, due to time limitations, these are not performed and analysed here. Instead, the Controlled Random Search algorithm will be used subsequently due to its overall best performance observed so far.

5.4 A Bottom-up Approach to the Definition of Control Strategies

5.4.1 Towards a Systematic Definition of Frameworks

It is obvious that the success (or failure) of a control strategy depends on its framework, *i.e.,* on specifying which control devices are to be operated and which sensor information is to be used for control of which control devices. Thus, definition of appropriate frameworks is essential for comparison of local and integrated control. This becomes particularly important when control more sophisticated than constant settings (as discussed in the previous section) is to be analysed. Ideally, comparison of local and integrated control has to ensure that "good", if not best, representatives of these two are compared against each other. Therefore an attempt is made to base the definition of strategy frameworks not just on intuition, but on a more sophisticated method .

The most obvious way of defining a strategy framework would be to define it in such a way that it resembles commonly applied and proven control. Another option would be to define a number of different frameworks (including those which would not be conventionally thought of), to find the optimum strategy parameter sets for each of those, and to finally select the one which performs best. Such an approach is attempted in this section.

Here, three controls are considered – the last pump in the sewer system (P7), the setting of the maximum inflow rate into the treatment plant (*PCINMX*) and, finally, the return activated sludge rate (*RASPCT*). The RAS rate will be assumed here to be set proportional to the inflow rate, with the proportionality factor to be determined and dependent on system state. Each of these three devices is assumed to be controlled either with constant settings or to depend on information from one of five sensors. Accordingly, eighteen strategy frameworks are defined as shown in Figure 5.14. The figure also shows the ranges of the strategy parameters assumed for the subsequent simulation runs. Obviously, a larger number of control devices and sensors could be included in this analysis, resulting in an increase in the number of runs. In order to keep the number of runs low, it will be assumed that within each framework only one control device is operated as explained, whilst for the other devices constant settings (according to the base case) are assumed.

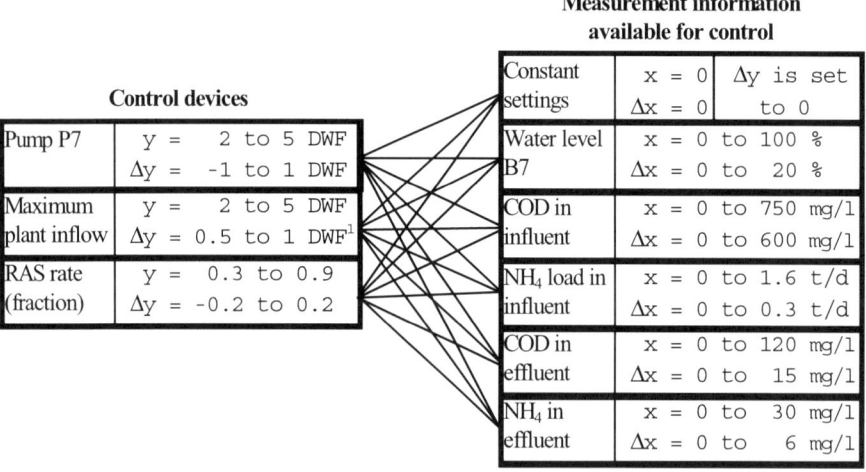

[1] For those frameworks, in which the maximum plant inflow rate is set dependent on influent or effluent loads or concentrations, a range for Δy of -1 to 1 is defined.

Figure 5.14. Definition of 18 frameworks representing various combinations of control devices and sensor information

For each of these frameworks, gridding runs are performed, varying all (*i.e.*, four, except in the case of constant settings, where it is just one) parameters over the ranges indicated. The resulting ranges of DO-DU values obtained for these frameworks are shown in Figures 5.15 to 5.17. As defined in Section 3.3.3, DO-DU denotes the duration for which the DO concentration in the river is below 4 mg/l. These values are shown here for convenience (since for the results obtained here, the F2 criterion, which is used in other sections of this work, would provide equivalent, but less tangible, information). For each framework, the range of DO-DU values obtained is shown by the lines; the boxes indicate the range [DO-DU$_{mean}$ - DO-DU$_{stddev}$; DO-DU$_{mean}$ + DO-DU$_{stddev}$][7], thus indicating the range in which the majority of the DO-DU values fall for each framework.

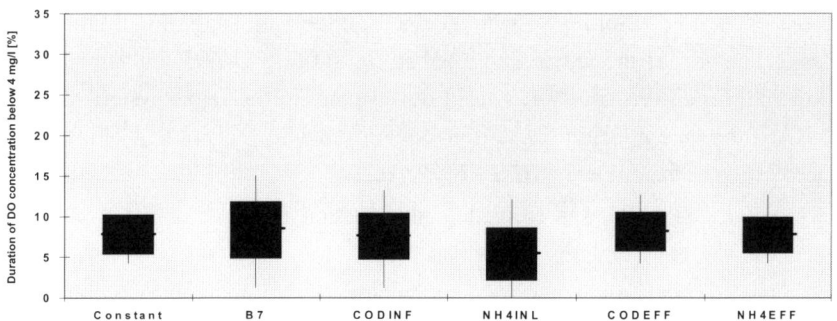

Figure 5.15. Control of pump P7 by various controllers – evaluation by duration of exceedance of DO threshold

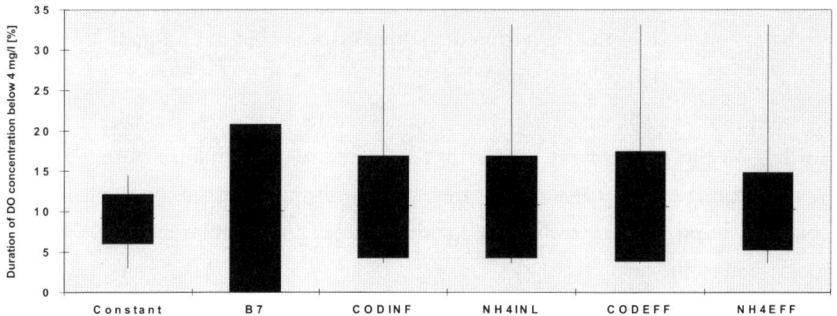

Figure 5.16. Control of maximum inflow rate by various controllers – evaluation by duration of exceedance of DO threshold

---

[7] DO-DU $_{mean}$, DO-DU $_{stddev}$: mean and standard deviation, respectively, of DO-DU values obtained in the gridding runs.

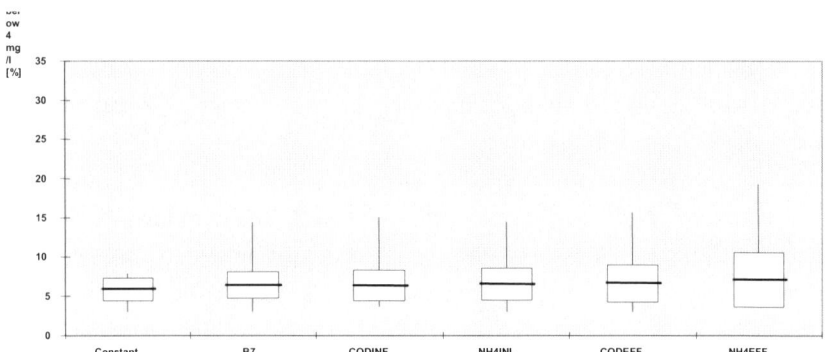

Figure 5.17. Control of return activated sludge rate by various controllers – evaluation by auration of exceedance of DO threshold

When attempting to draw conclusions from these results concerning which sensors to use to controlling of which devices, one might suggest to control Pump P7 as a function of the ammonium load in the influent (*NH4INL*), the maximum inflow rate dependent on a water level downstream in the sewer system (*B7*), and the RAS rate as a function of the ammonium load in the influent (*NH4INL*) or concentration in the effluent (*NH4EFF*).

However, the analysis presented in this section may be subjected to several criticisms. The most obvious one would be that not all the available sensors have been included. Secondly, since only one control device was considered at a time, interactions between synchronous settings and variations of several of those could not be covered. Lastly, the definition of ranges of the strategy parameters influences the results of the gridding procedure, in particular the minimum and maximum values obtained. Also statements made in the literature suggest that a successful systematic approach to the definition of frameworks has yet to be found (*cf.* Weijers *et al.*, 1995; Olsson and Jeppson, 1994).

Despite these valid criticisms, a combined strategy framework will be defined according to the conclusions made. Its parameters will be jointly optimised using the simulation and optimisation procedure described earlier on. Furthermore, several other frameworks are defined and subjected to a similar analysis. This will be described in the next section.

5.4.2 Analysis of Frameworks Involving Several Controllers

In this section, several strategy frameworks are defined which make use of the two-point controllers introduced in Section 5.1 for several control devices. Besides the framework suggested by the analysis of the previous section, several other, intuitively defined, frameworks are analysed. As in the previous sections, the

Control Random Search algorithm with a population size of 50 will be applied for the optimisation study. Once again the rainfall input time series covering one week as defined in Section 4.3 is used. Table 5.7 provides the definition of the frameworks studied here. Note that framework 3 was derived from the analyses described in the previous section.

Framework 4 is concerned solely with the settings of the pumps upstream in the sewer system. Comparing the corresponding results with those obtained for frameworks 5 (in which only treatment plant parameters are optimised) and 6 (a combination of frameworks 4 and 5) will allow assessment to be made as to whether more potential for improvement of the performance of the urban wastewater system can be gained by operation of the pumps in the sewer system or by utilising some of the control devices available in the treatment plant. Frameworks 7 and 8, finally, are defined almost identically to each other, except that in framework 7 most control decisions are based on COD measurements, whereas NH_4 is used for decisions on most of the control devices in framework 8.

The results obtained for these frameworks are shown in Appendix D. Figure 5.18 illustrates the results obtained for F2 (the criterion used as objective function in the optimisation) and also depicts the F3 values of the results obtained in the optimisation with regard to F2.

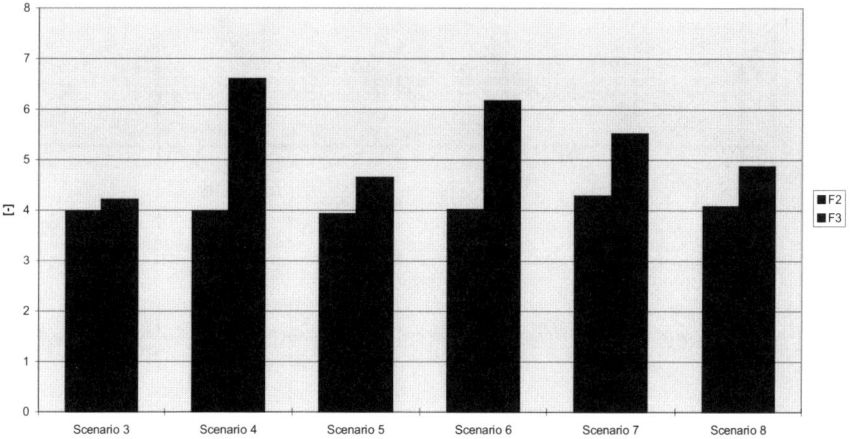

Figure 5.18. Results of optimisation of frameworks 3 to 8 with regard to F2

Table 5.7a. Strategy frameworks 3 to 5 involving several two-point controllers

Control device	Scenario 3 Controlled by	Scenario 3 Value	Scenario 4 Controlled by	Scenario 4 Value	Scenario 5 Controlled by	Scenario 5 Value
Pump P2 in the sewer system (maximum flow capacity)	(constant value)	5 DWF	B2	$x_0 \in [0;100]$ $y_0 \in [4;5]$ $\Delta x = 0$ $\Delta y = -1$	(constant value)	5 DWF
Pumps P4, P6 in the sewer system (maximum flow capacity)	(constant value)	5 DWF	B4/B6	$x_0 \in [0;100]$ $y_0 \in [4;5]$ $\Delta x = 0$ $\Delta y = -1$	(constant value)	5 DWF
Pump P7 in the sewer system (maximum flow capacity)	NH4INL	$x_0 \in [0;2.5]$ $y_0 \in [3;5]$ $\Delta x = 0$ $\Delta y \in [-1;0]$	B7	$x_0 \in [0;100]$ $y_0 \in [4;5]$ $\Delta x = 0$ $\Delta y = -1$	(constant value)	5 DWF
Maximum inflow rate into plant (PCINMX)	B7	$x_0 \in [0;100]$ $y_0 \in [2;5]$ $\Delta x = 20$ $\Delta y \in [0;1]$	(constant value)	3 DWF	NH4INL	$x_0 \in [0;2.5]$ $y_0 \in [2;5]$ $\Delta x = 0$ $\Delta y = -1$
Threshold: STTPTH	(constant value)	1000 m³/h	(constant value)	1000 m³/h	(constant value)	1000 m³/h
Emptying rate of storm tanks (STTPQ)	(constant value)	500 m³/h	(constant value)	500 m³/h	NH4INL	$x_0 \in [0;2.5]$ $y_0 \in [300;1000]$ $\Delta x = 0$ $\Delta y = -200$
Return activated sludge rate Scenarios 3: RASPCT Scenarios 4 and 5: RASABS	NH4EFF	$x_0 \in [0;30]$ $y_0 \in [0.3;0.9]$ $\Delta x = 0$ $\Delta y \in [-0.2;0.2]$	(constant value)	600 m³/h	QINTP	$x_0 \in [2;5]$ $y_0 \in [300;1000]$ $\Delta x = 0$ $\Delta y = 300$
Waste sludge rate (WASABS)	(constant value)	27.5 m³/h	(constant value)	27.5 m³/h	(constant value)	27.5 m³/h

n.b.: Those parameters the settings of which are to be optimised are highlighted.

Table 5.7b. Strategy frameworks 6 to 8 involving several two-point controllers (continued)

Control device	Scenario 6 Controlled by	Scenario 6 Value/Strategy parameters	Scenario 7 Controlled by	Scenario 7 Value/Strategy parameters	Scenario 8 Controlled by	Scenario 8 Value/Strategy parameters
Pump P2 in the sewer system (maximum flow capacity)	(constant value)	5 DWF	(constant value)	5 DWF	(constant value)	5 DWF
Pumps P4, P6 in the sewer system (maximum flow capacity)	$B4/B6$	$x_0 \in [0;100]$ $y_0 \in [3;5]$ $\Delta x = 0$ $\Delta y = -1$	(constant value)	5 DWF	(constant value)	5 DWF
Pump P7 in the sewer system (maximum flow capacity)	$B7$	$x_0 \in [0;100]$ $y_0 \in [3;5]$ $\Delta x = 0$ $\Delta y = -1$	$CODINL$	$x_0 \in [0;1000]$ $y_0 \in [4;5]$ $\Delta x = 0$ $\Delta y = -1$	$NH4INL$	$x_0 \in [0;2.5]$ $y_0 \in [4;5]$ $\Delta x = 0$ $\Delta y = -1$
Maximum inflow rate into plant ($PCINMX$)	$NH4INL$	$x_0 \in [0;2.5]$ $y_0 \in [3;5]$ $\Delta x = 0$ $\Delta y = -1$	$CODINL$	$x_0 \in [0;1000]$ $y_0 \in [2;5]$ $\Delta x = 0$ $\Delta y = -1$	$NH4INL$	$x_0 \in [0;2.5]$ $y_0 \in [2;5]$ $\Delta x = 0$ $\Delta y = -1$
Threshold: Emptying storm tank ($STTPTH$)	(constant value)	1000 m^3/h	(constant value)	1000 m^3/h	(constant value)	1000 m^3/h
Emptying rate of storm tanks ($STTPQ$)	$NH4INL$	$x_0 \in [0;2.5]$ $y_0 \in [300;1000]$ $\Delta x = 0$ $\Delta y = -200$	$CODINL$	$x_0 \in [0;1000]$ $y_0 \in [300;1000]$ $\Delta x = 0$ $\Delta y = -200$	$NH4INL$	$x_0 \in [0;2.5]$ $y_0 \in [300;1000]$ $\Delta x = 0$ $\Delta y = -200$
Return activated sludge rate (RASABS)	$QINTP$	$x_0 \in [2;5]$ $y_0 \in [300;1000]$ $\Delta y = 300$	$QINTP$	$x_0 \in [2;5]$ $y_0 \in [300;1000]$ $\Delta x = 300$ $\Delta y = 300$	$QINTP$	$x_0 \in [2;5]$ $y_0 \in [300;1000]$ $\Delta x = 0$ $\Delta y = 300$
Waste sludge rate ($WASABS$)	(constant value)	27.5 m^3/h	(constant value)	27.5 m^3/h	(constant value)	27.5 m^3/h

n.b.: Those parameters the settings of which are to be optimised are highlighted.

Comparing the results obtained for Scenario 4 (control only in the sewer system) with those of Scenario 5 (control only at the treatment plant) and the base case (*cf.* Figure 5.12), one can observe that prudent settings of the pumps in the sewer system can have significant impact on the performance of the entire wastewater system. It can be seen that framework 7 (using COD information for the control actions performed in the plant) performs better in terms of DO than framework 8, whereas the latter gives better results with regard to ammonium – even though the optimisation was carried out with regard to DO.

Compared to some of the other frameworks discussed here, Scenario 3, which was derived from the analyses of the previous section, gives slightly worse results in terms of DO. This demonstrates the difficulties of finding a "best" framework by means of a systematic study as opposed to intuition. However, since the approach chosen in the previous section for a systematic way to identify promising frameworks suffers from the weaknesses mentioned above, further studies would appear to be justified.

Although optimisation with regard to F2 (criterion based on DO, as was performed here) results only in slight improvements, fairly different results are obtained with regard to ammonium – here control of the treatment plant seems to have major effects. This observation suggests considering both (DO and ammonium) in the optimisation procedure (*cf.* also Section 5.6.2, where aspects of multiobjective optimisation are briefly discussed).

The frameworks discussed in this section would be considered to represent local control scenarios. Also frameworks 7 and 8, where information about treatment plant influent concentration is used to control a pump in the sewer system, are considered to be examples of local control scenarios, since the information transferred from the inlet of the treatment plant towards the pump furthest downstream in the sewer system, might be available in the sewer system itself as well. Further optimisation studies of other integrated scenarios could be carried out in the same manner. In order to conclude the analyses presented in Sections 5.3 and 5.4, the following section compares the results of the "best" strategies found so far.

5.5 Integrated Versus Local Control

This section summarises the findings of the previous two sections with regard to the driving question of this book ("Is integrated control superior to conventional control?"). According to the discussion above, Scenarios 1 and 3 to 8 could be considered to represent forms of local control, whereas Scenarios 2a and 2b exemplify integrated control. In order to compare local with integrated control, those strategies representing the best samples (of those analysed in this study) of each category are evaluated here. The selection of scenarios is done mainly according to the F2 values (which were used for the optimisation of strategy parameters). When two scenarios with similar F2 values are compared with each other, as a second selection criterion the F3 value (which is related to the ammonium level in the river) is chosen. This procedure leads to selection of Scenario 3 (representing the best local control scenario of those analysed so far) and Scenario 2a (as the best integrated scenario).

Figure 5.19 shows the values of the various criteria describing DO and ammonium levels in the river of the solutions found in the optimisation for Scenarios 2a and 3.

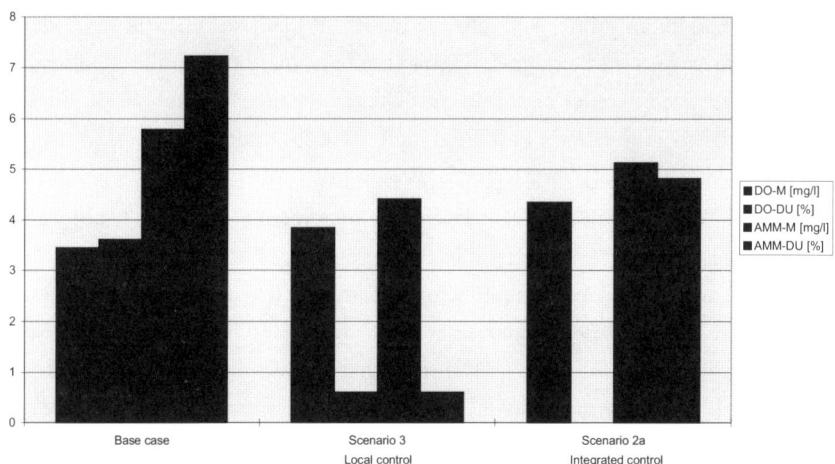

Figure 5.19. Base case and best local and integrated strategies – DO and ammonium criteria

It can be seen from these results that local control (with regard to oxygen) improves the oxygen levels in the river (increase of DO-M and decrease of DO-DU values). Here, an improvement of the ammonium levels is observed. Integrated control leads to further improvement of the DO levels (note that the duration of DO being below the critical threshold value is reduced to zero). However, this improvement is achieved at the expense of the ammonium criteria, which still assume better values than for the base case scenario.

Recalling the definition of the local control scenario considered here (Scenario 3) and the values found for the strategy parameters (*cf.* Appendix D), this scenario is characterised by the use of ammonium information for control of the pump downstream in the sewer system (reducing its pump rate at high ammonium loads in the plant influent) as well as for control of the RAS rate in the treatment plant, which is set according to the influent flow rate and the ammonium load in the effluent). The maximum inflow rate allowed to the plant is controlled by the water level in the last tank in the sewer system. Under normal conditions, the inflow to the treatment plant is limited in this scenario to 2.7 DWF. As soon as the basin downstream in the sewer system is filled more than a third (indicating increased water volumes in the sewer system), the plant influent rate is increased to 3.3 DWF.

The integrated control scenario found here to be best performing of those investigated (Scenario 2a) is defined by a simple two-level hierarchical approach (*cf.* Figure 5.11), where the settings of the pump downstream in the sewer system and of the maximum inflow rate into the treatment plant are influenced by the state of the entire system. For example, the maximum influent rate to the treatment plant is set to 2.75 DWF. Under certain conditions (as detailed in Table 5.3), this setting is increased to 3.75 DWF. Further details of this strategy are explained above (Section 5.3.1).

Overall, the conclusion may be drawn that integrated control can lead to superior performance of the urban wastewater system compared to local control. However, further research would be desirable to substantiate this finding. Further analyses appear to be necessary in particular with regard to an improvement of the optimisation procedure so as to ensure that within the given restrictions on computing time a solution is found which indeed can be considered to represent a global optimum. Furthermore, more work seems to be necessary on the method to determine best frameworks for control strategies (*cf.* Section 5.4). Various other aspects related to the suggested procedure to find optimum control strategies, which have not been discussed in this section, are briefly outlined in the next section.

5.6 Further Aspects

This section briefly discusses various aspects which are associated with the application of the simulation and optimisation procedure developed, outlined and applied in this book. Topics of this section include issues related to the sensitivity of the solutions found (Section 5.6.1), to the consideration of more than one criterion in the optimisation process (Section 5.6.2) and to the length of the input rainfall time series required for the procedure (Section 5.6.3). Finally, in Section 5.6.4, some remarks are made on the influence of variations in the definition of the case study site on the potential of local and integrated control.

5.6.1 Sensitivity of Solutions

When applying the simulation and optimisation procedure developed in this work, at least two questions might arise. One of these would be "How sensitive are the solutions suggested by the optimisation algorithm? And secondly, one might wonder whether a set of most sensitive strategy parameters could be selected prior to the application of the optimisation procedure and to perform the optimisation only with those parameters which are considered to be most important, thus reducing the dimension of the optimisation problem.

In Section 5.3.1, a strategy framework for an integrated control scenario was defined, involving seven parameters (Scenario 2b). This will be used here for the illustration of the questions raised above. Here, the results obtained (see Section 5.3.3) by optimisation of all seven parameters are compared with the results obtained by optimisation of only those parameters which are found in an *a priori* analysis to be most sensitive.

The determination of the sensitivity of the objective function chosen for the optimisation (*i.e.*, the F2 criterion) to each of the seven parameters is determined by calculating the sensitivity coefficients as defined by Lei (1996). Each strategy parameter is varied individually over the its range (assuming n values $x_1,..,x_n \in \Re$) as defined below whilst keeping all other strategy parameters at their optimum values determined earlier on, and then the simulation tool is run. The corresponding values for F2 obtained by the simulation (these are denoted $y_i \in \Re$ ($i=1,..,n$) here) are used for the calculation of the sensitivity coefficient SC for each strategy parameter as follows :

$$SC = \frac{CV_y}{CV_x}, \qquad (5.2)$$

where

$$CV_y = \frac{100}{\sqrt{n-1}} * \frac{s_y}{\bar{y}} \qquad \text{(relative coefficient of variation of } y\text{)}$$

$$\bar{y} = \frac{1}{n}\sum_{i=1}^{n} y_i \qquad \text{(mean value of y)}$$

$$s_y = \sqrt{\frac{1}{n-1}\sum_{i=1}^{n}(x_i - \bar{x})^2} = \sqrt{\frac{1}{n-1}\left[\sum_{i=1}^{n}y_i^2 - \frac{1}{n}\left(\sum_{i=1}^{n}y_i\right)^2\right]} \quad (5.3)$$

(standard deviation of y)

CV_x, $\bar{x}$, and s_x are defined in an analogous way, describing coefficient of variation, mean and standard deviation of the values assumed by that strategy parameter for which the sensitivity coefficient is calculated. Due to its numerical advantages, the rightmost term in (5.3) is used for calculation of the standard deviation (Sachs, 1992).

CV_x and CV_y represent the "relative coefficient of variation" (Sachs, 1992), which essentially is the quotient of standard deviation and the mean of the values considered. The sensitivity coefficient is then determined by dividing the coefficient of variation of the simulation output by the coefficient of variation of the simulation input, in which the parameter the significance of which is to be investigated, is varied. The sensitivity coefficient is considered to represent a measure of the sensitivity of the model output to a change in the parameter input (Lei, 1996).

In order to determine the *a priori* sensitivity of the F2 criterion to the seven strategy parameters investigated in Scenario 2b, these are varied over the entire range defined as the feasible region for the optimisation (see Table 5.2 in Section 5.3.1). For convenience to the reader, the definition of their ranges as well as the parameter settings found by optimisation (Table A.8.1) are repeated here (Rows 1 and 4, respectively, in Table 5.8). Graphical representations of the objective function in the neighbourhood of the solution found by the optimisation procedure (row 4) can be found in Appendix D.

Table 5.8 summarises the sensitivity analysis carried out. Row 1 gives the feasible range of the strategy parameters as defined in Section 5.3.1. Row 2 shows the sensitivity coefficients which were obtained acccording to Equations (5.2). These served for the selection of the most significant parameters, which are found to be z_1, z_2 and z_3 (the corresponding entries are shaded in Table 5.8). Row 3 shows the results of an optimisation of only these parameters (within their ranges defined in row 1) and the corresponding F2 value of the optimum solution found (here: 4.04).

The results of the optimisation of all strategy parameters over their full ranges (row 1) is shown in row 4 (*cf.* also Section 5.3). After the optimisation, the sensitivity of the F2 value to the parameters within the neighbourhood of the optimum is investigated. In order to do this, ranges were defined (spanning ± 10% of the original range around the solution just obtained; except for the case of P7, which is operated in discrete stages). These are shown in row 5.

Table 5.8. *a priori* and *a posteriori* sensitivity analysis of the strategy parameters of Scenario 2b

		z_1 P7	z_2 PCINMX	z_3 STTPTH	z_4 Sewer system	z_5 Storm tank	z_6 Tr. plant	z_7 River	F2
						Threshold			
1a	Range from	2	2	700	0	0	3	75	n/a
1b	to	5	5	1000	100	100	5	120	n/a
A priori sensitivity analysis									
2.	SC over this range	0.10	0.12	0.15	< 0.01	< 0.01	< 0.01	< 0.01	n/a
3.	Sol. (1)	5	3.10	855	n/a	n/a	n/a	n/a	4.04
Optimisation of all parameters (*cf.* Section 5.3.1)									
4.	Sol. (2)	4	2.75	1259	83	27	4.53	79	4.31
Sensitivity analysis of solution found by optimisation (*a posteriori*)									
5a	Range around solution	2	2.45	1199	73	17	4.33	74	n/a
5b	(2)	4	3.05	1319	93	37	4.63	84	n/a
6.	SC over this range	0.16	0.38	0.66	0.36	0.01	0.00	0.49	n/a
7.	Sol. (3)	4	2.79	1220	81	n/a	n/a	77	4.35

n.b.: The shaded entries in rows 2 and 6 mark the most sensitive parameters found in the *a priori* and *a posteriori* analyses, respectively.

Row 6 shows the resulting *a posteriori* sensitivity coefficients. It can clearly be seen that the sensitivity is more pronounced locally around the parameter set suggested by the optimisation than when calculated over the entire feasible domain. Also the parameters z_4 and z_7 now show sensitivity coefficients comparable to those of z_1 to z_3. Those parameters showing greatest sensitivity (*i.e.*, z_1 to z_3, z_7) are subjected to an additional call of the optimisation routine, the results of which are shown in row 7. This represents a "refinement" of the solution found earlier (row 4).

Analysing the F2 values obtained in the various optimisation runs, the following conclusions may be drawn. An attempt to reduce the dimension of the optimisation problem by considering only those strategy parameters which appear *a priori* to be the most significant ones (with regard to the objective function chosen) may lead to a suboptimum solution (compare the F2 value of 4.04 shown in row 3 with the one of 4.31 obtained when optimising all seven parameters). Therefore, particular care has to be taken when a sensitivity analysis is applied in an attempt to select a set of most relevant parameters for the optimisation and thus to reduce the dimensionality of the optimisation problem.

Performing an *a posteriori* sensitivity analysis (row 6) not only reveals the most significant parameters of the solution obtained, but also proves to be helpful for an additional run of the optimisation procedure, which is carried out to "refine" the solution obtained earlier on.

It is clear that the analysis of the sensitivity of the parameters found by the optimisation procedure would merit a much more detailed study. Such a study of the objective function in the vicinity of the global optimum could involve the computation of second-order approximations of the objective function and parameter covariance matrices (Gill *et al.*, 1991; Kuczera, 1997).

Overall, due to the close resemblance of the strategy optimisation problem to the problem of model parameter estimation (as pointed out in Section 2.5.1 and detailed by Schütze (1996c)), methods suggested for and potential problems (e.g. identifiability of the strategy parameters) of parameter appear to be of relevance in the present context. A comprehensive review of these issues is given by Beck (1987).

5.6.2 Multi-objective Optimisation

Most analyses performed in this work were carried out with regard just to the F2 criterion, which condenses information about minimum DO concentrations in the river and the duration of the DO concentration being below a threshold value of 4 mg/l. However, operation of the urban wastewater system generally has to pursue several objectives, which may be contradictory. The discussion in Section 5.2 showed (for a simple example) that optimisation with regard to ammonium may lead to different control strategies from a strategy found to be optimising DO levels. Often a variety of objectives are to be considered when defining control strategies (*cf.* Section 2.4). Due to the possibly contradictory nature of these objectives, no general solution can be found which is optimum with regard to each of the individual objectives. Therefore, optimisation with regard to several criteria (multiobjective optimisation) has to be concerned with the determination of the set (or, at least, an element of the set) of non-inferior solutions. A non-inferior solution is one in which an improvement in one objective results in deterioration with respect to another one.

The simplest approach to considering several objectives for the optimisation consists in weighting and summing the functions defining the individual objectives. Here, an immediate problem lies in the definition of the weights. Although this approach is widely applied (for example in the projects involving optimisation for RTC of sewer systems, see Section 2.4.2 and 2.4.3), it is not easy to define a set of appropriate weights. A suggestion to define the variety of objectives in terms of monetary costs is made by Vanrolleghem *et al.* (1996b).

Also various other approaches to finding non-inferior solutions are suggested in the literature on optimisation. Among the potential options would be to consider all but one objectives as constraints in the optimisation procedure. In this section, however, a simple and pragmatic approach ("preemptive goal programming"; Lobbrecht, 1997) is suggested which makes use of the features of the global optimisation algorithms discussed in earlier chapters of this book. Since these algorithms, in particular the Controlled Random Search, propose not only a global optimum but also maintain a set of best solutions found so far during the optimisation process, it seems plausible to evaluate this set (resulting in best values of the criterion used for optimisation) also with regard to the other criteria of interest. This is demonstrated here, using again the optimisation results of Scenario 2b (defined in Section 5.3.1) as an example, for the consideration not only of oxygen levels in the river (F2 criterion), but also of the ammonium levels (F3 criterion). Figure 5.20 shows the set of best solutions at termination of the algorithm plotted against their F2 and F3 values.

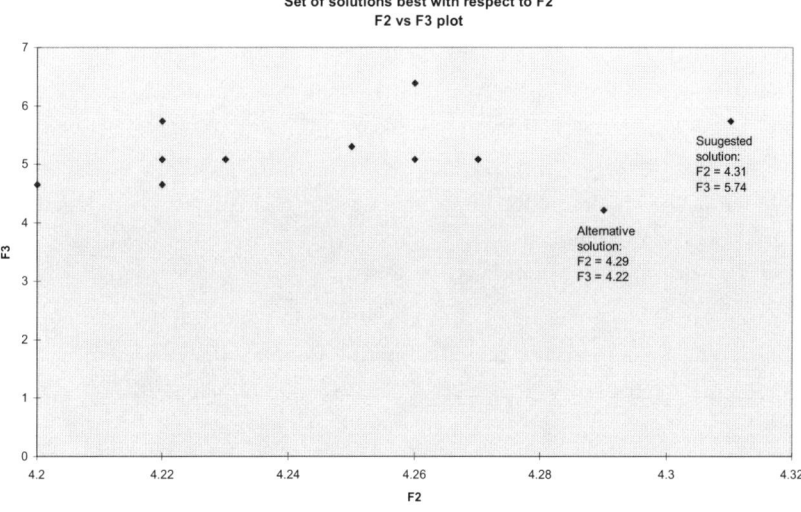

Figure 5.20. F2-F3 plot of those candidate control strategies with best results in terms of F2

As can be seen from this figure, the F3 values of those strategies resulting in largest (best) F2 values vary considerably. Those (F2;F3)-pairs expressing best performance in terms of DO and ammonium concentration in the river can be found in the lower right corner of the plot (*i.e.*, large F2 values, small F3 values). In this example, none of the (F2; F3)-pairs shown appears to represent an overall best solution. Thus, two solutions are suggested as shown in the figure. Depending on how DO and ammonium concentrations are weighted against each other, either one or the other would be suggested as the solution. Table 5.9 gives the values of the strategy parameters of these two solutions.

Despite its simplicity, this procedure appears to be of some interest as long as one of the objectives to be considered can be identified as the primary objective (here: F2), which serves as the objective function for the optimisation, whilst the remaining objective(s) assist in the final selection of the solution.

Table 5.9. Strategy parameters of the solutions suggested when considering F2 and F3 simultaneously

No.	z_1	z_2	z_3	z_4	z_5	z_6	z_7	F2	F3
1.	3	2.75	1259	83	27	4.53	79	4.31	4.29
2.	3	2.86	986	76	18	3.98	88	4.29	4.22

5.6.3 Simulation Period Required for Optimisation

Although the simulation and optimisation procedure developed in this book can be applied rainfall input time series of any length (besides the limitations in computing time; however here without pressure to find a solution within one control time step as is the case in on-line optimisation), most optimisation runs described were performed using a time series of only one week duration (as defined in Section 4.3) as rainfall input. This restriction to a time series of relatively short length was necessary since a large number of optimisation runs for test and evaluation of the different optimisation algorithms was carried out. Ideally, the evaluation of the impacts of a given control strategy (this constitutes the evaluation of the objective function) should be done over a long time period in order to take long-term effects of pollution into account as well as a great variety of rainfall patterns. However, this may not be practical, simply because of the computing time required for a large number of simulation runs covering a long time period. Therefore, the question may arise to what extent a control strategy found by optimisation over a relatively short period can lead to improved performance of the urban wastewater system when applied over a longer period.

An analysis has been conducted, in which the parameters of one particular strategy framework (again, Scenario 2b is used to serve as an example) are optimised. However, five different series of rain data of different length are used as input for the optimisation. The five control strategies defined by the resulting parameter sets are then evaluated by using them as the control input in a simulation of the longest of these five rainfall data series. From the results conclusions are drawn as to whether a control strategy derived from a shorter rainfall may be of similar quality to a parameter set which requires more simulation time for its determination. The rainfall series used here are characterised as follows: Series 1 comprises a continuous data set covering the period April to September 1982. The summer months have been chosen, since these are generally characterised by more rainfall (in Northern Germany) and water quality conditions being more critical for the oxygen levels as observed by Lammersen (1997a,b). Rainfall series 2 consists of half of the data, namely the first three months of series 1. A selection of events was carried out to obtain data set 3, which is characterised in Table 5.10. Every third event from a ranked list of all events (defined here by the settings of the event separation criteria as motivated by the findings of Section 4.5 of an interevent time of 4 days and a minimum depth of 0.3 mm) during the summer months of 1992 was selected. The ranking of rain events was performed according to their total rain depth. The rainfall series of one week length used throughout this study constitutes series No. 4 used in this analysis.

Table 5.10. Event selection defining rainfall series 3 (selected events are highlighted)

Event start date	Event start time	Event end date	Event end time	Event duration [days]	Total depth [mm]
11.06.1982	15:40	10.07.1982	22:15	29.28	118.8
05.08.1982	19:40	11.09.1982	13:35	36.75	99.9
28.04.1982	15:00	11.05.1982	16:10	13.05	30.8
06.04.1982	20:55	17.04.1982	20:00	10.97	21.2
17.05.1982	18:35	31.05.1982	23:40	14.21	20.0
23.04.1982	20:20	28.04.1982	06:30	4.43	9.6
15.07.1982	12:50	20.07.1982	20:15	5.31	6.1
20.09.1982	19:50	26.09.1982	00:35	5.20	4.0
26.09.1982	17:55	30.09.1982	00:55	3.30	4.0

Finally, optimisation was also carried out using dry-weather (zero rainfall) data as input. Since the definition of Scenario 2b implies that none of the settings of the control devices affected by these strategy parameters affects a system with no rainfall runoff, the resulting set of strategy parameters constitutes in fact a random sample from their feasible domain. As the last test case for the evaluation, the base case scenario of control (see Table 4.10) is tested (No. 6). Table 5.11 presents the parameter sets obtained from different rainfall series.

Table 5.11. Parameter sets obtained for Scenario 2b over various input rainfall series

No.	Rainfall data used for optimisation	Length [days]	z_1	z_2	z_3	z_4	z_5	z_6	z_7	F2
1.	April – September 1982	182	2	2.94	997	46	84	4.42	104	3.87
2.	April – June 1982	100	2	2.99	1009	27	69	4.92	95	3.91
3.	Selected events of 1.	63	3	3.55	1147	86	90	3.82	94	3.91
4.	1 week in 1977	7	3	2.75	1259	83	27	4.53	79	4.31
5.	Dry-weather flow	7	3	4.57	905	44	23	4.59	101	7.60
6.	Base case definition	n/a	5	3.00	1000	n/a	n/a	n/a	n/a	

Although the parameter sets obtained seem to be fairly different from each other, it can be observed that the values of each of them (probably less so the values for z_4 and z_5), are clustered in certain regions of their respective ranges (*e.g.*, the overall best value of z_1 seems to be either 2 or 3, z_2 seems to have a value of about 2.75 to 3.55 *etc.*). Furthermore, it can be observed from the results that optimisation of strategy parameters over a shorter time period gives better values of the criterion according to which the optimisation is performed. This corresponds to expectations since finding parameters over a shorter period of rainfall data allows the parameters to be better adopted to the specific characteristics of the particular rainfall input than this would be the case if they had to represent the optimum parameter set for a longer input time series.

Figure 5.21 shows the results of application of the different parameter sets of Table 5.11 to the simulation of the rainfall series covering April to September 1992, which was used to obtain parameter set 1.

Various conclusions can be drawn from the results shown in Figure 5.21. Use of a rain series of shorter length (Series 2) than the full six month series (Series 1) leads to strategy parameters giving almost the same F2 value. A similar statement holds true for even shorter series (such as Series 3 and 4), with the F2 values being slightly lower. As the example 4 shows, a parameter set obtained from rainfall input taken from a shorter series (taken from a different year) leads to reasonable performance. A parameter set obtained by random sampling can result in an F2 value far from satisfactory, as case 5 shows. This is reassuring, since this shows that the selection of the strategy parameters indeed has an influence on F2. Finally, it is confirmed again that the base case setting of the strategy parameters leads to fairly good results.

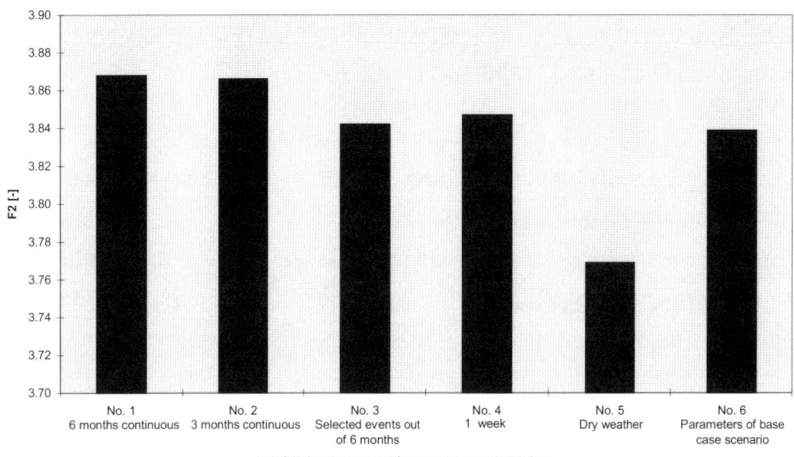

Figure 5.21. Results obtained for the F2 criterion obtained when applying the parameter sets shown in Table 5.11 to the April to September 1982 rain series

Even though this substudy is far from comprehensive, it suggests that the use of a relatively short time series for the rainfall input data for optimisation of strategy parameters may be justified. However, a more detailed study is considered to be necessary to confirm these findings.

5.6.4 Control Potential of Various Case Study Sites

As discussed in Section 5.5, integrated control can lead to better DO levels in the river. However, this improvement was not very significant. Therefore, the question arises which key factors have the greatest influence on the extent to which integrated control leads to better river water quality. Obviously the potential of integrated control as assessed with the methodology and tools developed in this work depends on

- characteristics of the case study site defined;
- type of integrated control investigated (definition of the framework; complexity of control actions considered for the analysis; definition of the ranges of the strategy parameters);
- methodological approach (*e.g.*, optimisation algorithms implemented).

Further research on each of these topics is recommended (*cf.* Chapter 6). However, a brief analysis of the first of these points is outlined in this section.

Five variations of the case study site defined in Chapter 4 and used throughout this book are investigated here. These include the following.

0. unmodified case study site (as defined in Section 4.1);
1. original case study site, but with reduced storage in the sewer system (tank volumes reduced by 50%);
2. original case study site, but with no restriction of inflow to the aeration tank;
3. original case study site, but with reduced river base flow (reduced from 1.5 m^3/s at all times to 0.75 m^3/s; thus the dilution ratio of dry-weather treatment plant discharges is altered from 4.7 to about 2.4);
4. original case study site, but with reduced river catchment rainfall runoff (reduced by 50%; resulting in less dilution in the river during rainfall events);
5. original case study site, but with no river catchment rainfall runoff at all (thus the upstream river flow has a constant value of 1.5 m^3/h at all times and is not influenced by rainfall).

For each of these variations of the case study site, the strategy parameters of the frameworks defined for Scenarios 1 ("local control") and 2b ("integrated control") (*cf.* Section 5.3.1) are optimised with regard to the F2 criterion (applying the Controlled Random Search algorithm as before). Also the F2 value corresponding to the base case (*cf.* Section 4.1.5) is determined for each of the sites defined above. The results are converted to duration of DO concentration in the river being below 4 mg/l and shown in Figure 5.22.

It can be seen from Figure 5.22 that for the unmodified case study site local control reduces the duration of critical DO levels in the river. Integrated control reduces the duration even further down to zero. As one would expect, the case study site with less storage in the system performs worse in terms of DO. Furthermore, the benefits gained by control are less pronounced. Waiving the restriction of inflow to the aeration tank (variation 2) does not give results which are significantly different in terms of control potential from those obtained for the original case study site. The catchment variations with reduced river base flow or reduced catchment runoff (3 to 5) are characterised by considerably longer periods of critical DO levels. However, local as well as integrated control seem to be of greater potential here than for the other case study site variations analysed. It is interesting to note that for the case study with reduced catchment runoff (4), integrated control can reduce the periods of critical DO concentrations to zero.

Analysis of Control Scenarios 275

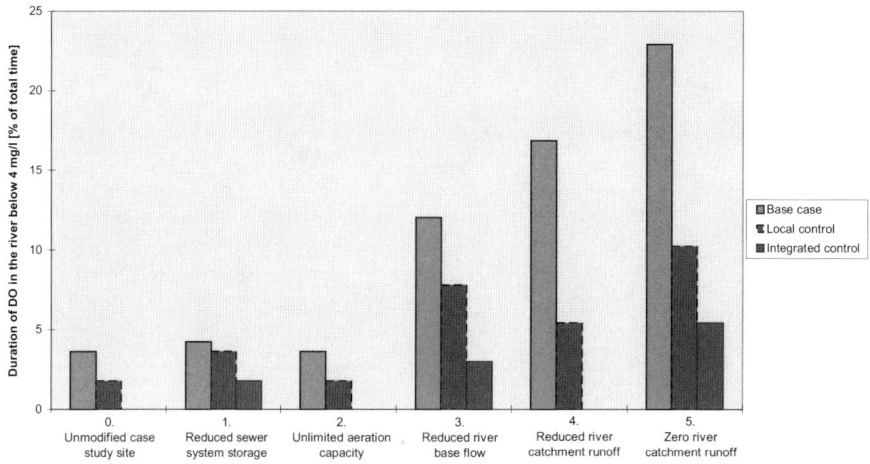

Figure 5.22. Variations of the case study site: DO-DU values obtained for Scenarios 1 and 2b

In summary, this brief analysis suggests that the effects of control (aiming at improvements in terms of DO levels in the river) appear to be less pronounced in systems with reduced storage in the sewer system, whereas control appears to be of particular benefit in case study sites characterised by low dilution conditions. A more detailed analysis of such topics can be found in Schütze *et al.* (in print).

Chapter 6
Conclusions and Further Research

6.1   Summary

In Chapter 1, the ultimate goal of this book has been defined as the assessment of the potential of integrated control of the urban wastewater system. This objective has been approached in two main steps. The first subgoal was the assembly of a simulation tool, which is capable of the simulation of the water flow and quality processes in sewer system, treatment plant and receiving river. After having achieved this subgoal, the tool developed was linked to a variety of optimisation procedures to form the program package SYNOPSIS ("software package for synchronous optimisation and simulation of the urban wastewater system") and subsequently used for the evaluation of control strategies.

The main achievements with regard to the first subgoal can be summarised as follows: a detailed review of the processes in the urban wastewater system as well as of the impacts of storm events on this system has been compiled. This review is complemented by an overview of currently available simulation tools for the components of the urban wastewater system. Since none of these nor any of the integrated simulation approaches pursued previously (cf. Section 2.3) met the particular requirements of this work, the most important of which is related to the ability of the model to consider information from all parts of the urban wastewater system for the determination of control actions in the system, a simulation tool was assembled by adapting and linking existing simulation packages. Various problems arising from the different characteristics of the constituent models (such as different sets of state variables) were discussed and solved in a pragmatic way. The package SYNOPSIS has been designed in such a way that the constituent submodels can be easily replaced by other models. This ensures that future developments in modelling can be incorporated in a straightforward manner in the simulation and optimisation environment developed in this book.

Since no field data covering an urban wastewater system in its entirety were available, a semi-hypothetical case study site had to be defined and modelled within this work. Test runs confirmed the plausibility of the results provided by the simulation tool. Thus, a tool for the simulation of water quantity and quality processes within the urban wastewater system has been established.

In order to develop and to analyse strategies for the operation of the urban wastewater system (the second subgoal of this research), a number of steps were carried out: control actions and procedures conventionally applied in each of the constituent parts of the system were reviewed. Particular emphasis was placed on the procedures applied for the determination of control actions. After provision of a comprehensive review of mathematical optimisation methods, a so-called "off-line optimisation approach" was defined and implemented for the determination of frameworks of control strategies and of the optimum settings of their parameters. The resulting optimisation problem is solved by several global and local optimisation techniques, which make only weak assumptions about the objective function. This constitutes a clear novelty compared to earlier applications of optimisation techniques for control of parts of the urban wastewater system, where the model used to describe processes in the system the performance of which was to be optimised had to be simplified considerably (or even linearised). The optimisation techniques used for the present approach, however, allow the use of basically any simulation model for the evaluation of the objective function of the optimisation problem.

The application of the optimisation procedures has been demonstrated for a test example and for the optimisation of the parameters of various control scenarios. Of the various optimisation algorithms implemented and evaluated within the context of this work, the Controlled Random Search algorithm due to Price (1979) proved to be the most successful one.

Finally, in order to address the driving question of this book ("Is integrated control superior to conventional control?"), a clear definition of the term "integrated control" was proposed, since no such definition could be found in the literature. In order to compare integrated control with conventional (non-integated) control and thus to assess its potential, a variety of control scenarios has been defined. The parameters of the corresponding strategies have been optimised by application of the tool and the methods defined in this research. As the most influential control devices, downstream pumps in the sewer system, the limitation of maximum flow to the treatment plant and (if longer time periods are considered) the sludge rates at the treatment plant have been identified. Results suggest that integrated control, which makes use of information from all parts of the urban wastewater system, can lead to improved performance of the urban wastewater system. The performance was assessed here by a criterion combining extreme pollutant concentrations and the duration for which critical concentrations were exceeded. Although, for the semi-hypothetical case study site analysed here the improvement in performance appears to be limited, quite marked improvements were noted under certain conditions (in particular when low-flow conditions in the river prevail).

To summarise this overview of the book, the main achievements and results can be listed as follows:

- review of the processes in the main components of the urban wastewater system, including the impacts caused by storm events;
- review of existing software packages and previous approaches to integrated modelling of the urban wastewater system;
- assembly of an integrated simulation and optimisation tool (SYNOPSIS) by adaptation and connection of existing software packages. Problems arising from the use of different sets of state variables in the constituent models have been solved by use of conversion factors. The novelty of this simulation tool consists not only in its integration in a variety of optimisation procedures (see below), but also in its ability to simulate the processes within sewer system and treatment plant simultaneously and in its provision to consider information from all parts of the urban wastewater system when taking control decisions;
- definition of a semi-hypothetical case study site since no appropriate data of a real site were available;
- discussion of the conventional notion of "event", concluding that consideration of individual events becomes problematic and does not lead to significant reduction of computation time when simulating the urban wastewater system in its entirety;
- review of mathematical optimisation methods and of real-time control of sewer systems, treatment plants and rivers;
- definition of the term "integrated control";
- implementation of various optimisation methods (Controlled Random Search, several variations of a genetic algorithm, local optimisation after Powell), which call the simulation tool as a means for the evaluation of the objective function. Of the algorithms implemented, the Controlled Random Search gives best results within the given limitations of computation time;
- definition of an off-line optimisation approach for the development of control strategies and optimisation of their parameters using the procedures implemented in SYNOPSIS;

280   Modelling, Simulation and Control of Urban Wastewater Systems

- application of this procedure to a variety of local and integrated control scenarios: optimisation of the parameters of the corresponding strategies. Results obtained suggest that integrated control, making use of information from all parts of the urban wastewater system, can lead to some improvement of the performance of the urban wastewater system. The extent of these improvements depends, among others, on the characteristics of the case study site.

6.2   Suggestions for Further Research

Due to the various assumptions made in the course of the work described in this book, some topics for further research evolve directly. To these, a number of issues, indicating further developments, have been added to form the following list.

Simulation of the urban wastewater system

- Naturally, the simulation modules implemented in this work could benefit from improvements of various kinds. These include consideration of some of the processes in the sewer system in a more detailed way (*e.g.*, surface flooding; as soon as appropriate data are available: sedimentation and washoff on the surface and within sewers; spatial distribution of rainfall). The treatment plant model could be extended to include denitrification and, possibly, phosphorus removal processes (*e.g.*, by implementation of the full IAWQ Activated Sludge Models). The secondary clarifier model may benefit from upgrading to the model of Takács *et al.* (1991). Further developments of the submodule simulating the river are suggested as follows: inclusion of a more detailed module for the simulation of catchment rainfall runoff processes; consideration of temperature variations, and of erosion, transport and deposition of sediments in the river. In order to allow a greater variety of integrated control strategies to be analysed, the river model would have to be implemented, similarly to the sewer system and treatment plant models, in a truly parallel way.
- Among the weakest parts of the simulation tool developed in this book might be the definition of conversion factors used for interfacing different state variables in different submodules. These could either be revised, *e.g.*, by making use of information available about the different fractions of organic matter, or made obsolete by further pursuing a reconciliation of different models (see also Section 2.3).

- In order to make further step towards model integration, the concept of simulating sewer system, treatment plant and receiving water in separate routines could be given up and this subsystem-based model structure replaced by an element-based structure, comprising of an overall set of equations governing the whole urban wastewater system as discussed by Rauch (1996a). Also the use of parallel processing hardware for the simulation of synchronous processes could be considered.

However, all modifications to the simulation tool could be in vain if they are not based on a comprehensive data set sampled from all parts of the urban wastewater system.

Control
- In this book only a small number of fairly simple control actions could be investigated. Thus, straightforward extensions would include the consideration of more sophisticated controllers, *e.g.,* those involving the use of more than one state variable for the control decision and/or of past and predicted future system states. Furthermore, additional control options, such as aeration control, step-feed control, control of nutrient removal processes and control actions in the river, could be included in the definition of control strategies.
- After implementation of the river module in a truly parallel way (see above), system states information also from downstream locations in the river could be considered for control decisions. Furthermore, control actions in the river and their coordination with other parts of the urban wastewater system (*cf.* Reda, 1996) could be included in the analysis of control in the urban wastewater system.
- The representation and implementation of control strategies could be extended. Potential options include the integration of a fuzzy controller into SYNOPSIS, the numerical parameters of which (*e.g.,* those involved in the definition of the membership functions) are optimised using the routines proposed in this work. Another option to extend the control module would consist in the inclusion of a rule interpreter, containing the control strategies, which are defined by rules and other forms of knowledge representation (Schütze, 1994b; Schütze *et al.*, 1994). The settings of the numerical parameters defining the control strategy could be optimised as described.

Optimisation

- Further research with regard to the optimisation procedures suggested in this work would, on the one hand, be related to the optimisation algorithms itself (*e.g.,* implementation of the Shuffled Complex Evolution algorithm by Duan *et al.* (1992, 1994)); a more detailed analysis of the fitness function and of the parameter settings employed in the genetic algorithm, the latter of which could be assisted by use of another genetic algorithm (Grefenstette, 1986); improved consideration of constraints in the local optimisation procedure). On the other hand, further improvements of the optimisation procedure (*e.g.,* by focusing the search on the most promising hyperellipsoids in the search space as suggested by Kuczera (1997)) are possible, which would probably result in solutions obtained with less computational effort.

- Additional global optimisation methods could be investigated, in particular those which assume Lipschitz-continuity of the objective function. Since optimisation is carried out to determine strategy parameters, which in practice will assume discrete values (under a sufficiently fine discretisation), Lipschitz constants could be defined for the objective function, thus opening up the use of dedicated global optimisation techniques such as those described by Zhigljavsky (1991) and Pintér (1996).

- A significant gain in computing speed is to be expected if the call of the full simulation model as a means to evaluate the objective function can be replaced by the call of simpler substitute models. Besides further analysis of the applicability of the Response Surface Methodology (with its drawbacks as a local optimisation method), also the use of neural network models of the processes in the urban wastewater system would represent important steps towards the application of the methods developed in this work with reduced computational effort.

- Obviously, the methodology developed here would benefit also from further developments in computing hardware.

- Optimisation of strategy parameters is this work was done solely with regard to criteria based on oxygen and ammonium concentrations in the receiving river. However, in practice, also a variety of other objectives has to be considered as well. This motivates a more detailed analysis of multiple objective optimisation techniques. For example, Folnseca and Fleming (1993), Cieniawski *et al.* (1995), To (1997) and Gupta *et al.* (1999) propose an optimisation technique involving a genetic algorithm which considers multiple objectives. Reference is also made to Section 5.6.2 of this book and to Schütze *et al.* (submitted).

- Also the other subsections of Section 5.6 contain suggestions for further research (related to sensitivity analysis and to the required length of the input rainfall time series).

- Since the ultimate goal of the application of optimisation techniques in the context of this research does not consist in the minimisation of certain numerical values, but in a comparative evaluation of control scenarios, techniques of ordinal optimisation (Ho *et al.*, 1992; Ho and Larson, 1995) appear to be of some interest. However, their applicability within the given context would have to be evaluated first.

- Lastly, the concepts of off-line optimisation (as developed in this book) and on-line optimisation (as applied in many real-time control projects) do not necessarily have to be seen as mutually exclusive approaches. Further research could aim at a reconciliation of both techniques, thus combining the advantages of both approaches. A general control framework developed by an off-line optimisation approach could be complemented by control decisions within this framework, which are determined by on-line optimisation at every time step.

Application to a real case study

- The most obvious continuation of the research reported in this book would be its application to a real case study site. It is believed that this is possible in a straightforward way as soon as appropriate data are available. Such an application would allow for a more sound appraisal of the potential of the procedure developed than s is possible when using a semi-hypothetical case study site.

- In a similar way, the SYNOPSIS package developed in this book may prove useful for a variety of other analyses, such as the assessment of the effects of water conservation and source control measures on the urban wastewater system in its entirety. Progress in this direction was made by Parkinson (1999), Parkinson *et al.* (2001).

In summary, it can be stated that the research reported in this book has developed a simulation tool and an optimisation procedure which have been shown to be very useful for studies concerning the potential of conventional and integrated real-time control of urban wastewater systems in their entirety. Among the novelties of this research are the development of an integrated simulation tool, which allows strategies for integrated control of urban wastewater systems to be simulated, and its application to the development, refinement and assessment of control strategies by off-line optimisation of its parameters, employing techniques of global optimisation. It could be shown (for a semi-hypothetical case study site) that integrated control of the urban wastewater system can lead to its improved performance. The extent of the potential improvement depends on the characteristics of the given case study site. It is hoped that this work will motivate further studies and developments in the area of integrated simulation and control, thus contributing to changes in urban wastewater management, resulting in a better use of existing wastewater infrastructure and to the environment as a part of God's creation.

Appendix A
Overview of Existing Software

This appendix provides a survey of some of the existing commercially available software products for simulation of the individual components of the urban wastewater system. Only a brief overview will be given here. This will be done in alphabetical order within each group of packages. Descriptions of the models used in the present work can be found in Section 3.2 of this book. Section 2.3 reviews previous (and current) approaches to integrated simulation of sewer system, treatment plant and rivers and will put the work described in this book into the context of similar projects. Parts of the survey presented in this section have been compiled in 1998. These have been complemented by some recent developments (2001).

A.1 Software for Simulation of Sewer Systems

This section outlines the main characteristics of some of the currently available sewer system models which include simulation of water quality parameters. More details about the individual models can be found in the references given.

An overview of the models HSPF, ILLUDAS, AUTO QI, STORM, SWMM, MOUSE, Wallingford as of 1993 is given by Maršalek *et al.* (1993). Other model reviews include those by Ashley and Goodison (1991) and House *et al.* (1993). It should be noted that some progress has been made since, in particular with regard to modelling water quality processes.

Figure A.1, taken from Ahyerre *et al.* (1998), gives an overview over various degrees of complexity found in common sewer models.

286   Appendix A

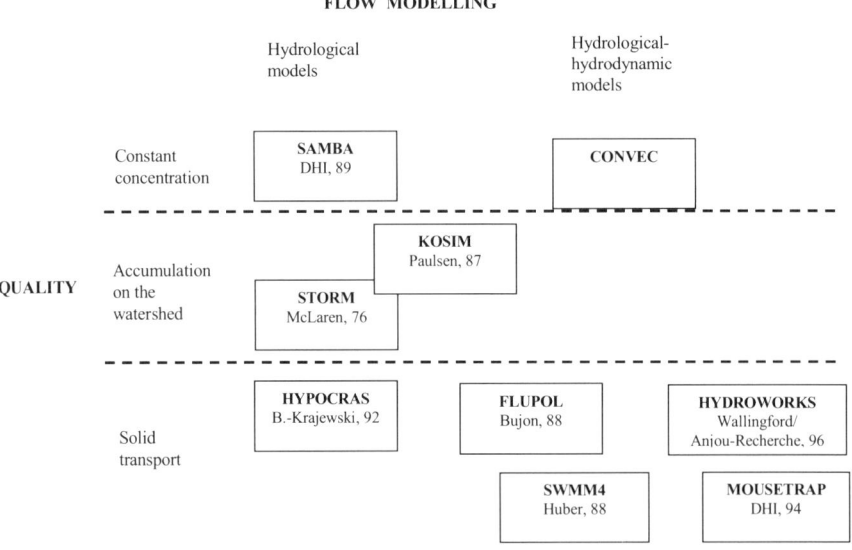

Figure A.1. Typology of existing sewer system flow and quality software (reproduced from Water Science and Technology, Vol. 37, No. 1, p. 207, with permission from the copyright holders, IAWQ)

FLUPOL

FLUPOL was developed by Agence de l'Eau Seine-Normandie (AESN), the Syndicat des Eaux d'Ile-de-France (SEDIF), and the Compagnie Générale des Eaux (CGE). Similarily to MOSQITO, the FLUPOL model simulates pollutant transport (without dispersion), sedimentation and erosion of sediments in sewers, aiming at producing hydrographs and pollutographs for various locations of the system. Pollutants modelled include TSS, COD, BOD and TKN. The hydraulic model uses three flow routing methods (Muskingum-Cunge, kinematic and diffusive wave). FLUPOL uses a less detailed description of the subcatchment characteristics than MOSQITO. For example, FLUPOL uses a global runoff coefficient whereas MOSQITO allows (or requires) a more specific definition of runoff characteristics. FLUPOL also models surface sedimentation and washoff (based on the linear reservoir approach), pollutant transport in sewers (advection only), sedimentation and suspension in sewers (determined by the Velikanov criteria, see Phan *et al.* (1994)), but no biochemical transformation processes. More details about FLUPOL can be found in Bujon *et al.* (1992), Phan *et al.* (1994) and Blanc *et al.* (1995). Modelling developments of MOSQITO and FLUPOL are being integrated in the quality module of HYDROWORKS.

HYDROWORKS

The HYDROWORKS package has been developed by Wallingford Software (UK) and Anjou Recherche (France). According to promotional material, it is designed as an integrated package, providing modules for sewer systems, treatment plants and rivers once the development process will have finished. As of today, modules for the sewer system and for the river (ISIS; see Section A.1.3) are available. Since development of HYDROWORKS was in its early stages when this work was started, it could not be considered in the model selection phase of this project. This section gives a brief characterisation of HYDROWORKS PM and DM. References describing HYDROWORKS and its applications include Michas (1995), Neylon (1995), Magne *et al.* (1996), Ashley *et al.* (1997), Udale *et al.* (1997) and Parkinson (2000).

The hydraulic model of HYDROWORKS is based on the full Saint Venant equations. Surcharge flows are modelled using the Preismann slot concept (see, for example, Havlik, 1996). Water quality modelling is based on the development of MOSQITO and FLUPOL (see above). It can be characterised as follows (Neylon, 1995): modelling of transport of suspended sediment and dissolved pollutants uses a mass-conservation approach. Dispersion of these two is assumed to be negligible. Pollutants modelled include BOD, COD, Total Suspended Solids, ammonia, TKN, total P, up to four user-defined dissolved pollutants and a single suspended sediment fraction. No physical or biochemical degradation of sediments and pollutants is modelled in HYDROWORKS.

A presentation by Udale *et al.* (1997) criticises HYDROWORKS for currently representing only one sediment fraction. Furthermore, these authors state that pollutants are assumed to be fully mixed at manholes, resulting in the fact that sedimentation in tanks cannot be modelled by HYDROWORKS as of yet. Magne *et al.* (1996) report successful HYDROWORKS applications for several catchments. Ashley *et al.* (1995) describe an application of HYDROWORKS, without, however, giving any details about the development stage of this model. In a later paper, Ashley *et al.* (1997) suggest that HYDROWORKS-QSIM does not account even for the limited knowledge that does exist in pollutant modelling.

More specifically, these authors state:

"it should be expected that for all systems in which sewer solids are known to play a significant part, the software will be unable to make a useful representation. This is because:

it models only one sediment fraction

it ignores bedload transport

it is relatively insensitive to quality parameter changes"

HYDROWORKS also includes a Real-time Control module, allowing for simple RTC strategies in the sewer system (such as logical rules, PID controllers and pumps with time-dependent settings) to be simulated. Applications of this module are described by Ashley *et al.* (1995) and Michas (1995). Despite the uncertainties associated to its quality module (as in many other packages), HYDROWORKS represents one of the most widespread packages in the UK.

KOSIM

This package is described in detail in Section 3.2.1.

MOSQITO

The sewer flow quality model MOSQITO was developed by Wallingford Software, based on research carried out by HR Wallingford and WRc. MOSQITO was designed as an add-on package to the flow model WALLRUS. Due to its simplifications (flow routing using the Muskingum-Cunge approach), WALLRUS cannot model looped networks. According to Crabtree *et al.* (1994a), MOSQITO was to be linked to the model SPIDA, which is capable of simulating looped networks as well (Price, 1994). However this development might have been dropped due to redesigning the MOSQITO engine as part of HYDROWORKS (Osborne, 1995).

MOSQITO simulates surface runoff, pollutant transport (dispersion is neglected), sedimentation, washoff and sediment transport. Pollutants are modelled in two forms (dissolved and sediment attached) and include TSS, BOD, COD and NH_4-N. Coarse and fine sediment fractions are distinguished. Three sediment type are considered: pipe sediment, surface sediment and foul flow sediment. MOSQITO does not simulate any biochemical interactions between pollutants nor are any degradation processes considered.

According to Jack (1995), simulation runs with MOSQITO require a significant amount of computation time. For a 1700 pipe catchment on a Sun workstation, this auther reports a simulation time of one-third of real time. Though a statement like this obviously depends on the specific characteristics of the casestudy site under consideration and of software and hardware being used, it may be concluded that such simulation times are not feasible for this project.

A critical appraisal of MOSQITO can be found in Crabtree *et al.* (1993). This report describes the Urban Pollution Managment Management procedure, and the applicability of MOSQITO as part of the model family MOSQITO-STOAT-MIKE11 (see also Section 2.3). Besides various data format problems, problems with selection of MOSQITO's timestep and problems with using MOSQITO output values as input for the treatment plant simulation with STOAT were reported.

Further details about MOSQITO can be found in Osborne and Payne (1990), Payne *et al.* (1990), Crabtree *et al.* (1993, 1994a), Gent *et al.* (1994, 1996), FWR (1994), Petrie and Jack (1994); the contributions of various UK research establishments to the development of this model are listed by Henderson and Moys (1987). A comparison with the French FLUPOL model is given by Phan *et al.* (1994) and Blanc *et al.* (1995) and is summarised in the FLUPOL section.

MOUSETRAP

A consortium was established in 1992 to specify, develop and test a sewer flow quality model based around the sewer model MOUSE (Crabtree *et al.*, 1994b). This consortium included representatives of Danish Hydraulic Institute (DHI), Water Quality Institute (VKI), WRc and others. The model developed, called MOUSETRAP, simulates the parameters DO, BOD, COD (as dissolved and sediment-attached parts), ammonia, nitrogen, phosphorus, non-uniform sediments, three types of bacteria and user-specified metals. More details on MOUSETRAP are described by Mark *et al.* (1993, 1996, 1998), Crabtree *et al.* (1994b, 1995), Gustafsson *et al.* (1994), Appelgren *et al.* (1995), Garsdal *et al.* (1995). According to Appelgren and Hernebring (1994), their application of MOUSETRAP for the Rya catchment in Göteborg was the first attempt to use a deterministic modelling approach of sediment transport on real catchment data. An overview of MOUSE and its different add-on programs is given by Gustafsson *et al.* (1994).

The program package consists of four modules:

- *Surface Runoff Quality module*

    This module models build-up and wash-off of sediments on the catchment surface and of dissolved pollutants in gully-pots. On the surface, fine and coarse sediment fractions are considered.

- *Sediment Transport module*

    This module describes the dynamic development of either uniform or graded, non-cohesive sediment deposits. The resistance from the sediment deposits gives a feedback (continuously updated Manning number) to the hydrodynamic flow module. Appelgren *et al.* (1995) stress the need for a well verified hydrodynamic model as the sediment transport is a highly nonlinear function of the hydrodynamic parameters. A detailed description of the Sediment Transport module is given by Mark *et al.* (1995).

- *Advection–Dispersion module*

    Advective transport (with mean flow) and dispersive transport (due to concentration gradients) of dissolved substances is described in this module, based on a one-dimensional advection–dispersion equation, which is solved by an implicit finite difference scheme. Linking of the Sediment Transport and Advection-Dispersion modules allows consideration of release of pollutant interstitial liquid in deposited sediments into the water phase.

- *Water Quality module*

    Finally, this module, which is linked to the Sediment Transport and Advection–Dispersion modules, models the reaction processes of DO, BOD, ammonia, nitrogen, phopshorus, metals, bacteria and temperature. These include degradation of organic matter, exchange of oxygen with the atmosphere, oxygen demand from eroded sediments, hydrolysis, growth of heterotrophic organisms and bacterial fate. Degradation of organic matter is assumed in MOUSETRAP to be carried out by heterotrophic micro-organisms. These are considered to exist in two environmental compartments within the pipe system: biofilms and sediments; and flowing sewage. Inert fractions of organic matter and particulate products resulting from decay are not included in MOUSETRAP since they are assumed not to generate BOD.

Sediments are grouped into three types in MOUSETRAP: Surface sediments and in-pipe sediments are assumed to consist of two fractions (coarse and fine, characterised by having a diameter > 0.5 mm and < 0.5 mm, respectively). The third type of sediments, deposited foul flow sediments, is modelled as just one (the fine) fraction.

Almeida (1999), providing a detailed evaluation of this model, states that the available information regarding the quality processes in MOUSETRAP is very limited; also some aspects of the nomenclature used in the literature about MOUSETRAP appear to be unclear, leading to some difficulties in evaluation of this model.

Crabtree *et al.* (1994b) claim that testing has shown that MOUSETRAP is a robust tool capable of producing reliable results. However, as is the case in many applications of sewer quality models, only a limited set of data for calibration and verification was available. Applications of MOUSETRAP include the sewer systems of Göteborg and Ljubljana (Appelgren et al., 1995; Mark et al., 1996). Fronteau et al. (1997a) provide a comparative evaluation of the modelling principles of MOUSETRAP with those of the IAWPRC Activated Sludge Model No. 1 (see Section 2.1.2.3).

SIMPOL

SIMPOL is not a simulation model *per se*, but a set of EXCEL procedures. It was developed during the Urban Pollution Management (UPM) project (FWR, 1994, 1998). These reference also include descriptions of SIMPOL; a summary is given by Dempsey et al. (1996).

Usage of SIMPOL is part of the UPM procedure (FWR, 1994). Its main purpose is to serve as a tool for selection of relevant events for integrated simulation studies. After calibration against detailed models (such as, for example, MOSQITO or HYDROWORKS), a variety of rainfall events is simulated using SIMPOL. Because of SIMPOL's short execution time, a large number of rainfall events can be used at this step. According to the simulation results, the most interesting events, for which a detailed analysis is considered to be necessary, are selected and simulated using detailed models.

SIMPOL models the main elements of a sewer system by a variety of tanks: surface tank, sewer tank, CSO tank, storm tank. It should be noted that treatment plant performance is not modelled by SIMPOL; the treatment plant effluent characteristics have to be supplied externally for SIMPOL studies.

The surface tank module models the rainfall runoff process by using a percentage relationship without any storage. BOD is assumed to be constant. Since SIMPOL is simulating only individual events of limited length, it does not take into account the detailed inter-event history of wetting and drying of the catchment surface (Udale et al., 1997).

Flow within sewer systems is modelled using a nonlinear relationship between volume of water in the tank and outflow rate. BOD is deposited and eroded during storm events (erosion is proportional to runoff quantity); the related model parameters are to be found by calibration against a detailed sewer flow quality model. A CSO tank is modelled as an on-line tank with a maximum pass-forward capacity; the discharge concentration is equal to the inflow concentration. In a similar way, a storm tank in SIMPOL is modelled as an off-line tank. The BOD concentration in the tank is time-variant: it is related to the residence time in the tank. The related model parameters are to be found by calibration against a detailed sewer flow quality model.

An upgraded version of SIMPOL has been published as Version 2 (FWR, 1998). New features include a simple (steady-state) river model, simulating BOD and ammonia decay and surface reaeration. This model includes DO, BOD and ammonia as state variables. Futhermore, direct evaluation against intermittent river water quality standards is included. Treatment plant discharges into the river still have to be supplied externally to the model.

SWMM

The program package SWMM ("Storm Water Management Model") was developed by the US Environmental Protection Agency. It models flow and pollutants in sewer systems. Quality simulations are based on SWMM's transport module, which does not allow for simulation of backwater effects and pressure flows to be considered. For simulation of these processes, an extended transport model (EXTRAN) is available. However, this extended model includes the flow processes only.

Modelling pollutant transport is done by advection and mixing in conduits and by plug flow or complete mixing in the storage tanks. Sedimentation and resuspension are included as well as decay processes, which are modelled by first-order expressions. Non-conservative pollutants are not linked with each other, and the decay of one has no effect on any other.

Further details about SWMM are given by Delleur (1996) and Almeida (1999).

A.2 Software for Simulation of Activated Sludge Wastewater Treatment Plants

AQUASIM

AQUASIM, developed at the Swiss Federal Institute for Environmental Science and Technology (EAWAG), is considered not to be the implementation of a particular model, but to represent an "environmental system identification tool" (Reichert, 1995). Besides simulation of a model to be specified by the user, it also provides functionalities for sensitivity analysis and parameter estimation. The sensitivity analysis module performs calculation of linear sensitivity functions of arbitrary variables with respect to each of the parameters included in the analysis. When provided with measured data and a model structure, AQUASIM performs an automatic parameter estimation, using the weighted least-squares technique.

When defining his/her own model, the user can choose from a variety of so-called "compartments" as building blocks of the model. Compartment types include "mixed reactor", "river section" (simulating flow by solving the kinematic or diffusive wave approximation of the Saint Venant equations and mass-transport by the Advection–Dispersion equation; also lateral inflows can be modelled), "advective–diffusive", "biofilm reactor" (including a one-dimensional biofilm model) and "saturated soil".

Within each compartment, an arbitrary number of state variables and transformation processes can be defined by the AQUASIM user. For example, Reichert (1994b) describes modelling of an activated sludge treatment plant with AQUASIM. Also an example for river water quality modelling is given in this reference.

AQUASIM is available on various hardware platforms and operating systems, including Unix and DOS/Windows. A list of applications of AQUASIM is given in Reichert *et al.* (1996). It should be noted that the long list of AQUASIM applications lists treatment plants, rivers, lakes, biofilm studies, but no sewer system implementation. In a recent paper, Holzer and Krebs (1998) describe the use of AQUASIM for the simulation of the tanks within the sewer system; for the simulation of the remaining parts of the system the MOUSETRAP package is used.

Further details about AQUASIM can be found in the references mentioned above as well as in Reichert (1994a) and Reichert *et al.* (1995).

ASIM

ASIM was developed by the the Swiss Federal Institute for Environmental Science and Technology (EAWAG) as a public-domain educational tool. It models the activated sludge process only (including related control options), *i.e.* it does not contain any module for primary clarification. However, very simple models for primary and secondary clarifier can be built using the reactor building blocks in ASIM (Gujer, 1995). It should be noted that the ASIM in its original version allows only for a limited number of timesteps to be simulated. Details of ASIM can be found in Gujer (1990, c.1990), Gujer and Henze (1991) and Gujer and Larsen (1995). Güven (1995) provides a detailed comparison of ASIM with STOAT and the treatment plant model developed by Lessard (1989).

ASIM allows for a free definition of the flow scheme and of the biokinetic model; a variety of biokinetic models is also supplied with the program (with/without nitrification, denitrification, phosphorus removal). Also models similar to the IAWPRC Activated Sludge Models Nos. 1 and 2 are implemented.

Furthermore, a number of controllers can be simulated in ASIM: return and waste sludges, internal recirculation, aeration (by setting the oxygen transfer coefficient $k_L a$) and a second influent flow rate can be controlled. Control can be performed either in a proportional way or by on/off control.

EFOR

EFOR was developed by Krüger A.S. in Søborg/Denmark in cooperation with the Water Quality Institute, Technical University of Denmark, Emolet Data and Cowi Consult (STOWA, 1995). Its activated sludge part is an extension of the IAWPRC Activated Sludge Model No. 1. Extensions include "an extra component for slowly degradable soluble substrate, changes in the description of the hydrolysis processes which allows for hydrolysis of the extra substrate component and for hydrolysis under anaerobic conditions, and components and processes which cover simultaneous precipitation." (Dupont and Sinkjaer, 1994). Dupont and Henze's (1992) secondary clarifier model serves as the base for the secondary clarifier module of EFOR. Up to 18 different wastewater components can be modelled.

A particular feature of EFOR is the variety of implemented control options. These include aeration, return and excess sludges, recirculation, chemical dosage, carbon dosage and weir control. Control is possible by on/off regulation, step regulation (involving more intermediate positions than on/off regulation), proportional and flow-proportional regulation (Pedersen, 1992).

Practical implementations include Damhusåen and Lynetten treatment plants in Denmark, which are large-scale treatment plants of 400000 and 800000 PE, respectively. Further details about EFOR can be found in the references mentioned above as well as in Pedersen and Sinkjaer (1992), Finnson (1993), Thornberg *et al.* (1994) and Nyberg *et al.* (1996).

GESIM

GESIM is a simple treatment plant model developed by itwh in Hannover/Germany (Kollatsch and Kenter, 1992). It can be linked to the KOSIM model for sewer systems (see Section 3.2.1) and assists in the evaluation of total emission discharges from CSOs and treatment plants into the receiving water. Because of its simplified model structure, it is intended more for statistical evaluation of long-term simulations rather than for detailed simulation of individual events.

Input data are read from a KOSIM generated file; these include inflow rate and BOD, COD, total nitrogen and P loads per time interval. Values for dry solids and NH_4-N and organic nitrogen are then computed from the input as these are required by the subsequent modules of the program package. Primary clarification is simulated by linear relationships between residence time and sedimentation efficiency (ATV A131, 1991). Since the model developers considered the IAWPRC Activated Sludge Model No. 1 as being too complex for long-term simulations they rearranged the formulae for treatment plant design developed by Kayser (1987) and Böhnke *et al.* (1989) in such a way that these can be applied for simulation. As opposed to the IAWPRC model, these formulae are based on BOD and NH_4-N as main pollutants. Nitrification and denitrification as well as filtration and phosphorus removal can be modelled as well. However, for the latter two processes a simple linear approach is employed. The secondary clarifier is modelled in ten layers, applying the flux theory.

An application of GESIM within a study of sewer system, treatment plant and river in Dresden in described by Fuchs *et al.* (1996).

GPS-X

GPS-X constitutes a modular multi-purpose modelling environment for wastewater treatment plants, developed by Hydromantis in Canada (STOWA, 1995). It can be applied to analysis, design and control of WWTPs. The model library of GPS-X includes models for primary and secondary clarification, aerobic and anaerobic biological treatment as well as modules for hydraulic components (equalisation basins, splitting devices, pumps), thus including almost all modelling approaches of the last 20 years (Hoen *et al.*, 1994). Modules for chlorination, filtration and chemical phosphorus removal are under development. GPS-X is applied to many large-scale treatment plants in several countries (Patry and Barnett, 1992).

A comparative evaluation of four commercially available treatment plant simulation tools (Deakin and Vickers, 1995), commissioned by Northumbrian Water Ltd, concluded that out of the four programs under investigation (Biowin, ESP, GPS-X and STOAT), GPS-X was the one most closely meeting the requirements defined by the water company. The water company's objectives were to select a model which assists in improvement of operational efficiency and performance of existing WWTPs as well as in the design of new plants. Furthermore, the model to be selected shold be able to be used in determining discharge pollutant loads as part of urban pollution management studies; another requirement was that the costs of the model should be not too excessive. Stokes *et al.* (1993) and Patry and Takács (1995) describe details as well as applications of GPS-X. Also Ashley *et al.* (1997) report a successful application of GPS-X within their project.

LESSARD

A detailed description of this program is given in Section 3.2.2.

SIMBA

The treatment plant modelling environment SIMBA (ifak, 2001) was developed by ifak Magdeburg e. V. in Barleben/Germany in cooperation with Otterpohl Wasserkonzepte. It is based on the MATLAB/SIMULINK simulation system. SIMBA allows the user to define his treatment plant from building blocks. However, SIMBA allows the user to freely modify and extend the building blocks of the model. Since SIMBA is based on MATLAB, all of MATLAB's features can be used for subsequent analyses. A recent version allows parameter estimation, sensitivity analysis, controller design to be carried out in a convenient way. Biochemical process models can be edited and defined easily by the user using an editor with built-in ASM-matrix-like notation.

Various models for the different parts of the treatment plant are implented in SIMBA; however, the user also can define his own models. Models implemented include a primary sedimentation model by Otterpohl, the IAWPRC Activated Sludge Models No. 1, 2, and 3 and various secondary clarifier model (including those by Otterpohl and Freund (1992) and Takács *et al.* (1991)). Additional modules serve to model of specific parts and configurations of the plant (*e.g.* biofilm, Sequencing Batch Reactors *etc.*).

A description of SIMBA can be found in ifak (2001), and an application of SIMBA within a study investigating aeration of treatment plants is described by Veersma *et al.* (1995). A survey conducted in the Netherlands (STOWA, 1995) recommends SIMBA out of a variety of various packages for the simulation of treatment plants.

SIMBA is particularly suited to on-line simulation and control applications and is applied at a number of plants for these purposes (see, for example, Alex *et al.* (2001), Alex and Tschepetzki (2001), Jumar *et al.* (2000) and Jumar and Tschepetzki (2001)).

Even though SIMBA has originally been designed as a treatment plant simulator, recent addition of modules ("SIMBA Sewer"; Alex *et al.*, 1999) allows SIMBA to simulate water flow and quality processes in the sewer system as well. In a similar way, also river water quality modelling is possible, thus making SIMBA suitable for studies involving integrated simulation and control.

STOAT

The STOAT package (STOAT = Sewage Treatment Operation and Analysis over Time) was developed by WRc in cooperation with Imperial College and several British water utilities; its first version was released in 1994 (Dudley and Chambers, 1995). During its development STOAT was extensively validated against treatment plant data, and the accuracy of the process models within STOAT was found to be comparable to the accuracy of the original data meaurements (Güven, 1995). Since STOAT is described and discussed in detail by Güven, only a very brief overview of the model is given here. Further details can also be found in Crabtree *et al.* (1993, 1994a), WRc (1994), FWR (1994), Dudley *et al.* (1994), Petrie and Jack (1994) and Dudley and Chambers (1995).

STOAT contains a variety of submodels. The submodels for storm tanks and primary clarifiers are based on the work conducted by Lessard and Beck (1988, 1991b) (*cf.* also Section 2.1.2). Various modifications of the activated sludge process can be modelled in STOAT.

For simulation of the Activated Sludge Process, either the IAWPRC Activated Sludge Models or a BOD-based model by Jones (1978) (see also Section 2.1.2) can be chosen within STOAT. The secondary clarifier is modelled according to Takács *et al.* (1991). Other model components included are for trickling filters, disinfection, phosphorus removal, sludge digestion and dewatering. Furthermore, a variety of control options is included. Pollutants included in the models are BOD (or COD), SS, NH_3-N, NO_3, DO. For simulation of biological phosphorus removal, soluble phosphorus and volatile fatty acids are required as determinands.

STOAT has been applied to many treatment plants, particularily in the UK (see references given above).

WEST++

This simulation package was developed recently by Hemmis n.v., University of Gent and Epas n.v. in Belgium. According to Hemmis (1998), WEST++ contains a variety of module libraries for the individual parts of the treatment plant (*e.g.* preliminary treatment models by Lessard and Beck, Otterpohl and Freund), the IAWQ activated sludge models, the secondary clarifier model by Takács *et al.*). Additional modules include modules for parameter estimation and sensitivity analysis. Particular emphasis was given to the performance of the simulations. Furthermore, the platform allows the user to specify his own models, using a model specification language. After addition of additional modules, WEST++ is now also being used for projects involving integrated simulation and control of the complete urban wastewater system (*cf.* Meirlaen *et al.* (2000a, 2000b, 2001)). WEST++ is being used in many installations, inside and outside Belgium.

A.3   Software for Simulation of Rivers

The following lists some of the currently available river water quality models. A recent review paper (Rauch *et al.*, 1998b) provides a summary overview of some river models. Another brief review of river water quality models can be found in LFU (1996); Reda (1996) provides a comparative evaluation of MIKE11, RATTS (Norreys, 1991; Norreys and Cluckie, 1996) and SPRAT (Crockett *et al.*, 1989, 1990). Relevant review work was also done by Reichert *et al* .(2001), Shanahan *et al.* (2001) and by Vanrolleghem *et al.* (2001). In particular due to the EU Water Framework Directive (CEC, 2000), river water quality simulation is gaining in importance.

AQUASIM

This simulation tool for aquatic systems can be used for the modelling of rivers as well. A description of its characteristics has already been given in the previous section.

Model by ATV

After model development and testing work carried out over several years, the German Association for Water Pollution Control (ATV) launched its "Commonly Available Water Quality Model" in 1998 (ATV 2.2.3, 1997). Müller (2001) provides an overview of the model, whilst Christoffels (2001) discusses its use for the example of river basin management of the Erft river.

The program package is designed for use on a PC platform in water authorities and consultancies. Because of its modular structure, the user can select processes and combine the appropriate modules in order to simulate those processes which are of interest for a particular study. The model consists of a flow module as well as of 17 quality modules, simulating various aspects of water quality (*e.g.* BOD/COD, oxygen balance, nitrogen, phosphorus, silica, several classes of algae, sediment processes, dissolved pollutants, conservative pollutants, heavy metals, temperature, irradiation, pH value). Flow (assumed to be one-dimensional) is modelled by solution of the Saint Venant equations, which are solved using an implicit method of characteristics. The equation describing mass transport is solved by an algorithm without damping.

DOSMO

The DOSMO ("Dissolved Oxygen Stream Model for combined sewer overflows") program family consists of three models with different complexity (Hvitved-Jacobsen and Schaarup-Jansen, 1991). These models, obtainable from the Danish MOUSE Service Centre, allow impacts of CSO discharges on the oxygen balance of the river to be assessed.

The model DOSMOSIM assumes flows to be steady and uniform; mass transport by dispersion is neglected. The DO equation, based on the classical Streeter–Phelps model, is solved analytically (Simonsen and Harremoës, 1978; Hvitved-Jacobsen and Schaarup-Jansen, 1991). Thus, simulation results can be obtained in very short computation time. DOSMO3.0 is able to model unsteady flow, applying the kinematic wave approach. DOSMO3.0 is available as an add-on to MOUSE; it calculates the 1-hour minima of DO stream concentrations for each CSO event, thus providing the data required for a statistical analysis of CSO events. Of this model family, DOSMO (DMSC, 1995) is the most complex model. It simulates unsteady flow and transport, taking also dispersion into account. The Saint Venant and Advection–Dispersion equations are solved by implicit schemes which are unconditionally stable and free of numerical dispersion (Hvitved-Jacobsen and Schaarup-Jansen, 1991). Biochemical transformations modelled include BOD decay, immediate and delayed oxygen demand, reaeration, respiration and photosynthesis.

DUFLOW

The program package DUFLOW was developed by various cooperating Dutch institutions: IHE/Delft; Rijkswaterstaat/The Hague; Delft Institute of Technology; Agricultural University of Wageningen; Stichting Toegepast Onderzoek Waterbeheer (STOWA)/Utrecht; Bureau Icim/Rijswijk. A full description of the program can be found in IHE (1992). Applications are discussed in van den Boomen *et al.* (1995), van Duin *et al.* (1995), Aalderink *et al.* (1995, 1996), Makkinga *et al.* (1998).

DUFLOW is a shell program which allows the user to define freely the biochemical water quality processes within the river. Modelling of flow and pollutant transport is implemented as fixed procedures (not to be modified by the user). Additionally, DUFLOW allows the definition of ancillary structures and control options within the river.

Dynamic unsteady flow in the river is simulated by solving the momentum and continuity equations using the four-point implicit Preissmann scheme. Mass transport is described by the Advection–Dispersion equation. The numerical method applied to solve the Advection–Dispersion equation was adapted from the model FLOWS. This method is reported to be unconditionally stable, showing little numerical dispersion.

DUFLOW allows the user to define state variables of the water phase as well as bottom variables. The latter ones can assist in modelling of interactions between the sediment and the overlying water column. They also showed to be useful for modelling other state variables that are not subject to horizontal transport (*e.g.* growth of macrophytes and fish) (Aalderink *et al.*, 1995).

DUFLOW allows its user to define his/her own quality model in a Pascal-type notation, describing the biochemical processes and interactions among the pollutants. Additionally, default models are supplied with the program, including two eutrophication models.

ISIS

ISIS is a software package developed by HR Wallingford and Sir William Halcrow and Partners Ltd. (Halcrow & Partners, 1995a,b; Leclerc, 1996). It consists of two main modules: ISIS Flow and ISIS Quality. Once the Flow module has been run, the Quality module can be called also several times without having to call the Flow module again. This can cut down computation time when various definitions of the quality processes are to be run. A similar feature is also found in the DUFLOW model (see above).

The Flow module allows "any sensible looped or branched network" to be modelled (Halcrow & Partners, 1995a). Free surface flows are represented by the Saint Venant equations for unsteady flows in open channels. The Saint Venant equations are solved using the four-point implicit Preissmann scheme (Preissmann, 1960). Various methods are available for computations of steady flow applications.

The Advection–Dispersion transport equation is solved by a finite difference approximation. The water quality module assumes three different layers in the river: the water column, a fluffy layer (a layer of mud lying on top of the bed; this layer is limited to a maximum thickness) and bed and pores. A variety of variables can be simulated, according to the specifications of the user. Variables simulated can include conservative and decaying pollutants, oxygen, sediment, phytoplankton, macrophytes, benthic algae, coliforms, salt, silicate, and temperature.

The DO module includes only the dissolved variables, *i.e.* fast and slow dissolved BOD, fast and slow organic nitrogen, ammoniacal nitrogen, nitrites, nitrates and DO.

MIKE11

The simulation package MIKE11 was developed at Danish Hydraulic Institute (DHI). It models water flow and quality in rivers, channels and estuaries. Besides its rainfall runoff part NAM, it consists of four main modules. Their names and functions are similar to those of the MOUSETRAP package for sewer systems (see Section A.1.1). The hydrodynamic module (HD) solves the full Saint Venant equations (solved by an implicit finite-difference scheme: "double sweep algorithm"), thus enabling branched and looped networks to be modelled. Alternatively to solving the full Saint Venant equations, the diffusive and kinematic wave approximations as well as a quasi-stationary approach can be chosen. Various hydraulic structures as well as control structures can be modelled.

Pollutant transport is modelled by the Advection–Dispersion (AD) module, solving the traditional Advection-Dispersion equation by an implicit finite difference scheme. Erosion and deposition of cohesive sediments are modelled as source–sink terms in the Advection–Dispersion equation. An add-on module provides an advanced description of cohesive sediment transport. For simulation of non-cohesive sediment transport four different models can be used. Water quality processes simulated by the water quality module (WQ) include BOD-DO relationships, nitrification, sedimentation and resuspension, influence from bottom vegetation, oxygen consumption from reduced chemicals as well as processes related to eutrophication and heavy metals.

MIKE 11 is applied internationally in various projects. Together with MOSQITO (sewer system modelling) and STOAT (treatment plant), MIKE 11 also constitutes the suite of models exemplifying modelling of the urban wastewater system in the UPM procedure. A project of DHI and WRc aims at linking MIKE 11 to MOUSETRAP and STOAT (see Section 2.3).

The following table, taken from Rauch *et al.* (1998b), shows (in matrix notation in a similar way to the common representation of the IAWPRC Activated Sludge Model No. 1) the processes which are modelled by MIKE11.

Table A.1. Biochemical and physical processes of MIKE11 in matrix notation (reproduced from Water Science and Technology, Vol. 38, No. 11, p. 242, with permission from the copyright holders, IAWQ)

	Component Process	1 DO	2 BODd	3 BODs	4 BODb	5 NH3	7 NO3	Process rate $[ML^{-3}T^{-1}]$
1	Reaeration	1						$K2\,(DO_{SAT}-DO)$
2a	BODd biodegration	-1	-1					Kd3 BODd
2b	BODs biodegradation	-1		-1				Ks3 BODs
2c	BODb biodegration	-1			-1			Kb3 BODb
3	BOD sedimentation			-1	+1			K5 BODs/d
4	BOD resuspension			1	-1			S1 BODb/d
5	Sediment DO demand	-1						B1
6	Nitrification	-Y1				-1	1	$K4\,NH3^{e4}$
7	Denitrification						-1	$K6\,NO3^{e6}$
8	Photosynthesis	1				-0.066		Pmax $\cos[2\pi(\tau/\alpha)]$
9	Respiration	-1				0.066		R

Where
BODd = dissolved BOD $[ML^{-3}]$; BODs = suspended BOD $[ML^{-3}]$;
BODb = settled BOD $[ML^{-3}]$; Kd3 = degradation rate constant for dissolved BOD $[T^{-1}]$;
Ks3 = degradation rate constant for suspended BOD $[T^{-1}]$;
Kb3 = degradation rate constant for settled BOD $[T^{-1}]$;
K5 = sedimentation rate for suspended BOD $[LT^{-1}]$;
d = mean river depth [L];
S1 = resuspension rate for sedimented BOD (zero, if the flow velocity or the concentration BODb are below critical values) $[LT^{-1}]$;
B1 = constant value of sediment oxygen demand (in addition to biodegradation of BODb) $[ML^{-3}T^{-1}]$;
Y1 = yield factor for oxygen consumed by nitrification [-];
K4 = nitrification rate constant $[T^{-1}$ or $M^{1/2}L^{-3/2}T^{-1}]$;
e4 = coefficient characterizing concentration dependence of nitrification (1 or 0.5) [-];
K6 = denitrification rate constant $[T^{-1}$ or $M^{1/2}L^{-3/2}T^{-1}]$;
e6 = coefficient characterizing concentration dependence of denitrification (1 or 0.5) [-];
Pmax = maximum production at noon (zero during the night) $[ML^{-3}T^{-1}]$;
τ = actual time of the day related to noon [-];
α = actual relative day length [-]; R = respiration rate $[ML^{-3}T^{-1}]$;

QUAL2E

According to Chapra (1997), the model QUAL2E (Brown and Barnwell, 1987), developed by Water Resources Engineers, Inc. (now Camp, Dresser and McKee) and maintained now by the US EPA, is presently the most widely used computer model for simulating streamwater quality. Shanahan *et al.* (1998) consider QUAL2E as being the current standard for river water quality modelling. It helps in understanding the limitations of QUAL2E to be aware of the purpose this model was developed for originally: the model is designed to be used for determination of wasteload allocations, according to US federal laws Shanahan *et al.* (1998). Such an allocation indicates the maximum amount of waste that can be discharged to the river whilst still meeting water quality standards under certain low-flow conditions. Thus simulations for this purpose make assumptions of constant low river flow and maximum permitted effluent discharge rate.

Hydraulic steady (*i.e.* time-invariant) and nonuniform (*i.e.* space-variant) flow conditions are assumed in QUAL2E. Mass transport is simulated by the Advection–Dispersion equation. Due to the assumption made about the flow conditions, its solution is considerably simplified. The dispersion coefficient is assumed to be a function of the channel's geometric properties.

State variables of the water quality module include DO, BOD, organic nitrogen, ammoniacal nitrogen, nitrate nitrogen, nitrite nitrogen, chlorophyll-a, organic phosphorus, dissolved phosphorus, coliform bacteria, temperature as well as user defined non-conservative and conservative constituents.

Table A.1.3 shows (in matrix notation drawn up in a similar way to the common representation of the IAWPRC Activated Sludge Model No. 1) the processes which are modelled by QUAL2E (Rauch *et al.*, 1998b). A comparison of the representation of the biochemical reactions in QUAL2E and in the IAWPRC Activated Sludge Model No. 1 was carried out by Masliev *et al.* (1995) (see also Section 2.3).

QUAL2E has two modes of simulation: steady-state (the model is run until a steady-state solution is obtained) and time-variable (at present, the latter is limited to diurnal simulations (Chapra, 1997)).

Ristenpart and Wittenberg (1991) extended the model QUAL2E to include non-steady flow conditions as well; numerical solution is obtained in their model by an implicit backward difference scheme. The model extension has been successfully applied to simulating BOD, DO, dissolved phosphorus and nitrate in North German wetland creeks. Another application of this model extension (called DYNAMO) was within an integrated study of sewer system, treatment plant and river in the city of Dresden (Fuchs and Gerighausen, 1995; Fuchs *et al.*, 1996).

Table A.2. Biochemical and physical processes of QUAL2E in matrix notation (reproduced from *Water Science and Technology*, Vol. 38, No. 11, p. 241, with permission from the copyright holders, IAWQ)

	Component	1	2	3	4	5	6	7	8	9	Process rate
	Process	DO	BOD	ABM	ORG-N	NH4	NO2	NO3	ORG-P	DIS-P	$[ML^{-3}T^{-1}]$
1	Reaeration	1									$K2 (DO_{sat}-DO)$
2	Bio-degradation	-1	-1								K1 BOD
3	BOD sedimentation		-1								K3 BOD
4	Sediment DO demand	-1									K4/d
5	Photo-synthesis	a3		1		-0.07 F_{NH4}		-0.07 $(1-F_{NH4})$		-0.01	μmax ABM f(L,N,P)
6	Respiration	-a4		-1		0.07				0.01	ρ ABM
7	Algae sedimentation			-1							$\sigma 1/d$ ABM
8	Nitrogen Hydrolysis				-1	1					$\beta 3$ ORG-N
9	Nitrification 1st step	-3.43				-1	1				$\beta 1$ NH4 f(nitr)
10	Nitrification 2nd step	-1.14					-1	1			$\beta 2$ NO2 f(nitr)
11	N sedimentation				-1						$\sigma 4$ NH4
12	N sediment release					1					$\sigma 3/d$
13	P hydrolysis								-1	1	$\beta 4$ ORG-P
14	P sedimentation								-1		$\sigma 5$ ORG-P
15	P sediment release									1	$\sigma 2/d$

Where DO = dissolved oxygen $[ML^{-3}]$; DO_{sat} = DO saturation concentration $[ML^{-3}]$;
BOD = biochemical oxygen demand of organic material $[ML^{-3}]$;
ABM = algal biomass $[ML^{-3}]$; ORG-N = organic nitrogen $[ML^{-3}]$;
NH4 = ammonia-N $[ML^{-3}]$; NO2 = nitrite-N $[ML^{-3}]$; NO3 = nitrate-N $[ML^{-3}]$;
ORG-P = organic phosphorus $[ML^{-3}]$; DIS-P = dissolved phosphorus $[ML^{-3}]$;
K2 = reaeration coefficient $[T^{-1}]$; K1 = deoxygenation coefficient $[T^{-1}]$;
K3 = BOD settling rate $[T^{-1}]$; K4 = sediment oxygen demand rate $[ML^{-2}T^{-1}]$;
d = mean stream depth [L]; μmax = maximum algal growth rate $[T^{-1}]$;
ρ = algal respiration rate $[T^{-1}]$; $\sigma 1$ = algal settling rate $[LT^{-1}]$;
$\sigma 2$ = benthos source rate for P $[ML^{-2}T^{-1}]$; $\sigma 3$ = benthos source rate for N $[ML^{-2}T^{-1}]$;
$\sigma 4$ = N settling rate $[T^{-1}]$; $\sigma 5$ = P settling rate $[T^{-1}]$;
$\beta 1$ = ammonia oxidation rate $[T^{-1}]$; $\beta 2$ = nitrite oxidation rate $[T^{-1}]$;
$\beta 3$ = N hydrolysis rate $[T^{-1}]$; $\beta 4$ = P hydrolysis rate $[T^{-1}]$;
a3 = stoichiometric coefficient gO/gABM [-]; f(L,N,P) = algal growth limitation factor;
f(nitr) = nitrification limitation factor; F_{NH4} = ammonia preference factor.

QUALSIM

The American model WQRRS (developed by the Hydraulic Engineering Centre of the US Army Corps of Engineers) was further developed into the package QUALSIM, which is now available from Hydrotec GmbH in Aachen/Germany (Hydrotec, 1992; Hydro-Ingenieure, 1993; LFU, 1996, Rieß-Dauer (1998).

Flow data can either be computed by an internal hydraulic module or these can be externally supplied to the water quality module. Mass transport is calculated according to the Advection–Dispersion equation. The user can choose from a variety of state variables to be modelled. These include oxygen, BOD, nitrogen in its various forms, phosphorus, sediment, inorganic matter, algae (sessile and suspended), user-defined conservative pollutants, coliform bacteria, benthic organisms, fish, temperature, pH value. Though this model models nitrification, no allowance is made for denitrification processes in the river. Particular emphasis is made on modelling of complete carbon, nitrogen and phosphorus cycles. Dead biomass is considered as detritus. Degradation of this detritus releases carbon as CO_2, nitrogen as ammonium and phosphorus as ortho-phosphate back to the aquatic system.

REDA

Reda's model was developed from a model for the Bedford-Ouse case study described by Beck and Finney (1987), simulating flow and several, interacting, water quality determinands determinands (BOD, DO, Amm-N, NO_3-N, and chlorophyll-a, representing algae - *cf.* Figure 2.8). Beck and Finney's model is based on data obtained from a daily measurement campaign. The CSTR approach, the main characteristics of which have already been described in Section 2.1.3.3, constitutes the core concept of the model. Reda (1996) (see also Beck and Reda, 1994) introduced two additional parameters for each reach, the "added depth" (h_a; which is added to the depth of the water level over the weir, thus permitting fine tuning of the oscillations of flow and solute concentrations) and the "dead fraction" (f_d; thus confining solute transport and mixing volume to a volume smaller than the volume available for flow), which allow for more accurate reprensentation of mass transport. Model development and calibration was carried out for a 10.2 km stretch of River Cam within and downstream of Cambridge. This stretch of the river is equipped with three gate and lock structures which can be used for flow control in the river. The river model was calibrated for water and solute transport based on hourly series on flow and ammoniacal nitrogen over a period of 13 days. The biochemical calibration used daily data series over a period of more than four months. Thus, development and calibration of this model was done on an extensive data set.

RIVER WATER QUALITY MODEL No. 1

This model was proposed in 2001 by the IWA Task Group on River Water Quality Modelling (Reichert *et al.*, 2001; Shanahan *et al.*, 2001; Vanrolleghem *et al.*, 2001; *cf.* also Reichert and Vanrolleghem, 2001). It represents the results of the efforts of the task group to define a unified, consistent model for river water quality processes. It is described in an ASM-like matrix notation and characterised by closed mass balances based on the elementary composition of organisms (rather than just on COD units). The model aims to include the most important processes for carbon, oxygen, nitrogen and phosphorus in the river under aerobic or anoxic conditions, assuming that nitrate is always available. If anaerobic processes in the water column or the river sediments are significant for the model variables, the model needs to be extended accordingly (Reichert *et al.*, 2001).

Processes considered in the full model include growth of heterotrophs, respiration of heterotrophs, growth and respiration of nitrifyers, algae, growth and death of consumers, hydrolysis and as well as adsorption and desorption of phosphate. The full model comprises 23 main processes and 24 variables. However, the authors suggest defining submodels according to the specific needs of the model application. For example, the Streeter–Phelps model and the QUAL2E model can be considered as special cases of the River Water Quality Model No. 1. Further documentation is provided in the references given as well as on EAWAG's web page (www.eawag.ch/~reichert).

Appendix B
Parameters of the Treatment Plant Model

Storm tank

Scouring parameter	0
Settling velocity for SS [m/h]	2
Settling velocity for VSS [m/h]	2
Settling velocity for CODp [m/h]	2
% of settleable matter of SS	75
% of settleable matter of VSS	75
% of settleable matter of CODp	60
% of tank volume influenced by sludge scrapers during drawing mechanism	25
% of tank volume influenced during filling	100
Reaction rate constant for removal of SS settleable fraction during filling [1/h]	0
Reaction rate constant for removal of VSS settleable fraction during filling [1/h]	0
Reaction rate constant for removal of CODt settleable fraction during filling [1/h]	0

Primary sedimentation

% of settleable SS for crude sewage	65
% of settleable VSS for crude sewage	65
% of settleable CODp for crude sewage	50
% of settleable SS for storm sewage	75
% of settleable VSS for storm sewage	75
% of settleable CODp for storm sewage	60
Number of CSTRs representing primary sedimentation	5
Scouring parameter	0
Sludge density	1.03
Sludge moisture	0.95
First-order decay rate for CODs [1/h]	0
Rate constant for generation of NH3 [1/h]	0
Ration of VSS to SS	0
SS settling velocity, crude sewage [m/h]	1
VSS settling velocity, crude sewage [m/h]	1
CODp settling velocity, crude sewage [m/h]	1

SS settling velocity, storm sewage [m/h]	2
VSS settling velocity, storm sewage [m/h]	2
CODp settling velocity, storm sewage [m/h]	2
Initial concentration of settleable fraction of SS in each CSTR [mg/l]	65
Initial concentration of non-settleable fraction of SS in each CSTR [mg/l]	100
Initial concentration of settleable fraction of VSS in each CSTR [mg/l]	55
Initial concentration of non-settleable fraction of VSS in each CSTR [mg/l]	80
Initial concentration of settleable fraction of CODp in each CSTR [mg/l]	40
Initial concentration of non-settleable fraction of CODp in each CSTR [mg/l]	100
Initial concentration of soluble COD in each CSTR [mg/l]	300
Initial concentration of ammonia in each CSTR [mg/l]	27
Initial concentration of nitrate in each CSTR [mg/l]	0
Initial mass of sludge solids in each CSTR [kg]	0

Activated sludge

Ratio of S_I to total COD [%]	10
Ratio of X_I to total COD [%]	18
Ratio of S_S to total COD [%]	29
Ratio of X_S to total COD [%]	43
Ratio of X_I to MLVSS, tail end of aerator [%]	15
Ratio of X_S to MLVSS, tail end of aerator [%]	1
Ratio of $X_{B,H}$ to MLVSS, tail end of aerator [%]	78
Ratio of X_P to MLVSS, tail end of aerator [%]	5
Ratio of $X_{B,A}$ to MLVSS, tail end of aerator [%]	0.1
μ_H [g/h]	0.24
Y_H [g VSS formed/g COD oxidised]	0.55
K_s [mg/l]	130
$K_{O,H}$ [mg/l]	0.2
b_H [1/h]	0.005
k_h [g slowly biodegradable COD/g COD]	0.125
K_X [g slowly biodegradable COD/g COD]	0.03
f_p	0.08
μ_A [1/h]	0.024
Y_A [g VSS formed / g NH4 oxidised]	0.06
K_{NH} [g/m^3]	0.6
$K_{O,A}$ [g/m^3]	0.5
b_A [1/h]	0.006
Dissolved oxygen concentration [g/m^3]	10
a, VSS settling parameter	4.7
b, VSS settling parameter	0.55
Height of sludge compaction zone [m]	0.3
Height of thickening [m]	2.0

Height of clarification zone [m]	0.9
Minimum SS in effluent [mg/l]	3
Proportionality constant for the effect of flow on effluent suspended solids	0.009
Proportionality constant for the COD concentration equivalent to the efflent total SS	1.0
Ratio of VSS to SS for the clarifier effluent	0.8
Proportionality constant for consumption of DO during oxidation of ammonia to nitrate	4.57
CSTR volumes as fractions of total aerator volume [%]	25,25,25,25
Ratio of total influent flow admitted to the CSTRs of the aerator [%]	100,0,0,0
Ratio of total air blown into the CSTRs [%]	50,25,15,10
Rate constant for aeration of the mixed liquors by the air blowers (K_La)	0.0008,0.00055,0.0004,0.00035
Initial S_I concentrations in the CSTRs [g/m^3]	40,40,40,40
Initial X_I concentrations in the CSTRs [g/m^3]	500,500,500,500
Initial S_S concentrations in the CSTRs [g/m^3]	20,5,2,1
Initial X_S concentrations in the CSTRs [g/m^3]	70,20,10,5
Initial $X_{B,H}$ concentrations in the CSTRs [g/m^3]	2500,2500,2500,2500
Initial X_P concentrations in the CSTRs [g/m^3]	100,100,100,100
Initial S_{NH} concentrations in the CSTRs [g/m^3]	6,2,1,0.3
Initial S_{NO} concentrations in the CSTRs [g/m^3]	22,25,27,27
Initial X_{BA} concentrations in the CSTRs [g/m^3]	15,15,15,15
Initial S_O concentrations in the CSTRs [g/m^3]	25,2.5,3,3
Initial S_I concentration in each CSTR of the clarifier [g/m^3]	40
Initial X_I concentration in each CSTR of the clarifier [g/m^3]	5
Initial S_{NH} concentration in each CSTR of the clarifier [g/m^3]	0.2
Initial S_{NO} concentration in each CSTR of the clarifier [g/m^3]	25

Appendix C
Rainfall Data Used in this Study

Figures C.1 and C.2 show the historical rainfall series used in these studies. These have been measured at Fuhrberg in Northern Germany.

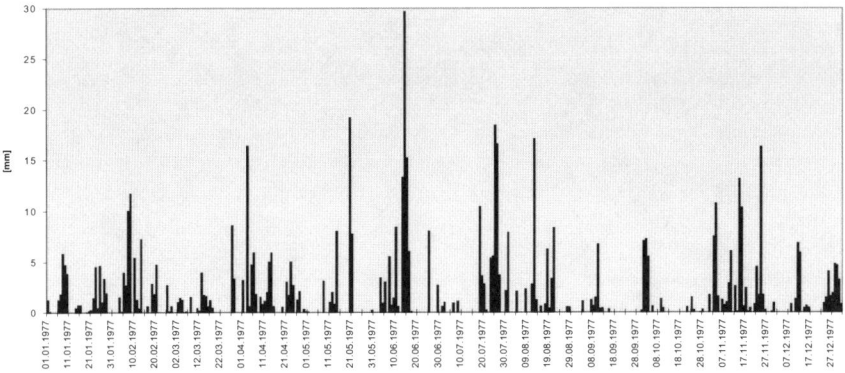

Figure C.1. Rainfall 1977: daily depth [mm] – Fuhrberg/Germany

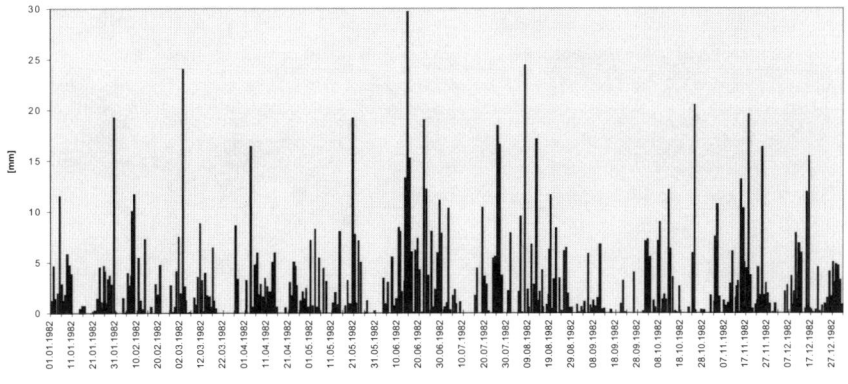

Figure C.2. Rainfall 1982: daily depth [mm] – Fuhrberg/Germany

Appendix D
Detailed Results of the Optimisation Runs Presented in Chapter 5

This section summarises some of the optimisation results obtained in Chapter 5.

Table D.1a. Results of optimisation runs for Scenarios 1 and 2: evaluation criteria

Criterion	Base case	Scenario 1	Scenario 2a	Scenario 2b
F2 (*)	3.86	3.93	4.35	4.31
F3	6.60	7.04	5.74	5.74
DO-M	3.45	3.74	4.35	4.31
DO-DU	3.61	1.81	0.00	0.00
DO-E	4.44	5.52	5.07	5.05
AMM-M	5.78	6.20	5.26	5.25
AMM-DU	7.23	8.43	4.82	4.82
AMM-E	1.77	1.86	1.81	1.81
BOD-M	21.11	19.56	18.39	18.43
BOD-E	16.81	9.98	12.68	12.51
CODtot	22556	22761	22633	22767
Qoverf	26781	23184	24516	26444
CSODur	3.01	4.82	4.22	4.22
QCSO	3214	11594	9776	9776
QST	23567	11590	14740	16668
(*): used as objective function for the optimisation				

Table D.1b. Results of optimisation runs for Scenarios 1 and 2: strategy parameters

Strategy parameter	Results			
1.	z_1 (P7) 5	z_1 (P7) 3	z_1 (P7) 3	z_1 (P7) 3
2.	z_2 (PCINMX) 3.00	z_2 (PCINMX) 3.09	z_2 (PCINMX) 2.75	z_2 (PCINMX) 2.75
3.	z_3 (STTPTH) 1000	z_3 (STTPTH) 1117	z_3 (STTPTH) 1259	z_3 (STTPTH) 1259
4.			z_4 83	z_4 83
5.			z_5 27	z_5 27
6.			z_6 4.53	z_6 4.53
7.			z_7 4.61	z_7 79

Table D.2. Results of optimisation runs for Scenarios 3 to 8

Criterion	Scenario 3	Scenario 4	Scenario 5	Scenario 6	Scenario 7	Scenario 8
F2 (*)	3.98	3.98	3.93	4.02	4.29	4.08
F3	4.22	6.60	4.65	6.17	5.52	4.87
DO-M	3.85	3.85	3.76	4.02	4.29	4.08
DO-DU	0.60	0.60	1.81	0.00	0.00	0.00
DO-E	5.08	5.22	4.74	5.65	5.24	5.22
AMM-M	4.41	5.88	5.33	5.58	5.17	4.87
AMM-DU	0.60	7.23	1.81	6.02	4.22	2.41
AMM-E	1.27	1.70	1.47	1.48	1.53	1.51
BOD-M	20.66	19.30	20.42	18.86	19.56	20.05
BOD-E	11.78	11.40	14.90	10.09	11.36	11.71
CODtot	25359	22677	24108	23211	22639	22669
Qoverf	22300	24680	13845	22446	22684	26893
CSODur	4.82	4.82	3.01	4.819	4.82	4.82
QCSO	10544	10864	3214	10864	12740	12740
QST	11756	13816	10631	11582	9944	14153

(*): used as objective function for the optimisation

Strategy parameter	Results					
1.	$P7\text{-}x_0$ 1.25	$P4\text{-}x_0$ 6	$PCINMX\text{-}x_0$ 4.72	$P4\text{-}x_0$ 23	$P7\text{-}x_0$ 25	$P7\text{-}x_0$ 1.02
2.	$P7\text{-}y_0$ 5	$P4\text{-}y_0$ 4	$PCINMX\text{-}y_0$ 4.22	$P4\text{-}y_0$ 4	$P7\text{-}y_0$ 4	$P7\text{-}y_0$ 4
3.	$P7\text{-}\Delta y$ -1	$P6\text{-}x_0$ 29	$STTPQ\text{-}x_0$ 1.42	$P6\text{-}x_0$ 34	$PCINMX\text{-}x_0$ 621	$PCINMX\text{-}x_0$ 1.73
4.	$PCINMX\text{-}x_0$ 37	$P6\text{-}y_0$ 4	$STTPQ\text{-}dy$ 465	$P6\text{-}y_0$ 4	$PCINMX\text{-}y_0$ 2.91	$PCINMX\text{-}y_0$ 2.76
5.	$PCINMX\text{-}y_0$ 2.70	$P7\text{-}x_0$ 95	$RASABS\text{-}x_0$ 4.61	$P7\text{-}x_0$ 25	$STTPQ\text{-}x_0$ 580	$STTPQ\text{-}x_0$ 1.52
6.	$PCINMX\text{-}\Delta y$ 0.58	$P7\text{-}y_0$ 4	$RASABS\text{-}y_0$ 933	$P7\text{-}y_0$ 4	$STTPQ\text{-}y_0$ 608	$STTPQ\text{-}y_0$ 879
7.	$RASPCT\text{-}x_0$ 24.0	$P2\text{-}x_0$ 17		$PCINMX\text{-}x_0$ 0.78	$RASABS\text{-}x_0$ 3.52	$RASABS\text{-}x_0$ 2.94
8.	$RASPCT\text{-}y_0$ 0.71	$P2\text{-}y_0$ 5		$PCINMX\text{-}y_0$ 3.10	$RASABS\text{-}y_0$ 794	$RASABS\text{-}y_0$ 751.4
9.	$RASPCT\text{-}\Delta y$ -0.15			$STTPQ\text{-}x_0$ 1.13		
10.				$STTPQ\text{-}y_0$ 659		
11.				$RASABS\text{-}x_0$ 2.86		
12.				$RASABS\text{-}y_0$ 879		

References

Aalderink RH (1990) Estimation of storm water quality characteristics and overflow loads from treatment plant influent data. Water Sci. Technol. 22,10/11,77-85

Aalderink RH, Klaver N, Noorman R (1995) DUFLOW V 2.0 Micro-Computer package for the simulation of 1-dimensional flow and water quality in a network of open water courses. Water Quality Modelling. Proc. Int. Conf. on Water Quality Modelling. Heatwole C (ed.). ASAE. Orlando 1995. *pp.* 416-426

Aalderink RH, Zoeteman A, Jovin R (1996) Effect of input uncertainties upon scenario predictions for river Vecht. Water Sci. Technol. 33,2,107-118

Abraham, CHM (1996) Erfahrungen der Hamburger Stadtentwässerung mit der Kanalnetzsteuerung. In Hahn HH, Trauth R (eds.) Zehnte Karlsruher Flockungstage. Wechselwirkung zwischen Einzugsgebiet und Kläranlage. Report No. 78. Institut für Siedlungswasserwirtschaft. Universität Karlsruhe. Oldenbourg München. *pp.* 157-162

Ackers P, White WR (1973) Sediment transport new approach and analysis. J. Hydr. Div. ASCE. 99,2041-2060

Ackley DH (1987) An empirical study of bit vector function optimization. In Davis (1987). *pp.* 170-204

Acton FS (1970) Numerical Methods That Work. 1990. Corrected edition. Mathematical Association of America, Washington, *pp.* 464-467

Ademoroti CMA (1986) Model to predict BOD from COD values. Effluent Wat. Treatment. 26,3,80-84

Ahyerre M, Chebbo G, Tassin B, Gaume E (1998) Storm water quality modelling, an ambitious objective? Water Sci. Technol. 37,1,205-213

Ainger ChM, Armstrong RJ, Butler D (1996) Characterisation of dry weather flow in sewers. Funders Report/IP/7, prepared under contract to CIRIA by Montgomery Watson in association with Imperial College, London, December 1996

Alabaster JS, Lloyd R (1980) Water Quality Criteria for Freshwater Fish. Published for the UN FAO. Butterworth & Co., London

Alarie RL, McBean EA, Farquahr GJ (1980) Simulation modelling of primary clarifiers. J. Env. Eng. Div. ASCE. 106,2,293-309

Albuquerque AJC (1993) Application of an Activated Sludge Model to Treatment of Swine Waste and Sensitivity Analysis. MSc dissertation. Department of Civil Engineering. Imperial College of Science, Technology and Medicine, London

Alex J, Risholt LP, Schilling W (1999) Integrated modeling system for simulation and optimization of wastewater systems. Eighth International Conference on Urban Storm Drainage. Sydney/Australia. 30 August – 3 September 1999. *pp.* 1553-1561.

Alex J, Tschepetzki R (2001) Prädiktive Steuerung und Regelung der Stickstoffelimination in biologischen Kläranlagen. *at – automatisierungstechnik*. 49,10,436-448

Alex J, To TB, Hartwig P (2001) Improved design and optimization of aeration control for WWTPs by dynamic simulation. *IWA Instrumentation, Control and Automation Conference*, Malmö, June 2001, *pp.* 499-506

Allitt R, Nelen F (1994) Real Time Control in The Hague, WaPUG Autumn Meeting. Blackpool

Almeida M (1992) Derivation of IF..THEN..ELSE operating rules from optimization of combined sewer systems under real time control. In Hartong, H., Lobbrecht, A. (eds.) Applications of Operations Research to Real Time Control of Water Resources Systems. Third European Junior Scientist Workshop. Formerum/Terschelling. The Netherlands. 20-25 September 1991. PREDICT 1992. *pp.* 109-118

Almeida M (1994) Real time control in small urban drainage systems - The Fehraltorf project. final Report. Laboratório Nacional de Engenharia Civil, Lisbon

Almeida M (1999a) Pollutant transformation processes in sewers under aerobic dry weather flow conditions. I&D Tese Hidráulica. Laboratório Nacional de Engenharia Civil. Lisboa. Imperial College of Science, Technology and Medicine. London

Almeida M (1999b) At-source domestic wastewater quality. Urban Water 1,1,49-56

Almeida M, Schilling W (1993) Derivation of if-then-else rules from optimised strategies for sewer systems under real time control. 6th International Conference on Urban Storm Drainage, Niagara Falls, *pp.* 1525-1530

Amandes CB, Bedient PB (1980) Stormwater detention in developing watersheds. J. Env. Eng. Div. ASCE. 106,2,403-419

Ambrose RB Jr, Wool TA, Connolly JP, Schanz RW (1988) WASP4, A Hydrodynamic and Water Quality Model--Model Theory, User's Manual and Programmer's Guide. Report EPA 600/3-87/039. US EPA. Athens, Georgia

Andersen NK, Sørensen S (1994) The Copenhagen RTC system. International User-Group Meeting "Computer aided analysis and operation in sewage transport and treatment technology". IUGM94. Chalmers Institute of Technology, Göteborg, *pp.* 235-245

Andersen NK, Harremoës P, Sørensen S, Andersen HS (1996) Monitoring and real time control in a trunk sewer. 7th International Conference on Urban Storm Drainage, Hannover, *pp.* 1157-1162

Andrews JF (1974) Review paper: Dynamic models and control strategies for wastewater treatment processes. Wat.Res.8, 261-289

Andrews JF (1993) Modeling and simulation of wastewater treatment processes. Water Sci. Technol. 28,11-12,141-150

Andrews JF (1994) Dynamic control of wastewater treatment plants. Environmental Science and Technology. 28,9,434A-440A

Anonymous (1994) The Wordsworth Dictionary of Biography. Wordsworth Reference. Ware

APHA American Public Health Association (1994) Standard Methods for the Examination of Water and Wastewater. 18th edition, Washington D.C.

Appelgren C, Hernebring C (1994) A pilot application of the MouseTRAP Sediment transport model in Göteborg. International User-Group Meeting "Computer aided analysis and operation in sewage transport and treatment technology" IUGM94. Chalmers Institute of Technology, Göteborg, *pp.* 151-158

Appelgren C, Hernebring C, Mark O (1995) Validation of MouseTRAP. First DHI Software Users Conference, Copenhagen

Ardern E, Lockett WT (1914) Experiments on the oxidation of sewage without the aid of filters. J. Soc. Chem. Ind. 33, *p.* 523

Arthur S, Ashley RA (1997) Near bed solids transport rate prediction in a combined sewer network. Water Sci. Technol. 36,8-9,129-134

Arthur S, Ashley RA (1998) The influence of near bed solids transport on first foul flush in combined sewers. Water Sci. Technol. 37,1,131-138

Ashley R, Budge F, Fleming R (1995) HYDROWORKS RTC modelling for Aberdeen. WaPUG Autumn 1995, Blackpool

Ashley R, Goodison MJ (1991) The development of best optimum solutions to drainage problems. Water Sci. Technol. 24,6,89-100

Ashley R, Verbanck M, Bertrand-Krajewski JL, Hvitved-Jacobsen T, Nalluri C, Perrusquia G, Pitt R, Ristenpart E, Saul A (1996) Solids in sewers - The state of the art. 7th International Conference on Urban Storm Drainage. Hannover, *pp.* 1771-1776

Ashley RM, Wotherspoon DJJ, Petrie M, Jack A (1997) Modelling to assist with the evaluation of the holistic performance of sewers and wastewater treatment plants. CIWEM Conference on Integrated Modelling '97. Ballyonnell, Ireland, October 1997

ATV (Abwassertechnische Vereinigung e. V.) (1989) Beeinflussung der Beschaffenheit von Fließ-gewässern durch Mischwassereinleitungen. Ergebnis einer Anhörung von Fachleuten am 18. und 19. April 1988 in Essen. Korrespondenz Abwasser, 36,7,755-760

ATV (Abwassertechnische Vereinigung e. V.) (1996) Allgemein verfügbares Gewässergütemodell. Projektabschlußbericht 02WA9104/4. Abwassertechnische Vereinigung, Hennef/Sieg

ATV A128 (Abwassertechnische Vereinigung e. V.) (1992) Richtlinien für die Bemessung und Gestaltung von Regenentlastungsanlagen in Mischwasserkanälen. ATV-Arbeitsblatt A128. Gesellschaft zur Förderung der Abwassertechnik, St. Augustin

ATV A131 (Abwassertechnische Vereinigung e. V.) (1991) Bemessung von einstufigen Belebungsanlagen ab 5000 Einwohnerwerten. ATV-Arbeitsblatt A131. Gesellschaft zur Förderung der Abwassertechnik, St. Augustin

ATV-DVWK A131 (Deutsche Vereinigung für Wasserwirtschaft, Abwasser und Abfall e. V.) (2000) Bemessung von einstufigen Belebungsanlagen. ATV-DVWK-Arbeitsblatt A131. Gesellschaft zur Förderung der Abwassertechnik, St. Augustin

ATV 1.2.4 (Abwassertechnische Vereinigung e. V.) (1991) Leitfaden Abflußsteuerung. Planung, Entwurf und Betrieb. ATV-Arbeitsgruppe 1.2.4 "Abflußsteuerung in Kanalnetzen". Dokumentation und Schriftenreihe der ATV aus Wissenschaft und Praxis, St. Augustin

ATV 1.2.4 (Abwassertechnische Vereinigung e. V.) (1995) Untersuchung zum Steuerungspotential von Kanalnetzen. 5. Arbeitsbericht der ATV-Arbeitsgruppe 1.2.4 "Abflußsteuerung in Kanalnetzen". Korrespondenz Abwasser, 42,1,103-108

ATV 1.2.6 (Abwassertechnische Vereinigung e. V.) (1986) Die Berechnung des Oberflächenabflusses in Kanalnetzmodellen - Teil 1 Abflußbildung. Arbeitsbericht der ATV-Arbeitsgruppe 1.2.6 "Hydrologie der Stadtentwässerung" gemeinsam mit dem DVWK. Korrespondenz Abwasser, 33,2,157-162

ATV 1.2.6 (Abwassertechnische Vereinigung e. V.) (1987) Die Berechnung des Oberflächenabflusses in Kanalnetzmodellen - Teil 2 Abflußkonzentration. Arbeitsbericht der ATV-Arbeitsgruppe 1.2.6 "Hydrologie der Stadtentwässerung" gemeinsam mit dem DVWK. Korrespondenz Abwasser, 34,3,263-269

ATV 1.9.3 (Abwassertechnische Vereinigung e. V.) (1992) Anwendung von Schmutzfrachtberechnungsmethoden im Sinne des neuen ATV-Arbeitsblattes A128. 6. Arbeitsbericht der ATV-Arbeitsgruppe 1.9.3 "Schmutzfrachtberechnung". Korrespondenz Abwasser, Vol. 39. No. 5. 1992. *pp.* 727-738

ATV 2.2.3 (Abwassertechnische Vereinigung e. V.) (1997) Einführung des ATV-Gewässergütemodells. Arbeitsbericht der ATV-Arbeitsgruppe 2.2.3 "Erstellung eines allgemein verfügbaren Gewässergütemodells". Korrespondenz Abwasser, 44,11,2058-2061

ATV 2.11.4 (Abwassertechnische Vereinigung e. V.) (1997) Simulation von Kläranlagen. 1. Arbeitsbericht der ATV-Arbeitsgruppe 2.11.4 "Simulation von Kläranlagen" im ATV-

Fachausschuß 2.11 "Entwurf und Bau von Kläranlagen". Korrespondenz Abwasser, 44,11,2064-2074

ATV 2.12.1 (Abwassertechnische Vereinigung e. V.) (1997) Auswirkungen der Mischwasserbehandlung auf den Betrieb von Kläranlagen. Arbeitsbericht der ATV-Arbeitsgruppe 2.12.1 "Auswirkungen der Regenwasserbehandlung auf Kläranlagen" im ATV-Fachausschuß 2.12 "Betrieb von Kläranlagen". Korrespondenz Abwasser, 44,8,1419-1428

Ayesa E, Goya B, Larrea A, Larrea L, Rivas A (1998) Selection of operational strategies in activated sludge processes based on optimization algorithms. Water Sci. Technol. 37,12,327-334

Babovic V (1991) Applied Hydroinformatics - a control and advisory system for real-time applications. IHE Report Series No. 26. International Institute for Hydraulic and Environmental Engineering, Delft

Balslev P, Nickelsen C, Lynggaard-Jensen A (1993) On-line flux-theory based control of secondary clarifiers, Water Sci. Technol. 30,2,209-218

Bandemer H, Bellmann A (1976) Statistische Versuchsplanung. MINÖA Series. Vol. 19. No 2. BSB B.G. Teubner Verlagsgesellschaft, Leipzig

Bardoel T (1995) Optimalisering van het peilbeheer van de Amstelveense poel. Doctoraalverslag afstudeervak. Vakgroep Wiskunde. Sectie operationale Analyse. Wageningen Agricultural University

Barovic U (1995) Einfluß der Kalibrierung eines hydrologischen Modells auf die Ergebnisse der Simulation am Beispiel des Mischsystems der Stadt Hildesheim. Diplomarbeit. Institut für Wasserwirtschaft, Hydrologie und landwirtschaftlichen Wasserbau, Universität Hannover

Bauwens W, Vanrolleghem P, Fronteau C, Smeets M (1995) An integrated methodology for the impact assessment of the design and operation of the sewer-waste water treatment plant system on the receiving water quality. Mededelingen Faculteit Landbouwkundige en toegepaste biologische Wetenschappen. Universiteit Gent. 60,4b,2447-2450

Bauwens W, Vanrolleghem P, Smeets,M (1996) An evaluation of the efficiency of the combined sewer - waste water treatment system under transient conditions, Water Sci. Technol. 33,2,199-208

Bayar A (1993) Identification of algal population dynamics in a model of river water quality. MSc dissertation. Department of Civil Engineering, Imperial College of Science, Technology and Medicine, London

Beasley D, Bull DR, Martin RR (1993a) An Overview of Genetic Algorithms Part 1, Fundamentals. University Computing. 15,2,58-69

Beasley D, Bull DR, Martin RR (1993b) An Overview of Genetic Algorithms Part 2, Research topics. University Computing. 15,4,170-181

Bechmann H (1999) Modelling of Wastewater Systems. Dissertation. Technische Universität Dänemark. Lyngby

Beck MB (1973) The application of control and systems theory to problems of river pollution. PhD thesis, University of Cambridge

Beck MB (1976) Dynamic modelling and control applications in water quality maintenance. Water Res. 10,575-595

Beck MB (1978) Real-time control of water quality and quantity. Research Memorandum RM-78-19. International Institute for Applied Systems Analysis, Laxenburg, Austria

Beck MB (1981) Operational water quality management beyond planning and design. Executive Report 7. International Institute for Applied Systems Analysis, Laxenburg, Austria

Beck MB (1983) Uncertainty, system identification and the prediction of water quality. In Beck MB, van Straten G (eds.) Uncertainty and Forecasting of Water Quality. Springer, Berlin, *pp.* 3-68

Beck MB (1984) Modelling and control studies of the activated sludge process at Norwich Sewage Works. Trans. Inst. .Measurement Control. 6, 3,117-131

Beck MB (1986) Identification, estimation and control of biological wastewater treatment processes. IEEE Proc. 133,254-264

Beck MB (1987) Water quality modelling. A review of the analysis of uncertainty. Water Resources Res., 23,8,1393-1442

Beck MB (1989) System identification and control. In Patry G, Chapman,D (eds.) Dynamic Modelling and Expert Systems in Wastewater Engineering. Lewis Publishers. Chelsea, Michigan, *pp.* 261-323

Beck MB (1991) Principles of modelling. Water Sci. Technol. 24,6,1-8

Beck MB (1996) Transient pollution events - acute risks to the aquatic environment. Water Sci. Technol. 33,2,1-15

Beck MB (1997) Applying systems analysis in managing the water environment: Towards a new agenda. Water Sci. Technol. 36,5,1-17

Beck MB, Young PC (1975) A dynamic model for DO-BOD relationships in a non-tidal stream. Water Res. 9,769-776

Beck MB, Finney BA (1987) Operational water quality management. Water Resources Res. 23,11,2030-2042

Beck MB, Finney BA, Lessard P (1987) Operational Water Quality Management. A sense of perspective. In Beck MB (ed.) Systems Analysis in Water Quality Management. Pergamon Press. Oxford. *pp.* 357-368

Beck MB, Adeloye AJ, Lessard P, Finney BA, Simon L (1989) Stormwater overflows modelling impacts on the receiving waters and the treatment plant. In J. Ellis (ed.) Urban Discharges and Receiving Water Quality Impacts. Pergamon Press, Oxford

Beck MB, Lumbers JP, Mackenzie HEC, Jowitt PW (1990) Un prototype de système expert pour le contrôle d'un procédé de boues activées. Sciences et techniques de l'eau. 23,2,161-167

Beck MB, Adeloye AJ, Finney BA, Lessard P (1991) Operational water quality management: transient events and seasonal variability. Water Sci. Technol. 24,6,257-265

Beck MB, Chen J (1994) System identification, parameter estimation and the analysis of uncertainty. A review. Paper 11. Preprints Second IAWQ Specialised Seminar on Modelling and Control of Activated Sludge Processes, Copenhagen, 22-24 August 1994

Beck MB, Chen J, Saul AJ, Butler D (1994) Urban drainage in the 21st century: assessment of new technology on the basis of global material flows. Water Sci. Technol. 30,2,1-12

Beck MB, Reda A (1994) Identification and Application of a dynamic model for operational management of water quality. Water Sci. Technol. 30,2,31-42

Beck MB, Watts JB, Winkler S (1997) An environmental process control laboratory The interface between instrumentation and model development. 7th IAWQ Workshop on Instrumentation, Control and Automation of Water and Wastewater Treatment and Transport Systems. Brighton, 6 - 9 July 1997

Becker M, Bischofsberger W, Brummer J, Geiger WF (1992) Vergleich des Abflußverhaltens bei verschiedenen Oberflächen und Kanalnetzstrukturen und bei belastungsabhängigen Steuerungsmaßnahmen. In Zielke *et al.* (1992), *pp.* 63-81

Beeneken T, Fuchs L, Scheffer C, Spönemann P (1994) Anwendung der Fuzzy-Logik in der Abflußsteuerung. Zeitschrift für Stadtentwässerung und Gewässerschutz. 26,65-127

Bellman R, Dreyfus SE (1962) Applied dynamic programming. Princeton University Press, Princeton NJ

Bennett JP, Rathbun RE (1972) Reaeration in open-channel flow. Geological Survey Professional Paper. 737. US Govern. Printing Office, Washington D.C.
Bente S, Schilling W, (1996) An object-oriented concept for an urban hydrology simulation system. 7th International Conference on Urban Storm Drainage. Hannover, September 1996. *pp.* 1777-1782
Berthouex PM, Fan R (1986) Evaluation of treatment plant performance causes, frequency and duration of upsets. J. Water Pollut. Control Fed. 58,368-375
Berthouex PM, Box GE (1996) Time series models for forecasting wastewater treatment plant performance. Water Res. 30,8,1865-1875
Bertrand-Krajewski, JL (1994) How to operate a combined sewer system, a wastewater treatment plant and its storage. TECHWARE workshop Environmental motivation for the abatement of the impact of combined sewer overflows on receiving waters. Molenheide/Belgium. 21-22 September
Bertrand-Krajewski JL, Briat P, Scrivener O (1993) Sewer sediment production and transport modelling: A literature review. J. Hydraulic Res. 31,4,435-460
Bertrand-Krajewski, JL, Lefebvre M, Lefai B, Audic JM (1995) Flow and pollutant measurements in a combined sewer system to operate a wastewater treatment plant and its storage tank during storm events. Water Sci. Technol. 31,7,1-12
Blanc D, Kellagher R, Phan L, Price R (1995) FLUPOL-MOSQITO, models, simulations, critical analysis and development. Water Sci. Technol. 32,1,185-192
Böhnke B, *et al.* (1989) Bemessung der Stickstoffelimination in der Abwasserreinigung (HSG-Ansatz). Korrespondenz Abwasser. 36,9,1046-1061
van den Boomen RM, Salverda AP, Uunk EJB, Roos C (1995) Modellering van het Reggesysteem met DUFLOW. H2O. 28,4,107-111
Borchardt D (1992) Wirkungen stoßartiger Belastungen auf ausgewählte Fließgewässerorganismen. Wasser - Abwasser - Abfall. Schriftenreihe des Fachgebietes Siedlungswasserwirtschaft. Uni-GH Kassel. Nr. 10
Bowie GL, Mills WB, Porcella,DB, Campbell CL, Pagenkopf JR, Rupp GL, Johnson KM, Chan PWH, Gherini SA, Chamberlin CE (1985) Rates, Constants, and Kinetic Formulations in Surface Water Quality Modeling. U.S. Environmental Protection Agency. Athens Georgia, EPA/600/3-85/040
Box GEP, Wilson KB (1951) On the experimental attainment of optimum conditions (with discussion). J. Roy. Statist. Soc. B13,1-45
Bradford BH (1977) Optimal storage in a combined sewer system. J. Water Res. Planning Man. Div. ASCE. 103,1-15
Brandt T, Drechsel U, Jacobi D, Zaiß H (1989) Schmutzfrachtsimulationsmodell SMUSI - Modellkonzeption. "Umweltplanung, Arbeits- und Umweltschutz". Schriftenreihe des Hessischen Landesamtes für Umwelt. No. 68
Brent RP (1973) Algorithms for Minimization Without Derivatives. Prentice Hall, Englewood Cliffs NJ
Brown LC, Barnwell TO Jr. (1987) The Enhanced Stream Water Quality Models QUAL2E and QUAL2E-UNCAS. Documentation and User Manual. Environmental Research Laboratory. Office of Research and Development. U.S. EPA. Athens, Georgia
Bryson AE, Ho YC (1969) Applied Optimal Control. Blaisdell, Waltham
BS8005 British Standard No. 8005. Sewerage
    Part 0 Introduction and guide to data sources and documentation. 1987
    Part 1 Guide to new sewerage construction. 1987
    Part 2 Guide to pumping stations and pumping mains. 1987
    Part 3 Guide to planning and construction of sewers in tunnel. 1989
    Part 4 Guide to design and construction of outfalls. 1987
    Part 5 Guide to rehabiliation of sewers. 1990. British Standards Institution

Bujon G, Herremans L, Phan L (1992) FLUPOL: a forecasting model for flow and pollutant discharge from sewerage systems during rainfall events. Water Sci. Technol. 25,8,207-215

Busby JB, Andrews JF (1975) Dynamic modeling and control strategies for the activated sludge process. J. Wat. Pollut. Contr. Fed. 47,1055-1080

Butler D (1991) A small scale study of wastewater discharges from domestic appliances. J. Inst. Water Env. Man. 5,178-185

Butler D (1993) The influence of dwelling occupancy and day of the week on domestic appliance wastewater discharges. Building and Environment. 28,1,73-79

Butler D (1994) Lecture on Urban Drainage. Lecture notes. Autumn Term 1994. Department of Civil Engineering. Imperial College of Science, Technology and Medicine, London

Butler D, Friedler E, Gatt K (1994) Characterising the quantity and quality of domestic wastewater inflows. Water Sci. Technol. 31,7,13-24

Butler D, Clark P (1995) Sediment management in urban drainage catchments. Construction Industry Research Information Association (CIRIA) Report 134, London

Butler D, Graham NJD (1995) Modeling dry weather wastewater flow in sewer networks. J. Env. Eng. 121,2,161-173

Butler D, Xiao Y, Karunaratne SHPG, Thedchanamoorthy S (1995) The gully pot as a physical, chemical and biological reactor. Water Sci. Technol. 31,7,219-228

Butler D, Gatt K (1996) Synthesising dry weather flow input hydrographs A Maltese case study. Water Sci. Technol. 34,3/4,55-62

Butler D, Davies J (2000) Urban Drainage. E&FN Spon. London

Butler, D, Schütze M (submitted) Integrating simulation models with a view to optimal control of urban wastewater systems. Workshop on Vulnerability of water quality in intensively developing urban watersheds. Making the case for High-performance integrated control. Athens, Georgia, USA. 14 -16 May 2001. Submitted to Urban Water.

Camacho, LA (2001) Development of a hierarchical modelling framework for solute transport under unsteady flow conditions in rivers. PhD thesis. Department of Civil and Environmental Engineering, Imperial College of Science, Technology and Medicine. London

Capodaglio A (1994a) Integrated control requirements for sewerage systems. Water Sci. Technol. 30,1,131-138

Capodaglio A (1994b) Evaluation of modelling techniques for wastewater treatment plant automation. Water Sci. Technol. 30,2,149-156

Capodaglio A (1994c) Transfer function modelling of urban drainage system, and potential uses in real-time control applications. Water Sci. Technol. 29,1-2,409-417

Carroll, DL (1996) Genetic algorithm. Version 1.6.2. University of Illinois

Carstensen J, Madsen H, Poulsen NK, Nielsen MK (1994) Identification of wastewater treatment processes for nutrient removal on a full-scale WWTP by statistical methods. Water Res. 28,10,2055-2066

Carstensen J, Nielsen MK, Harremoës P (1996) Predictive control of sewer systems by means of grey-box models. Water Sci. Technol. 34,3/4,189-194

CEC - Council of the European Communities (1991) Directive Concerning Urban Waste Water Treatment. EC Directive No. 91/271/EEC

CEC - Council of the European Communities (2000) Directive of Establishing a framework for Community action in the field of water policy. EC Directive No. 2000/60/EEC of 23 October 2000

Censor Y (1977) Pareto optimality in multiobjective problems. Appl. Math. Optimization. 4,41-59

Cerco CF, Cole T (1995) User's guide to the CE-QUAL-ICM three dimensional eutrophication model, release version 1.0. Technical Report EL-95-15. US Army Eng. Waterways Experiment Station Vicksburg, MS

Chang NB, Chen WC (1997) Combined genetic algorithm and neural network logic for optimal control of wastewater treatment plants. 7th IAWQ Workshop on Instrumentation, Control and Automation of Water and Wastewater Treatment and Transport Systems. Brighton, 6 - 9 July 1997. *pp.* 573-580

Chapra SC (1997) Surface Water-Quality Modelling. McGraw Hill, New York

Chebbo G, Saget A. (1995) Pollution of Urban Wet Weather Discharges. Encyclopedia of Environmental Biology. Volume 3. Academic Press Inc., San Diego, *pp.* 171-182

Chen J (1993) Modelling and control of the activated sludge process towards a systematic framework. PhD thesis. Department of Civil Engineering. Imperial College of Science, Technology and Medicine, London

Chen J, Beck MB (1993) Modelling, control and offline estimation of activated sludge bulking. Water Sci. Technol. 28,11-12,249-256

Chen J, Beck MB (1997) Towards designing sustainable urban wastewater infrastructures A screening analysis. Water Sci. Technol. 35,9,99-112

Christoffels E (2001) Ein Instrument zur Unterstützung wasserwirtschaftlicher Planungsaufgaben am Beispiel der Erft - ATV-Gewässergütemodell. Korrespondenz Abwasser 48,7,968-972

Churchill MA, Elmore HL, Buckingham RA (1962) Prediction of stream reaeration rates. J. San. Eng. Div. ASCE. SA41. Proc. Paper 3199

Cieniawski SE, Eheart JW, Ranjithan S (1995) Using genetic algorithms to solve a multiobjective groundwater monitoring problem. Water Resources Res. 31,2,399-409

CIRIA (1973) Cost-effective sewage treatment - the creation of an optimising model. CIRIA Report No. 46, London

Clemens FHLR (2001) Hydrodynamic models in urban drainage: Application and Calibration. PhD thesis. TU Delft

Clifforde IT. (1998) Personal communication. February 1998

Clifforde IT, Murell KN (1993) Urban Pollution Management Review of Products and Implementation. Foundation for Water Research report FR0405, Marlow

Clifforde I, Nielsen JB (1995) Integrated modelling - a vision for the future. Second International Users Group Meeting. Blackpool, November 1995

Cluckie ID, Han D, Tilford KA,. Lin K, Tyson J (1996) Real-time control of urban drainage systems using a Multiple Attribute Radar System (MARS). 7th International Conference on Urban Storm Drainage. Hannover. *pp.* 953-958

Collatz L, Wetterling W (1971) Optimierungsaufgaben. Springer. Berlin Heidelberg New York

Cooper VA, Nguyen VTV, Nicell JA (1997) Evaluation of global optimization methods for conceptual rainfall-runoff model calibration. Water Sci. Technol. 36,5,53-62

COST682 (1996) Integrated Wastewater Management. COST682. Reports from Working Group Meetings. Budapest, 10-12 January 1996, European Commission. DG XII. Unit XII-B-1

Côté M, Grandjean BPA, Lessard P, Thibault J (1995) Dynamic Modelling of the activated sludge process improving prediction using neural networks. Water Res. 29,4,995-1004

Couillard D, Zhu S (1992) Control strategy for the activated sludge process under shock loading. Water Res. 26,649-655

Covar AP (1976) Selecting the Proper Reaeration Coefficient for Use in Water Quality Models. Presented at the U.S. EPA Conference on Environmental Simulation and Modeling. 19-22 April, Cincinnati
Crabtree R, Dempsey P, Becker M, Gent R, Simpson K, Bryan D (1993) UPM applications methodology: Final report. Foundation for Water Research. Report FR 0384, Marlow
Crabtree RW, Becker M, Bryan D, Gent RJ, Threlfall JL (1994a) Review of UPM Modelling Tools. Foundation for Water Research. Report FR0442, Marlow
Crabtree R, Garsdal H, Gent R, Mark O, Dørge J (1994b) MOUSETRAP - A deterministic sewer flow quality model. Water Sci. Technol. 30,1,107-115
Crabtree R, Earp W, Whalley P (1994c) The Derby UPM Demonstration Project. WaPUG, Autumn Meeting, Blackpool
Crabtree RW, Ashley R, Gent R (1995) MOUSETRAP Modelling of real sewer sediment characteristics and attached pollutants. Water Sci. Technol. 31,7,43-50
Crabtree R, Earp W, Whalley P (1996) A demonstration of the benefits of integrated wastewater planning for controlling transient pollution. Water Sci. Technol. 33,2,209-218
Crockett CP, Crabtree RW, Markland HR (1989) SPRAT - a simple river quality impact model for intermittent discharges. Water Sci. Technol. 21,12,1793-1796
Crockett CP, Hutchings CJ, Crabtree RW (1990) The development of a dynamic water quality model (SPRAT). WRc report. FR0108. WRc Engineering, Swindon

Dannen J (1996) Methoden der linearen und nichtlinearen Optimierung zur Verbundsteuerung von Entwässerungssystemen. Diplomarbeit. Insitut für Angewandte Mathematik. Institut für Wasserwirtschaft, Universität Hannover
Dantzig GB (1963) Linear Programming and Extensions. Princeton University Press. Princeton, New Jersey
Dauber L, Novak B (1992) Quellen und Mengen der Schmutzstoffe in Regenabflüssen einer städtischen Mischkanalisation. EAWAG Separatum Nr. 927. Eidgenössische Anstalt für Wasserversorgung, Abwasserreinigung und Gewässerschutz, Dübendorf
Davies OL (ed.) (1978) The Design and Analysis of Industrial Experiments. Longman, London New York
Davis L (ed.) (1987) Genetic Algorithms and Simulated Annealing. Pitman, London
Davis L (1991) Handbook of Genetic Algorithms. Van Nostrand Reinhold, New York
Deakin PT, Vickers S (1995) An appraisal of four different sewage treatment works models. WaPuG Meeting, Blackpool
Debebe A (1996) Personal communication, December 1996
Decker J (1995) Auswirkungen von Abwasserinhaltsstoffen auf Kanalnetze und Möglichkeiten ihrer technischen Nutzung. Korrespondenz Abwasser. 42,6,904-916
Degrémont (1991) Water treatment handbook. 6th edition. Reuil-Malmaison, Degrémont
Delleur JW (1996) Water Quality Modeling in Sewer Networks. NATO-ASI on Hydroinformatics in Planning, Design, Operation and Rehabilitation of Sewer Systems. Harrachov, 16 - 29 June 1996
Dempsey P, Eadon A, Morris G (1996) SIMPOL - a simplified urban pollution modelling tool. 7th International Conference on Urban Storm Drainage. Hannover. *pp.* 1365-1370
Demuynck C, Mespreuve M, Bauwens W (1993) Application of a continuous simulation model on the sewer network of Brussels. 6th International Conference on Urban Storm Drainage. Niagara Falls. *pp.* 1472-1477
Dettmar J, Cassar A (1996) Components for the real time control of an urban drainage system. 7th International Conference on Urban Storm Drainage. Hannover. 815-820

Deyda, S. (1991) Parameteridentifikation für Simulationsmodelle in der Hydrologie. Diplomarbeit am Institut für Angewandte Mathematik und am Institut für Wasserwirtschaft, Hydrologie und landwirtschaftlichen Wasserbau. Universität Hannover

Deyda S, Khelil A, Siekmann M (1993) Erweiterung des Kanalnetzberechnungsmodells KMROUT auf die Simulation von Fließvorgängen im Einstau- bzw. Überstau-Bereich. Zeitschrift für Stadtentwässerung und Gewässerschutz. 22,3-57

DHI (1990) MOUSE 3.0. Users Guide and Technical Reference. Danish Hydraulic Institute, Hørsholm

DHI (1992) Danish Hydraulic Institute. MIKE 11, User Manual

DHI (1998a) Danish Hydraulic Institute. WRc plc Webpage
http//www.wrcplc.co.uk/tabs/rschnet/twp/index.htm, consulted on 31 January 1998

DHI (1998b) Danish Hydraulic Institute. WRc plc: CD-ROM describing the Technology Validation Project (TVP)

Dick RY (1970) Role of activated sludge final settling tanks. J. San. Eng. Div. ASCE. 96,423-436

Dick RI, Young KW (1972) Analysis of thickening performance of final settling tanks. Proceedings of 27th Industrial Waste Conference (ed. Bell, J.M.). Purdue University, N.C.

Dierickx M, Van Assel J, Heip L (1998) Implementation of UPM-Procedure in Flanders. Abstract submitted for the 3rd Hydroinformatics Conference. Copenhagen, 24-26 August 1998

DMSC Dansk MOUSE Service Center (1995) Iltsvindsmodellen DOSMO Version 3.1. Dokumentation. Charlottenlund, June 1995

Dochain D, Vanrolleghem PA, van Daele M (1995a) Structural identifiability of biokinetic models of activated sludge respiration. Water Res. 29,11,2571-2578

Dochain D, Vanrolleghem P, Henze M (1995b) Optimizing the design and operation of biological wastewater treatment plants through the use of computer programs based on a dynamic modelling of the process. COST682. Report 1992-1995. European Commission. Directorate-General XII. Science, Research and Development. Environment research programme. ISBN 92-827-4344-6

Dosztányi I (1987) Gabčikovo-Nagymaros. Umwelt und Staustufe. Aqua Kiadó, Budapest

Draper N, Smith H (1981) Applied Regression Analysis. Second edition, Wiley

Dreyfus SE, Averill M L (1977) The Art and Theory of Dynamic Programming. Academic Press, New York, London

Duan Q, Sorooshian S, Gupta V (1992) Effective and efficient global optimisation for conceptual rainfall-runoff models. Water Resources Res. 28,4,1015-1031

Duan Q, Sorooshian S, Gupta V (1994) Optimal use of SEC-UA global optimization methnod for calibrating watershed models. J. Hydrology, 158,265-284

Dudley J (1995) Comparison of wastewater treatment plant models. Personal communication to D. Butler. 27 January 1995

Dudley JWO, Bryan DA, Chambers B (1994) STOAT - Development and application of a fully dynamic sewage treatment works model. International User-Group Meeting "Computer aided analysis and operation in sewage transport and treatment technology" IUGM94. Chalmers Institute of Technology, Göteborg, 81-98

Dudley J, Chambers B (1995) Dynamic Modelling of wastewater treatment processes using STOAT. WaPUG Meeting. Blackpool, November 1995

van Duin EHS, Portielje R, Aalderink RH (1995) Modelling Water quality and flow in river Vecht using DUFLOW. Water Quality Modelling. Proc. of the Int. Conf. on Water Quality Modelling. (ed. C. Heatwole). Orlando, ASAE. 313-324.

Dupont R. Henze M (1992) Modelling of the secondary clarifier combined with the activated sludge model no. 1. Water Sci. Technol. 25,6,285-300
Dupont R, Sinkjaer O (1994) Optimisation of Wastewater Treatment plants by means of computer models. Water Sci. Technol. 30,4,181-190
Dupont R, Dahl C (1995) A one-dimensional model for a secondary settling tank including density current and short-circuiting. Water Sci. Technol. 31,2, 215-224
Durchschlag A (1989) Bemessung von Mischwasserspeichern im Nachweisverfahren unter Berücksichtigung der Gesamtemission von Mischwasserentlastung und Kläranlagenablauf. Schriftenreihe für Stadtentwässerung und Gewässerschutz, 3, SuG Verlagsgesellschaft, Hannover
Durchschlag A (1990a) Gesamtemissionen als Planungsgröße von Mischwasserentwässerungssystemen. Korrespondenz Abwasser. 37,8,889-893
Durchschlag A (1990b) Long-term simulation of pollutant loads in treatment plant effluents and combined sewer overflows. Water Sci. Technol. 22,10/11,69-76
Durchschlag A (1991) Zusammenhang zwischen Mischwasserspeicherung und Kläranlagenablauf. Zeitschrift für Stadtentwässerung und Gewässerschutz, 4
Durchschlag A, Schilling W (1990) The total pollution discharge as a design criterion for combined sewer systems. 5th International Conference on Urban Storm Drainage. Osaka. *pp.* 1077-1082
Durchschlag A, Härtel L, Hartwig P, Kaselow M, Kollatsch D, Schwentner G (1991) Total emissions from combined sewer overflow and wastewater treatment plants. Eur. Water Pollution Control. 1,6,13-23
Durchschlag A, Härtel L, Hartwig P, Kaselow M, Kollatsch D, Schwentner G (1992) Joint consideration of combined sewerage and wastewater treatment plants. Water Sci. Technol. 26,5/6,1125-1134

Einfalt T (1993) FITASIM - A simulator for the real-time control of urban drainage systems. 6th International Conference on Urban Storm Drainage. Niagara Falls, *pp.* 1514-1518
Einfalt T, Wolf-Schumann U (1992) Training real time control on the FITASIM simulator. In Hartong H, Lobbrecht A (eds.) Applications of Operations Research to Real Time Control of Water Resources Systems. Third European Junior Scientist Workshop. Formerum/Terschelling. The Netherlands. 20-25 September 1991. PREDICT. 1992, *pp.* 71-74
Einfalt T, Huyskens R, Schilling, W (1993) The impact of spatial and temporal uncertainties from point rainfall measurements for RTC of urban drainage systems. 6th International Conference on Urban Storm Drainage. Niagara Falls, *pp.* 1496-1501
Einfalt T, Hatzfeld F, Hüsemann W (1994) Real-time application of rainfall measurement and forecasting in the Altenau catchment. International User-Group Meeting "Computer aided analysis and operation in sewage transport and treatment technology" IUGM94. Chalmers Institute of Technology, Göteborg, *pp.*263-270
Einfalt T, Semke M (1994) Tools for assessing the real time control potential of urban drainage systems. In Vervey, Minns, Babovic, Maksimović (eds.) Hydroinformatics '94. Balkema, Rotterdam, *pp.* 281-286
Elgerd OI (1967) Control Systems Theory. Mc Graw Hill, New York
Ellis JB (1986) Pollutional aspects of urban runoff. in Torno, H.C.. Maršalek, J.. Desbordes, M. (eds.) Urban Runoff Pollution. NATO ASI Series G10. Springer. *pp.* 1-38
Ellis JB (1989a) The management and control of urban runoff quality. J. Inst. Water Env. Man. April 1989. 3,2,116-125
Ellis JB (ed.) (1989b) Urban Discharges and Receiving Water Quality Impacts. Pergamon Press. Oxford

Ellis JB, Hvitved-Jacobsen T (1996) Urban drainage impacts on receiving waters. J. Hydraulic Res. 34,6,771-783
Elmore HL, Hayes TW (1960) Solubility of atmospheric oxygen in water. J. San. Eng. Div. ASCE. 86, 41-53
von der Emde W (1998) Geschichte des Belebungsverfahrens. Korrespondenz Abwasser. 45,6,1086-1101
Entem S, Lahoud A, Yde L, Bendsen B (1998) Real time control of the sewer system of Boulogne Billancourt - A contribution to improving the water quality of the Seine. Water Sci. Technol. 37,1,327-332
Esat V, Hall MJ (1994) Water resources system optimisation using genetic algorithms. Proceedings of the Hydroinformatics'94 Conference. *pp.* 225-231
Euler G, Jacobi D (1986) Vergleichende Darstellung von Schmutzfracht-Berechnungsmethoden. Korrespondenz Abwasser. 1,20-24

Fair GM, Geyer JC, Okun DA (1966) Water and Wastewater Engineering. Vol. 1 Water Supply and Wastewater Removal. Wiley, New York
Fair GM, Geyer JC, Okun DA (1968) Water and Wastewater Engineering. Vol. 2 Water purification and waste-water treatment and disposal. Wiley, New York
Fehlberg E (1960) Neuere genauere Runge-Kutta-Formeln für Differentialgleichungen zweiter Ordnung bzw. n-ter Ordnung. ZAMM. 40,252-259. 449-455
Fehlberg E (1966) New high-order Runge-Kutta-formulas with an arbitrarily small truncation error. ZAMM. 46,1-16
Fehlberg E (1969) Klassische Runge-Kutta-Formel fünfter und siebenter Ordnung mit Schrittweitenkontrolle. Computing. 4,93-103
Fehlberg E (1970) Klassische Runge-Kutta-Formel vierter und niedrigerer Ordnungen mit Schrittweitenkontrolle und ihre Anwendung auf Wärmeleitungsprobleme. Computing. 6, 61-71
Ferrara RA, Hildick-Smith A (1982) A modelling approach for storm water quantity and quality control via detentioin basin. Water Ressources Bull. 18,975-981
Fiacco AV, McCormick GP (1968) Nonlinear Programming Sequential Unconstrained Minimization Techniques. Wiley, New York Toronto
Finney DJ (1963) An Introduction to the Theory of Experimental Design. The University of Chicago Press, Chicago London
Finnson A (1993) Simulation of a strategy to start up nitrification at Bromma sewage plant using a model based on the IAWPRC model No. 1. Water Sci. Technol. 28,11-12,185-195
Folnseca, C.M. Fleming, P.J.H. (1993) Genetic algorithms for multiobjective optimization. Formulation, discussion and generalization. Proc. 5th International Conference on Genetic Algorithms. *pp.* 416-423
Freund M, Otterpohl R, Dohmann M (1993) Dynamische mathematische Modelle von Nachklärbecken - Übersicht und Vergleich. Korrespondenz Abwasser. 5,738-746
Freund M, Rolfs T (1995) Kläranlagenbemessung mittels statischer Bemessungsansätze - Möglichkeiten und Grenzen. awt - abwassertechnik. 1,22-25
Friedler E, Brown DM, Butler D (1996) A study of WC derived sewer solids. Water Sci. Technol. 33,9,17-24
Friedler E, Butler D (1996) Quantifying the inherent uncertainty in the quantity and quality of domestic wastewater. Water Sci. Technol. 33,2,65-78
Fries J (1996) Pollution discharge control by integrated modelling of the sewage treatment plant and the sewer system in Sweden. In Schütze M (ed.) Impact of Urban Runoff on Wastewater Treatment Plants and Receiving Waters. Proceedings of the Ninth

European "Junior" Scientist Workshop. Kilve/UK, April 1996. Foundation for Water Res. Marlow. *pp.* 111-114

Fronteau C (1999) Water Quality Management of River Basins and Evaluation of the Impact of Combined Sewer Overflows Using an Integrated Modelling Approach. Dissertation. Freie Universität Brüssel

Fronteau C, Bauwens W, Smeets M, Vanrolleghem P (1995) Een evaluatie van de efficientie van het rioolstelsel-RWZI-rivier systeem onder dynamische omstandigheden. Water 84. September/October 1995. *pp.* 203-210

Fronteau C, Bauwens W, Vanrolleghem P, Smeets M (1996) An immission based evaluation of the efficiency of the Sewer-WWTP-river system under transient conditions. 7th International Conference on Urban Storm Drainage. Hannover, 467-472

Fronteau C, Bauwens W, Vanrolleghem P (1997a) Towards a solution for the reconciliation of sewer and waste water treatment plant models. Second International Conference 'The Sewer as a Physical, Chemical and Biological Reactor. Aalborg, 1997

Fronteau C, Bauwens W, Vanrolleghem P (1997b) Integrated modelling comparison of state variables, processes and parameters in sewer and wastewater treatment models. Water Sci. Technol. 36,5,373-380

Fuchs L (1996) Hydrology or Urban Catchments. NATO-ASI on Hydroinformatics in Planning, Design, Operation and Rehabilitation of Sewer Systems. Harrachov, 16 - 29 June 1996

Fuchs S (1996) Abflußretention unter Berücksichtigung gewässerökologischer Grundsätze. In Hahn HH, Trauth R (eds.) 10. Karlsruher Flockungstage. Wechselwirkung zwischen Einzugsgebiet und Kläranlage. Report No. 78. Institut für Siedlungswasserwirtschaft. Universität Karlsruhe. Oldenbourg. München. *pp.* 163-179

Fuchs S (1997) Wasserwirtschaftliche Konzepte und ihre Bedeutung für die Ökologie kleiner Fließgewässer - Aufgezeigt am Beispiel der Mischwasserbehandlung. Institut für Siedlungswasserwirtschaft. Universität Karlsruhe, Oldenbourg München

Fuchs L, Müller D, Neumann A (1987) Learning production system for the control of urban sewer systems. In Beck MB (ed.) Systems Analysis in Water Quality Management. Pergamon Press. Oxford. *pp.* 411-421

Fuchs L, Hurlebusch R (1994) Steuerung von Regenrückhaltebecken im Bereich der Stadt Flensburg. Zeitschrift für Stadtentwässerung und Gewässerschutz. 26,65-127

Fuchs L, Beeneken T, Scheffer C, Spönemann P (1994a) Modelling of Real-time Control with Fuzzy Logic - The Flensburg Case Study. International User-Group Meeting "Computer aided analysis and operation in sewage transport and treatment technology" IUGM94. Chalmers Institute of Technology, Göteborg, *pp.* 247-262

Fuchs L, Scheffer C, Verworn HR (1994b) Mikrocomputer in der Stadtentwässerung. Kanalnetzberechnung. Modellbeschreibung HYSTEM-EXTRAN. Version 5.1

Fuchs L, Gerighausen D (1995) Untersuchung von Sanierungskonzepten basierend auf Gesamtemissionsbetrachtungen für das Kanalnetz der Stadt Dresden. Zeitschrift für Stadtentwässerung und Gewässerschutz. 30,27-74

Fuchs L, Scheffer C (1995) Zur Entwicklung der Modelltechnik in der Stadtentwässerung - Rückblick und zukünftige Tendenzen. Korrespondenz Abwasser. 42,10,1826-1834

Fuchs L, Gerighausen D, Schneider S (1996) Emission-Immission based design of combined sewer overflows and treatment plant - the Dresden case study. 7th International Conference on Urban Storm Drainage. Hannover, *pp.* 1121-1126

Fuchs L, Beeneken T, Spönemann P, Scheffer C (1997) Model based real-time control of sewer system using fuzzy-logic. Water Sci. Technol. 36,8-9,343-347

Fujie K, Urano,K, Ohtake M, Kubota H (1990) Mathematical model simulation and evaluation of the MLSS control strategies in an activated sludge process. In Briggs R (ed.) Instrumentation, Control and Automation of Water and Wastewater

Treatment and Transport Systems. Proc. 5th Workshop. Kyoto, Yokohama. Pergamon Oxford, *pp.* 539-544

FWR (1994) Urban Pollution Management Manual. Foundation for Water Res. Marlow

FWR (1998) Urban Pollution Management Manual. Second Edition. Foundation for Water Res. Marlow

Gall RAB, Patry GG (1989) Knowledge-Based System for the Diagnosis of an Activated Sludge Plant. In Patry GG, Chapman D (eds.) Dynamic Modeling and Expert Systems in Wastewater Engineering. Lewis Publishers Inc. Chelsea, Michigan. *pp.* 193-240

Gammeter S (1996) Einflüsse der Siedlungsentwässerung auf die Invertebraten-Zönose kleiner Fliessgewässer. PhD thesis ETH 11673. Eidgenössische Technische Hochschule, Zürich

Gan TY, Dlamini EM, Biftu GF (1997) Effects of model complexity and structure, data quality, and objective functions on hydrologic modeling. J. Hydrology. 192,1/4,81-103

Ganzevles PPG, van Luijtelaar H (1994) Sewer design on dynamic principles. Eur. Water Pollution Control. 4,5,18-23

Garsdal H, Mark O, Dørge J, Jepsen S (1995) MOUSETRAP Modelling of water quality processes and the interaction of sediments and pollutants in sewers. Wat. Sci, Tech. 31,7,33-41

Geldof GD (1995) Adaptive water management integrated water management on the edge of chaos. Novatech '95. Lyon, *pp.* 539-546

Gent R (1992) Use of foul in MOSQITO. Personal communication to D. Butler. 7 December 1992

Gent RJ, Crabtree RW, Eperson M (1994) Improved MOSQITO Applications Procedures. Foundation for Water Research report FR0443. Marlow

Gent R, Crabtree B, Ashley R (1996) A review of model developments based on sewer sediemtns research in the UK. Water Sci. Technol. 33,9,1-7

Gill EJ, Bryan DA (1994) The Development and use of the urban pollution database. Foundation for Water Research report FR 0441, Marlow

Gill PE, Murray W (eds.) (1974) Numerical Methods for Constrained Optimization. Academic Press, London

Gill PE, Murray W, Wright MH (1981) Practical Optimization. Academic Press, London

Gillblad T, Olsson O (1977) Computer control of a medium-sized activated sludge plant. Prog. Water Technol. 1977. 9,5/6,427-434

Giordana A, Neri F. (1996) Genetic algorithms in machine learning. AI Comm. 9,1,21-26

Goforth GF, Heaney JP, Huber WC (1983) Comparison of basin performance modeling techniques. J. Env. Eng. Div. ASCE. 109,1082-1098

Goldberg DE (1989) Genetic Algorithms in Search, Optimization and Machine Learning. Addison-Wesley, Reading/Mass.

Gonwa W, Capodaglio AG, Novotny V (1993) New tools for implementing real time control in sewers. 6th International Conference on Urban Storm Drainage. Niagara Falls, *pp.* 1375-1380

Grady CPL Jr (1989) Dynamic modelling of suspended growth biological wastewater treatment processes. In Patry G, Chapman D (eds.) Dynamic Modelling and Expert Systems in Wastewater Engineering. Lewis publishers, Chelsea Michigan, 1-38

Grau P, Sutton PM, Henze M, Elmaleh S, Grady CP, Gujer W, Koller J (1987) Report: Notation for use in the description of wastewater treatment processes. Water Res. 21,2,135-139

Greenstadt J (1972) A auasi-Newton method with no derivatives. Math. Computat. 26,117,145-166

Grefenstette JJ (1986) Optimization of control parameters for genetic algorithms. IEEE Transactions on Systems, Man, and Cybernetics. 16,1,12-128

Grijspeerdt K, Vanrolleghem P, Verstraete W (1995) Selection of one-dimensional sedimentation: Models for on-line use. Water Sci. Technol. 31,2,193-204

Grotehusmann D, Semke M (1990) KMROUT, ein hydrologisches detailliertes Kanalnetzberechnungsmodell. Zeitschrift für Stadtentwässerung und Gewässerschutz. 13

Grosche G, Ziegler V, Ziegler D (1984) Ergänzende Kapitel zu Bronstein-Semendjajew Taschenbuch der Mathematik. Teubner, Leipzig

Guderian J, Durchschlag A, Bever J (1998) Evaluation of total emissions from treatment plants and combined sewer overflows. Water Sci. Technol. 37,1,333-340

Gujer W (1977) Design of a nitrifying activated sludge process with the aid of dynamic simulation. Prog. Water Technol. 9,323-336

Gujer W (c. 1990) Biologische Abwasserreinigung. Nachdiplomstudium Siedlungswasserbau und Gewässerschutz. Studienrichtung Umweltingenieur, ETH Zürich.

Gujer W (1990) Mathematische Beschreibung von technischen Systemen. Nachdiplomstudium Siedlungswasserbau und Gewässerschutz. Studienrichtung Umweltingenieur, ETH Zürich

Gujer W (1995) ASIM Activated Sludge SIMulation Program. Version 3.0. Program description

Gujer W (1999) Siedlungswasserwirtschaft. Springer. Berlin Heidelberg New York

Gujer W, Henze M (1991) Activated sludge modelling and simulation. Water Sci. Technol. 23,1011-1023

Gujer W, Henze M, Mino T, Matsuo T, Wentzel MC, Marais GvR (1995) The Activated Sludge Model No. 2 Biological Phosphorus removal. Water Sci. Technol. 31,2,1-11

Gujer W, Larsen TA (1995) The implementation of biokinetics and conservation principles in ASIM. Water Sci. Technol. 31,2,257-266

Gülen S (1995) Operation and control of wastewater treatment plant - A literature review. MSc dissertation. Department of Civil Engineering. Imperial College of Science, Technology and Medicine, London

Gupta K, Saul AJ (1996) Suspended solids in combined sewer flows. Water Sci. Technol. 33,9,93-99

Gupta VK, Sorooshian S (1983) Uniqueness and observability of conceptual rainfall-runoff parameters. The percolation process examined. Water Resources Res. 19,1,269-276

Gupta VK, Bastidas LA, Sorooshian S, Shuttleworth WJ, Yang ZL (1999) Parameter estimation of a land surface scheme using multicriteria methods. J. Geophys. Res. 104(D16), 19491-19503

Gustafsson LG, Lumley DJ, Lindeborg C, Haraldsson J (1993a) Integrating a catchment simulator into wastewater treatment plant operation. Water Sci. Technol. 28,11/12,45-54

Gustafsson LG, Lumley DJ, Persson B, Lindeborg C (1993b) Development of a catchment simulator as an on-line tool for operating a wastewater treatment plant. 6th International Conference on Urban Storm Drainage. Niagara Falls, *pp.* 1508-1513

Gustafsson LG. Andréasson M, Winberg S (1994) From planning tools to on-line modelling systems encapsulating the complete urban aquatic environment. International User-Group Meeting "Computer aided analysis and operation in sewage transport and treatment technology" IUGM94. Chalmers Institute of Technology, Göteborg, *pp.* 279-289

Güven O (1995) Comparison of wastewater treatment plant models. MSc dissertation. Department of Civil Engineering. Imperial College of Science, Technology and Medicine, London

Haarsma GJ, Keesman K (1995) Robust model predictive dissolved oxygen control. Mededelingen Faculteit Landbouwkundige en toegepaste biologische Wetenschappen. Universiteit Gent, 60/4b,2415-2432
Halcrow & Partners Ltd. HR Wallingford Ltd (1995a) ISIS Flow User Manual
Halcrow & Partners Ltd. HR Wallingford Ltd (1995b) ISIS Quality User Manual
Hammer MJ (1996) Water and Wastewater Technology, third edition. Prentice Hall, Englewood Cliffs NJ
Hansen J (1997) Der Einsatz von Fuzzy Control für Regelungsaufgaben im Bereich der Nährstoffelimation in kommunalen Kläranlagen. Schriftenreihe des Fachgebietes Siedlungswasserwirtschaft. Universität Kaiserslautern, Vol. 10
Hansen J, Krauss M, Buchholz B (1994) Initial experience with a fuzzy logic control system for optimizing nitrogen removal at a municipal sewage treatment plant. Abwassertechnik. 45,4,35-38
Hansen OB, Pedersen J (1994) Integrated Planning of Improvements of sewer system and treatment plant for suburbs of Copenhagen. Water Sci. Technol. 30,1,157-166
Hansen OB, Jacobsen C, Harremoës P, Nielsen PS (1993) Model studies of storm water loading of a treatment plant. 6th International Conference on Urban Storm Drainage. Niagara Falls, *pp.* 1938-1943
Harms R, Kenter G (1990) Mischwasserentlastungen - KOSIM Version III.0. Programmdokumentation. Institut für technisch-wissenschaftliche Hydrologie, Hannover
Harremoës P (1982) Immediate and delayed oxygen depletion in rivers. Water Res. 16,1093-1098
Harremoës P (1988a) Stochastic Models for Estimation of extreme pollution from urban runoff. Water Sci. Technol. 22,8,1017-1026
Harremoës P (1988b) Überlauf von Mischwasser und Sauerstoffinhalt im Vorfluter. Korrespondenz Abwasser. 35,11,1168-1173
Harremoës P (1989) Overflow quantity, quality and receiving water impact. In Ellis, J.B. (ed.) Urban Discharges and Receiving Water Quality Impacts. Pergamon. *pp.* 9-16
Harremoës P (1991) Real time control - in context. In Maksimović, Č (ed.) New Technologies in Urban Drainage - UDT 91. Elsevier, London
Harremoës P (1994) Overview of Basic Principles for Combined Sewer Overflows, Legislation, Desing standards and Methods in EU Member States. Seminar Combined Sewer Overflow - An European Perspective. Delft, 24 March 1994
Harremoës P, Capadaglio AG, Hellström BB, Henze M, Jensen KN, Lynggaard-Jensen A, Otterpohl R, Søeberg H (1993) Wastewater treatment plants under transient loading - performance, modelling and control. Interurba'92 workshop. Water Sci. Technol. 27,12,71-115
Harremoës P, Hvitved-Jacobsen T, Lynggaard-Jensen A, Nielsen B (1994) Municipal Wastewater Systems, Integrated Approach to Design, Monitoring and Control. Water Sci. Technol. 29,1-2,419-426
Harremoës P, Rauch W (1996) Integrated design and analysis of drainage systems, including sewers, treatment plants and receiving waters. J. Hydraulic Res. 34,6,815-826
Härtel L (1990) Modellansätze zur dynamischen Simulation des Belebtschlammverfahrens. PhD thesis. Technische Hochschule Darmstadt, WAR-Schriftenreihe 47
Härtel L, Pöpel HJ (1992) A dynamic secondary clarifier model including processes of sludge thickening. Water Sci. Technol. 25,6,267-284
Härtel L, Durchschlag A, Hartwig P, Kaselow M, Kaiser C, Kollatsch D, Otterpohl R (1995) Kläranlagensimulation im Vergleich - Zweiter Arbeitsbericht der Gruppe Gesamtemissionen. Korrespondenz Abwasser. 42,6,970-980

Hartwig P (1993) Beitrag zur Bemessung von Belebungsanlagen mit Stickstoff- und Phosphorelimination. Veröffentlichungen des Institutes für Siedlungswasserwirtschaft und Abfalltechnik der Universität Hannover, 84

Harvey A (1989) Forecasting Structural Time Series Models and the Kalman Filter. Cambridge University Press

Häßlein M, Seyfried CF (1995) IAWQ Model No. 1 - always sufficient for description of nitrogen removal? Medelingen Faculteit Landbouwkundige en toegepaste biologische Wetenschappen. Universiteit Gent, 60,4b,2395-2402

Havlik V (1996) Computational Hydraulic Modelling I. NATO-ASI on Hydroinformatics in Planning, Design, Operation and Rehabilitation of Sewer Systems. Harrachov, 16 - 29 June 1996

HEC (1986) HEC-5 Simlation of Flood Control and Conservation Systems, Appendix on Water Quality Analysis. Report CPD-5Q. Hydrologic Engineering Center. U.S. Army Corps of Engineers, Davis, CA

Heinemann A (1992) Untersuchung von Klassifikationsverfahren im Rahmen eines selbstadaptierenden regelbasierten Systems. Diplomarbeit. Insitut für Informatik, Institut für Wasserwirtschaft, Universität Hannover

Hemmis n.v. (1998) WEST - Waste-water treatment plant engine for simulation and training. Promotional leaflet. Hemmis n.v. Kortrijk, January 1998

Henderson R (1995) Stormwater pollution control in EU member states - An Overview. Second International Modelling Users' Group Meeting. Blackpool, November 1995

Henderson RJ, Moys GD (1987) Development of a sewer flow quality model for the United Kingdom. 4th International Conference on Urban Storm Drainage. Lausanne, 201-207

Hendrickson JD, Sorooshian S, Brazil LE (1988) Comparison of Newton-type and direct search algorithms for calibration of conceptual rainfall-runoff models. Water Resources Res. 24,5,691-700

Henze M (1987) Storm water handling in wastewater treatment plants. EWPRC Symposium '87. 7. Europäisches Abwasser- und Abfallsymposium, München, 19 - 22 May 1987

Henze M (1992) Characterisation of wastewater for modelling of activated sludge processes. Water Sci. Technol. 25,6,1-15

Henze M, Grady CPL Jr, Gujer W, Marais GvR, Matsuo T (1986) Activated Sludge Model No. 1. IAWPRC Task Group on Mathematical Modelling for Design and Operation of Biological Wastewater Treatment IAWPRC. London, Report July 1986

Henze M, Gujer W, Mino T, Matsuo T, Wentzel MC, Marais GvR (1995a) Activated Sludge Model No. 2. IAWQ Scientific and Technical Reports, No. 3. IAWQ London

Henze M, Gujer W, Mino T, Matsuo T, Wentzel MC, Marais GvR (1995b) Wastewater and biomass characterisation for the Activated Sludge model No. 2 Biological Phosphorus removal. Water Sci. Technol. 31,2,13-23

Henze M, Gujer W, Mino T, Matsuo T, Wentzel MC, Marais GvR., van Loosdrecht (1999) Activated Sludge Model No. 2d. Water Sci. Technol. 39,1,165-182

Henze M, Gujer W, Mino T, van Loosdrecht M (2000) Activated Sludge Models ASM1, ASM2, ASM2d and ASM3. Edited by IWA Task Group on Mathematical Modelling for Design and Operation of Biological Wastewater Treatment. IWA Scientific and Technical Reports, No. 9. IWA London

Ho YC (1994) Heuristics, Rules of Thumb, and the 80/20 Proposition. IEEE Trans. Automatic Control. 39,5,1025-1027

Ho YC, Sreenivas RS, Vakili P (1992) Ordinal optimization of DEDS. J. Discrete Event Dynamic Syst. 2,2,61-68

Ho YC, Larson ME (1995) Ordinal optimization approach to rare event probability problems. Discrete Event Dynamic Systems. Theory and Applications. 5,281-301

Hoen K, Schuhen M, Köhne M (1994) Dynamische Simulation von Kläranlagen - Ein Hilfsmittel für den planenden Ingenieur? Korrespondenz Abwasser. 41,5,760-771

Holland JH (1975) Adaptation in Natural and Artificial Systems. University of Michigan Press, Ann Arbor

Holthausen E (1995) Numerische Simulation in Belebung und Nachklärung - Eine Methode zur Optimierung. Korrespondenz Abwasser. 42,10,1812-1819

Holzer P, Krebs P (1998) Total ammonia impact of CSO and WWTP effluent. Water Sci. Technol. 38,10,31-39

Holzhausen V, Krusbersky J, Reich D, Meßmer A (1990) Mischwasserbehandlung im Kanalnetz Frankfurt am Main West - Effektivere Speichernutzung durch Verbundsteuerung. Korrespondenz Abwasser. 37,3,242-248

Hornberger GM, Spear RC (1980) Eutrophication in Peel Inlet, I. Problem-defining behaviour and a mathematical model for the Phosphorus scenario. Water Res. 14,29-42

Hornberger GM, Spear RC (1981) An approach to the preliminary analysis of environmental systems. J. Env. Management. 12,1,7-18

Horst R, Pardalos PM (eds.) (1985) Handbook of global optimization. Kluwer, Dordrecht

HouseM.A, Newson DH (1989) Water quality indices for the management of surface water quality. Water Sci. Technol. 21,10/11,1137-1148

House MA, Ellis JB, Herricks EE, Hvitved-Jacobsen T, Seager HJ, Lijklema L, Aalderink H, Clifforde I (1993) Urban drainage - impacts on receiving water quality. Water Sci. Technol. 27,12,117-158

Howard CDD. (1993) Sewer and treatment plant operations control for receiving water protection. 6th International Conference on Urban Storm Drainage. Niagara Falls, 1933-1937

Huisman JL, Krebs P, Gujer W (2000) Aerobic degradation in the sewer system. Dresdner Berichte 16,109-126

Hvitved-Jacobsen T (1982) The impact of combined sewer overflows on the dissolved oxygen concentration of a river. Water Res. 16,1099-1105

Hvitved-Jacobsen T (1986) Conventional pollutant impacts on receiving waters, a review paper. In Torno HC, Maršalek J, Desbordes M (eds.) Urban runoff pollution. Proceedings NATO workshop . Montpellier. 25-30 August 1995. NATO ASI Series. Series G. Ecological Sciences. Vol. 10. Springer. Heidelberg. *pp.* 345-378

Hvitved-Jacobsen T (ed.)(1998) The sewer as a physical, chemical and biological reactor II. Selected Proceedings of the 2nd IAWQ International Specialised Conference on The Sewer as a Physical, Chemical and Biological Reactor. Aalborg. 25-28 May 1997. Water Sci. Technol. 37,1

Hvitved-Jacobsen T, Harremoës P (1981) Impact of combined sewer overflows on dissolved oxygen in receiving streams. 2nd International Conference on Urban Storm Drainage. Urbana, *pp.* 226-235

Hvitved-Jacobsen T, Schaarup-Jensen K (1990) Analysis of combined sewer overflow impact on the dissolved oxygen concentration of receiving streams. 5th International Conference on Urban Storm Drainage, Osaka, *pp.*517-522

Hvitved-Jacobsen T, Schaarup-Jensen K (1991) Pollution from urban runoff - oxygen depletion in streams and rivers, TRITON Training Course. Integrated Urban Runoff. EC programme COMETT II, 1991/92

Hvitved-Jacobsen T, Nielsen PH, Larsen T, Jensen AA (eds.)(1995) The sewer as a physical, chemical and biological reactor. Selected Proceedings of the International Specialised Conference on The Sewer as a Physical, Chemical and Biological Reactor. Aalborg, 16-18 May 1994, Water Sci. Technol. 31,7

Hydro-Ingenieure (1993) Gewässergütemodell QUALSIM. Kurzbeschreibung. Hydro-Ingenieure, Düsseldorf, March
Hydrotec (1992) Gewässergütesimulation mit dem Programm QUALSIM. Promotional material. Hydrotec GmbH, Aachen

Ibbitt RP (1970) Systematic parameter fitting for conceptual models of catchment hydrology. PhD thesis. Department of Civil Engineering. Imperial College of Science, Technology and Medicine, London
ifak (2001) SIMBA 4.0, SIMBA-Sewer 4.0. Manual. Institut für Automation und Kommunikation ifak Magdeburg e. V., Barleben
IHE (1992) DUFLOW - A Micro-Computer Package for the Simulation of One-dimensional Unsteady Flow and Water Quality in Open Channel Systems - Manual. Version 2.0. IHE Delft. Rijkswaterstaat, The Hague. Delft Institute of Technology. Agricultural University of Wageningen. Stichting Toegepast Onderzoek Waterbeheer (STOWA), Utrecht. Bureau Icim, Rijswijk
Imhoff K, Imhoff, K. (1990) Taschenbuch der Stadtentwässerung. 27th edition, Oldenbourg München Wien
Ingeduld P, Zeman E (1996) Computational hydraulic modelling II - Simulation of aquatic systems in open channels 1D and 2D. NATO-ASI on Hydroinformatics in Planning, Design, Operation and Rehabilitation of Sewer Systems. Harrachov, 16 - 29 June 1996
Isaacs S, Thornberg D (1997) A comparison between model and rule-based control of a periodic activated sludge process. 7th IAWQ Workshop on Instrumentation, Control and Automation of Water and Wastewater Treatment and Transport Systems. Brighton. 6 - 9 July 1997, *pp.*335-342
itwh (1995) Mikrocomputer in der Stadtentwässerung - Mischwasserentlastungen. Teil I. KOSIM. Version 4.1. Programmdokumentation. Institut für technisch-wissenschaftliche Hydrologie, Hannover
Ivanov P, Masliev I, De Marchi C, Somlyódy L (1996) DESERT - Decision Support System for Evaluating River Basin Strategies. User's Manual. International Institute for Applied Systems Analysis (IIASA). Laxenburg/Austria

Jack A.G (1995) Personal communication. January 1995
Jack AG, Petrie MM, Ashley RM (1995) Integrated Catchment Modelling - A Sustainable Approach for the City of Perth. 3rd International Conference. Water Pollution '95 - Modelling, Measuring and Prediction. Porto Carras, 25-28 April 1995
Jack AG, Petrie MM, Ashley RM (1996) The diversity of sewer sediments and the consequences for sewer flow quality modelling. Water Sci. Technol. 33,9,207-214
Jacobi D (1990) Evaluation of pollutant-load-calculation-methods by measurement - comparison and evaluation of the methods. 5th International Conference on Urban Storm Drainage. Osaka, *pp.*371-376
Jacobsen BN (1990) Wastewater Quality standards in European countries - Effluents from municipal wastewater treatment plants. European Water Pollution Control Association
Jakeman AJ, Hornberger GM (1993) How much complexity is warranted in a rainfall-runoff model? Water Resources Res. 29,8,2637-2649
Jakobsen C, Andersen NK, Harremoës P, Nielsen PS (1993a) Real Time Control in a pipe to minimise the CSO. 6th International Conference on Urban Storm Drainage. Niagara Falls, *pp.*1531-1536
Jakobsen C, Hansen OB, Harremoës P (1993b) Development and application of a general simulator for rule based control of combined sewer systems. 6th International Conference on Urban Storm Drainage. Niagara Falls, *pp.*1357-1362

James A (ed.) (1993) An introduction to Water Quality Modelling, 2nd edition. Wiley, Chichester

Jamsa K (1988) Using OS/2. McGraw Hill, Berkeley, California

Janikow CZ, Michalewicz Z, (1991) An experimental comparison of binary and floating point representations in genetic algorithm. In Belew RK, Booker LB (eds.) Proceedings of the Fourth International Conference on Genetic Algorithms. Morgan Kaufman, San Mateo, *pp.*31-36

Jeppson U, Olsson G (1993) Reduced order models for on-line parameter identification of the activated sludge process. Wat Sci. Tech. 28,11/12,173-182

Jeppson U, Diehl S (1995) Validation of a robust dynamic model of continuous sedimentation. Mededelingen Faculteit Landbouwkundige en toegepaste biologische Wetenschappen. Universiteit Gent. 60,4b,2403-2414

Jeppson U, Diehl S (1996) An evaluation of a dynamic model of the secondary clarifier. Water Sci. Technol. 34,5/6,19-26

Ji Z, Brown C, Vitasovic Z, Osborne M (1996a) Computer modelling study of Liverpool's real time control system. 7th International Conference on Urban Storm Drainage, Hannover, *pp.*1031-1036

Ji Z, McCorquodale J, Zhou S, Vitasovic Z (1996b) A dynamic solids inventory model for activated sludge systems. Wat. Env. Res., 68,3,329-337

Johann G, Verworn HR (1996) Requirements for radar rainfall data in urban catchment modelling and control. 7th International Conference on Urban Storm Drainage. Hannover, *pp.* 223-228

Johansen NB (1985) Discharge to receiving waters from sewer systems during rain. PhD thesis. Department of Environmental Engineering. Technical University of Denmark, Lyngby

Johansen NB, Harremoës P, Jensen M (1984) Methods for calculation of annual and extreme overflow events from combined sewer systems. Water Sci. Technol. 16,8/9,311-325

Jones GL (1978) A mathematical model for bacterial growth and substrate utilization in the activated sludge process. In James, A. (ed.) Mathematical Models in Water Pollution Control. Wiley. Chichester. *pp.* 265-280

Jørgensen M, Schilling W, Harremoës P (1995) General assessment of potential CSO reduction by means of real time control. Water Sci. Technol. 32,1,249-257

Jumar U, Alex J, Seick I, Tschepetzki R. (2000) Betriebsunterstützung im Klärwerk Magdeburg-Gerwisch durch Online-Simulation im Leitsystem. *gwf Wasser/Abwasser* 141,15,30-38

Jumar U, Tschepetzki R (2001). Implementation of a WWTP operation support tool based on on-line simulation. *IWA Instrumentation, Control and Automation Conference*, Malmö, June 2001, *pp.* 711-718

Jumar U, To TB (2001) Evolutionäre Optimierung im Entwurfsprozess von Kläranlagen *at – automatisierungstechnik* 49,10,449-461

Kaemaki P-St (1991) Modelling impacts of acid deposition and afforestation on catchment hydrochemistry. PhD thesis. Faculty of Engineering, University of London

Kaijun W, Zeeman G, Lettinga G (1995) Alteration in sewage characteristics upon aging. Water Sci. Technol. 31,7,191-200

Kappeler J, Gujer W (1994a) Development of a mathematical model for "Aerobic Bulking". Water Res. 28,2,311-322

Kappeler J, Gujer W (1994b) Verification and aplications of a mathematical model for "Aerobic bulking". Water Res. 28,2,311-322

Kappeler J, Gujer W (1994c) Influences of wastewater composition and operating conditions on activated sludge bulking and scum formation. Water Sci. Technol. 30,11,181-189

Kayser R (1987) Biologische Stickstoff- und Phosphorelimination in Abwasserreinigungsanlagen. Veröffentlichungen des Institutes für Stadtbauwesen der Technischen Universität Braunschweig, 42

Kayser R, Langendörfer H, Hofmann M. Ladiges G (1992) Expertensysteme als Hilfsmittel beim Kläranlagenbetrieb. Research report. BMFT-Forschungsvorhaben 02WA8828. Institut für Siedlungswasserwirtschaft. Insitut für Betriebssysteme und Rechnerverbund, Technische Universität Braunschweig

Karacs I (1997) Judges wash their hands of the Danube. The Independent, 26 September 1997, 11

Keinath TM (1985) Operational dynamics and control of secondary clarifiers. J. Water Pollution Control Fed. 57,7,770-776

Keinath TM (1990) Diagram for designing and operating secondary clarifiers according to the thickening criterion. Research J. Pollution Control Contr. Fed. 62,3,254-258

Khelil A (1990a) Steuerung eines Mischsystems zur Verbesserung der Gewässergüte und zur Verminderung der Betriebskosten - wissenschaftliche Erarbeitung von Steuerungskonzepten, Research report. Forschungsvorhaben 02-WA 86470. Institut für Wasserwirtschaft, Hydrologie und landwirtschaftlichen Wasserbau. Universität Hannover

Khelil A (1990b) Experience with real time control of an urban drainage system. First European Junior Scientist Workshop "Application of Operations Research to Real Time Control of Water Resources Systems". Kastanienbaum. Schriftenreihe der EAWAG. 3,199-204

Khelil A (1992a) Methodology for the development and improvement of control strategies in an urban drainage system (UDS). In Hartong H, Lobbrecht A (eds.) Applications of Operations Research to Real Time Control of Water Resources Systems. Third European Junior Scientist Workshop. Formerum/Terschelling. The Netherlands. 20-25 September 1991. PREDICT. *pp.* 119-128

Khelil A (1992b) Überarbeitung des Steuerungskonzepts in Bremen-Links-der-Weser. Final report. Institut für Wasserwirtschaft, Hydrologie und landwirtschaftlichen Wasserbau. Universität Hannover, September 1992

Khelil A (1998) personal communication. June 1998

Khelil A. Grottker M, Semke M (1990) Adaptation of an expert-system for the real time control of a sewerage network Case of Bremen left side of the Weser. 5th International Conference on Urban Storm Drainage. Osaka, *pp.*1329-1334

Khelil A, Schneider S (1991) Development of a control strategy to reduce combined sewerage overflows the case of Bremen-left of the Weser. Water Sci. Technol. 24,6,201-208

Khelil A, Heinemann A, Müller D (1993a) Learning algoritms in a rule based system for control of UDS. 6th International Conference on Urban Storm Drainage. Niagara Falls, *pp.* 1401-1408

Khelil A, Knemeyer B, Dehnhardt J (1993b) Comparison of optimisation algorithms to determine control strategies in UDS. 6th International Conference on Urban Storm Drainage. Niagara Falls, *pp.* 1395-1400

Khelil A, Eberl H, Schaad P, Wilderer P (1995) Reduction of Combined Sewer Discharges in Germany Planning Methodology for Implementing RTC in the AOI-Project. Proceedings 12th International Conference on Case Method Research and Case Method Application. Maribor, 13-15 November

Khelil A, Achatz S, Anton H-J, Eberl H, Schaad P, Wilderer P (1997) Online Prediction of Rainfall Inflows into Drainage Systems - Modeling, Calibration, Verification. Proceedings 1st International Conference on "Treatment of solid Waste and Wastewaters". Narbonne, April

Khuri AI, Cornell JA (1996) Response Surfaces - Designs and Analyses. Second Edition. Marcel Dekker, New York Basel Hong Kong

Kido Y, Sueishi T (1990) Real time control strategy for urban storm drainage using a decision matrix searching method. 5th International Conference on Urban Storm Drainage. Osaka, *pp.* 1293-1298

Kirkpatrick S, Gelatt CD, Vecchi MP (1983) Optimization by simulated annealing. Science. 220,671-680

Kjaer J, Wilson G, Mark O (1996) Model-based real-time predictive control for urban drainage systems. 7th International Conference on Urban Storm Drainage. Hannover, *pp.* 935-940

Klapwijk A, Spanjers H, Temmink H (1993) Control of activated sludge plants based on measurement of respiration rates. Water Sci. Technol. 28,11-12,369-376

Kleijnen JPC (1974) Statistical Techniques in Simulation. Parts I and II. Marcel Dekker Inc, New York

Kleijnen JPC (1982) Statistical aspects on simulation an update. Statistica Neerlandica. 36,4,165-186

Kleijnen JPC (1987) Statistical Tools for Simulation Practitioners. Marcel Dekker, New York

Kleissen FM (1990) Uncertainty and Identifiability in conceptual models of surface water acidification. PhD thesis. Department of Civil Engineering. Imperial College of Science, Technology and Medicine, London

Kleissen FM, Beck MB, Wheater HS (1990) The identifiability of conceptual hydrochemical models. Water Resources Res. 26,12,2979-2992

Klepper O, Scholten H, van de Kamer JPG (1991) Prediction uncertainty in an ecological model of the Oosterschelde Estuary. J. Forecasting. 10,1/2,191-209

Knemeyer B (1992) Optimierungsalgorithmen zur Entwicklung von Steuerungsstrategien in städtischen Kanalnetzen. Diplomarbeit. Institut für Wasserwirtschaft, Hydrologie und landwirtschaftlichen Wasserbau, Universität Hannover

Knollmann C (1993) Optimierte Kanalnetzsteuerung mit kombinierter hydrodynamischer Abflußsimulation am Beispiel. Diplomarbeit. Institut für Wasserwirtschaft, Hydrologie und landwirtschaftlichen Wasserbau, Universität Hannover

Kolbinger A (1996) Data processing concept for real-time control of a combined sewer system. In Schütze M (ed.) Impact of Urban Runoff on Wastewater Treatment Plants and Receiving Waters. Proceedings of the Ninth European "Junior" Scientist Workshop. Kilve/UK, April 1996. Foundation for Water Res. Marlow, *pp.* 127-132

Kollatsch D (1992a) Real time control of sewer systems and the influences on the wastewater treatment plant. In Hartong H, Lobbrecht A (eds.) Applications of Operations Research to Real Time Control of Water Resources Systems. Third European Junior Scientist Workshop, Formerum/Terschelling, The Netherlands, 20-25 September 1991, PREDICT. 1992, pp17-25

Kollatsch D (1992b) Total discharges taken into account for comprehensive planning of urban drainage and waste water treatment. Water Sci. Technol. 26,9-11,2609-2612

Kollatsch D (1993) Comprehensive planing of urban drainage and wastewater treatment. Water Sci. Technol. 27,12,205-208

Kollatsch D (1995) Übergreifende Planung von Kanalnetz und Kläranlage unter Berücksichtigung kombinierter Kanalnetze. Schriftenreihe für Stadtentwässerung und Gewässerschutz. Vol. 13. SuG Verlagsgesellschaft, Hannover

Kollatsch D, Schilling W (1990) Control strategies of sanitary sewage detention tanks to reduce combined sewer overflow pollution loads. 5th International Conference on Urban Storm Drainage. Osaka, 1365-1370

Kollatsch D, Kenter G (1992) Mikrocomputer in der Stadtentwässerung - Gesamtemission aus Mischsystemen. GESIM Version II.0. Institut für technisch-wissenschaftliche Hydrologie, Hannover, 1992

Koskinen K, Viitasaari M (1990) Rule based expert system for the control of the activated sludge process. In Briggs R (ed.) Instrumentation, Control and Automation of Water and Wastewater Treatment and Transport Systems. Proc. 5th Workshop, Kyoto Yokohama. Pergamon, Oxford, *pp.* 243-249

Kraft M (1996) Untersuchungen zum Steuerungspotential. In Schilling (1996c), *pp.* 17-28

Krauth Kh, Stolz G (1992) Belastung der Gewässer durch Kläranlagenabläufe und Regenwasserbehandlungsanlagen der Mischkanalisation. In Zielke *et al.* (1992). *pp.* 109-120

Krauth Kh, Müller JR (1996) Optimale Regenbecken- und Kläranlagenbewirtschaftung zum verbesserten Schutz der Vorfluter. In Hahn HH, Trauth R (eds.) 10. Karlsruher Flockungstage. Wechselwirkung zwischen Einzugsgebiet und Kläranlage. Report No. 78. Institut für Siedlungswasserwirtschaft. Universität Karlsruhe, Oldenbourg München, *pp.* 141-155

Krebs P (1991) The hydraulics of final settling tanks. Water Sci. Technol. 23,1037-1046

Krebs P (1993) Success and shortcoming of clarifier modelling. Water Sci. Technol. 28,11-12,141-150

Krebs P (1995) Success and shortcomings of clarifier modelling. Water Sci. Technol. 31,2,181-191

Krebs P (1997) Nutzung von Modellen für Gestaltung und Betrieb von Nachklärbecken. Wiener Mitteilungen, 137,159-190

Krebs P, Krejči V, Fankhauser R, Siegrist H (1996) Reducing the overall impact on a small receiving water. 7th International Conference on Urban Storm Drainage. Hannover, *pp.* 365-370

Krejči V, Fankhauser R, Gammeter S, Grottker M, Harmuth B, Merz P, Schilling W (1994a) Integrierte Siedlungsentwässerung - Fallstudie Fehraltorf. Schriftenreihe der EAWAG, Nr. 8, Dübendorf

Krejči V, Schilling W, Gammeter S (1994b) Receiving water protection during wet weather. Water Sci. Technol. 29,1-2,219-229

Kreutzberger WA, Race RA, Meinholz TL, Harper M, Ibach J (1980) Impact of sediments on dissolved oxygen concentrations following combined sewer overflows. J. Water Pollution Control Fed. 52,1,192-201

Krishnakumar K (1989) Micro-genetic algorithms for stationary and non-stationary function optimization. SPIE Intell. Control and Adaptive Syst. 1196, 289-296

Kroiss H (1994) Operation and monitoring at treatment plants. Proceedings EWPCA International Conference on Integrated Wastewater Management - Collection, Treatment and Reuse, Lisbon, 10-12 October. *pp.* 305-320

Kuczera G (1997) Efficient subspace probablistic parameter optimization for catchment models. Water Resources Res. 33,1,177-185

Künzi HP, Krelle W, von Randow R (1979) Nichtlineare Programmierung. 2nd edition. Springer, Berlin Heidelberg New York

Labadie JW, Morrow DM, Chen YH (1980) Optimal control of unsteady combined sewer flow. J. Water Res. Plann. Man. Div. ASCE. 106,205-223

Ladiges G, Kayser R (1993) On-line and off-line expert system for the operation of waste water treatment plants. Water Sci. Technol. 28,11-12,315-323

Ladiges G, Kayser R (1994) Applied off-line expert system for effluent, operational and technical problems of waste-water treatment plants. Water Sci. Technol. 30,2,157-164

Lammersen R (1997a) Die Auswirkung der Stadtentwässerung auf den Stoffhaushalt von Fließgewässern. Schriftenreihe für Stadtentwässerung und Gewässerschutz. Vol. 15. SuG Verlagsgesellschaft, Hannover

Lammersen R (1997b) Immissionsbezogene Anforderungen an die Stadtentwässerung. Zeitschrift für Stadtentwässerung und Gewässerschutz. Vol. 38, February, *pp.* 23-94

Lange J, Otterpohl R (1997) Abwasser: Handbuch zu einer zukunftsfähigen Wasserwirtschaft. Mall-Beton Verlag, Donaueschingen-Pfohren

Langendörfer H, Schönwälder J, Hofmann M (1990) Wissensbasierte Fehlerdiagnose von Kläranlagen. Fünfes Symposium 'Informatik für den Umweltschutz'. Wien, September 1990, Springer Informatik Fachberichte 256,197-206

Larsen TA, Gujer W (1996) Separate Management of Anthropogenic Nutrient Solutions (Human Urine)". Water Sci. Technol. 34,3-4,87-94

Larsen T, Broch K, Andersen MR (1998) First flush effects in an urban catchment area in Aalborg. Water Sci. Technol. 37,1,251-258

Larsson J (1995) Urban Drainage Modelling from an European user perspective - the past, the present and the future. Second International Users Group Meeting. Blackpool, November

Leclerc C (1996) River Water Quality Modelling. Imperial College International Diploma thesis. Department of Civil Engineering. Imperial College, London. Ecole Nationale Supérieure d'Hydraulique et de Mécanique de Grenoble

Le Dren IC (1996) Modelling and control of nutrient removal in the activated sludge process. PhD thesis. Department of Civil Engineering. Imperial College of Science, Technology and Medicine, London

Lei JH (1996) Uncertainty analysis of urban rainfall-runoff modelling. PhD thesis. Department of Hydraulic and Environmental Engineering. Norwegian University of Science and Technology, Trondheim

Lessard P (1989) Operational water quality management: Control of stormwater discharges. PhD thesis. Department of Civil Engineering. Imperial College of Science, Technology and Medicine, London

Lessard P, Beck MB (1988) Dynamic simulation of primary sedimentation. J. Env. Eng. Div. Proc. ASCE. 114,4,753-769

Lessard P, Beck MB (1990) Operational water quality management: Control of storm sewage at a wastewater treatment plant. Res. J. Water Pollution Control Fed. 62,6,810-819

Lessard P, Beck MB (1991a) Dynamic modelling of wastewater treatment processes - its current status. Env. Sci. Technol. 25,1,30-39

Lessard P, Beck MB (1991b) Dynamic simulation of storm tanks. Water Res. 25,4,375-391

Lessard P, Beck MB (1993) Dynamic Modelling of the activated sludge process: a case study. Water Res. 27,6,963-978

Lester JN (1990) Sewage and sewage sludge treatment. In Harrison RM (ed.) Pollution Causes, Effects and Control. Royal Society of Chemistry, 2nd edition, *pp.* 33-62

Lettenmaier DP, Burges SJ (1975) Dynamic water quality management strategies. J. Wat. Pollution Contr. Fed. 47,12,2809-2819

LfU (1996) Landesanstalt für Umweltschutz Baden-Württemberg: Untersuchung und Bewertung heute weltweit verfügbarer Gewässergütemodelle. Studie im Auftrag des Zentralen Fachdienstes Wasser, Boden, Abfall, Altlasten bei der LfU. Bearbeitet vom Institut für Siedlungswasserwirtschaft der Universität Karlsruhe, Karlsruhe

Lijklema L (1993) Integrated management of urban waters - a preface. Interurba 92. Water Sci. Technol. 27,12,vii-x

Lijklema L, Roijackers RMM, Cupper JGM (1989) Biological assessment of effects of CSOs and stormwater discharges. In Ellis JB (ed.) Urban Discharges and Receiving Water Quality Impacts. Pergamon, Oxford, *pp.* 37-46

Lijklema L, Tyson JM, Lesouef A (1993) Interactions between sewers, treatment plants and receiving waters in urban areas: a summary of the Interurba'92 workshop conclusions. Water Sci. Technol. 27,12,1-29

Lijklema L, Aalderink RH, de Ruiter H (1996) Procesbeschrijvingen DUFLOW Zuurstofhuishouding in stromende en stagnante Watersystemen. Landbouwuniversiteit Wageningen, June

Lindberg S, Nielsen JB, Green MJ (1993) A European concept for real time control of sewer systems. 6th International Conference on Urban Storm Drainage. Niagara Falls, *pp.* 1363-1368

Lindberg CF, Carlsson B (1996) Nonlinear and set-point control of the dissolved oxygen concentration in an activated sludge process. Water Sci. Technol. 34,3/4,135-142

Linde-Jensen JJ (1993) Real time control in part of Copenhagen. Water Sci. Technol. 27,12,209-212

Liong SY, Chan WT, ShreeRam J (1995) Peak-flow Forecasting with Genetic Algorithm and SWMM. J. Hydraulic Eng. 121,8,613-617

Liu R (2000) Monitoring, Modelling, and control of nutrient removal in the activated sludge process. PhD thesis. University of Georgia. Athens, Georgia, USA

Lobbrecht AH (1994) A network flow model for optimized control of rural water systems. Proceedings of the Hydroinformatics'94 conference. *pp.* 161-167

Lobbrecht AH (1997) Dynamic Water-System Control - Design and Operation of Regional Water-Resources Systems. PhD thesis. Technical University Delft

Lohmann J, Schlegel S (1981) Measurement and control of the MLSS concentration in activated plants. Water Sci. Technol. 13,9,217-224

Loke E, Warnaars EA, Jacobsen P, Nelen F, do Céu Almeida M (1996a) Artificial neural networks as a tool in urban storm drainage. 7th International Conference on Urban Storm Drainage. Hannover, *pp.* 1557-1562

Loke E, Warnaars EA, Jacobsen P, Nelen F, do Céu Almeida M (1996b) Problems in urban storm drainage addressed by artificial neural networks. 7th International Conference on Urban Storm Drainage. Hannover, *pp.* 1581-1586

Londong J (1994) Consequences of the behaviour of activated sludge plants with combined sewage inflows. Water Sci. Technol. 30,1,139-146

Londong J, Wachtl, P (1995) Chemisch-biologische Untersuchungen als Grundlage für die Anwendung dynamischer Simulationsmodelle für Belebungsanlagen. awt-abwassertechnik. February, *pp.* 51-54

Lumley DJ, Gustafsson LG, Haraldsson J, Balmér P (1995) Development and implementation of model-based real time control of the Rya wastewater treatment plant catchment. Novatech 95, Lyon, *pp.* 269-276

Lynggaard-Jensen A, Harremoës P (1993) Advanced integrated monitoring and control of municipal waste water systems. 6th International Conference on Urban Storm Drainage. Niagara Falls, *pp.* 1927-1932

Magne G, Phan L, Price R, Wixcey J (1996) Validation of HYDROWORKS-DM, a water quality model for urban drainage. 7th International Conference on Urban Storm Drainage. Hannover, *pp.* 1359-1364

Makkinga AA, Nieuwenhuis RA, Aalderink RH, Uunk EJB (1998) Modelling dissolved oxygen in the River Regge system using Duflow. Eur. Water Management. 1,1,37-42

Mance G (1981) The quality of urban storm discharges - a review. WRc Report ER 192-M

Mance G, Harman MM (1978) The Quality of Urban Storm-water Runoff. Urban Storm Drainage. Pentach Press, London, Plymouth, *pp.* 603-617

Mark O, Larsson J, Larsen T, (1993) A sediment transport model for sewers. 6th International Conference on Urban Storm Drainage. Niagara Falls, *pp.* 638-643

Mark OL, Appelgren C, Larsen T (1995) Principles and approaches for numerical modelling of sediment transport in sewers. Water Sci. Technol. 31,7,107-115

Mark OL, Cerar U, Perrusquía G (1996) Prediction of Locations with sediment deposits in sewers. Water Sci. Technol. 33,9,147-154

Mark O, Hernebring C, Magnusson P (1998) Optimisation and control of the inflow to a wastewater treatment plant using integrated modelling tools. Water Sci. Technol. 37,1,348-354

Maršalek J, Barnwell TO, Geiger,W, Grottker M, Huber WC, Saul AJ, Schilling W, Torno HC (1993) Urban drainage systems design and operation. Water Sci. Technol. 27,12,31-70

Maryns F (1996) Application of Activated Sludge Model No. 1 in riverine conditions. MSc dissertation. Interuniversity Programme in Water Resources Engineering. Katholieke Universiteit Leuven, Vrije Universiteit Brussel, University of Newcastle, September

Maryns F, Bauwens W (1997) The application of the Activated Sludge Model No. 1 to a river environment. Water Sci. Technol. 36,5,201-208

Marsili-Libelli S (1995) Operational wastewater treatment control in the context of flexible standards. Mededelingen Faculteit Landbouwkundige en toegepaste biologische Wetenschappen. Universiteit Gent, 60,4b,2503-2506

Masliev I, Somlyódy L, Koncsos L (1995) On Reconciliation of Traditional Water Quality Models and Activated Sludge Models. Working Paper WP-95-18. International Institute for Applied Systems Analysis (IIASA). Laxenburg/Austria

Masri F, Bekey GA, Safford FB (1980) A global optimization algorithm using adaptive random search. Appl. Math. Comp. 7,353-375

McCorquodale JA, Zhou S (1993) Effects of hydraulic and solids loading on clarifier performance. J. Hydraulic Res. 31,4, 461-478

Medina MA (1986) State-of-the-art, physically-based and statistically-based water quality modelling. In Torno HC, Maršalek J, Desbordes M (eds.) Urban Runoff Pollution. Proc. of a Workshop on Urban Runoff Pollution. Montpellier/France. 25-30 August 1985. NATO ASI Series G Ecological Sciences. Vol. 10. Springer. *pp.* 499-586

Medina MA, Huber WC, Heaney JP (1981a) Modelling stormwater storage/treatment transients: Theory. J. Env. Eng. Div. ASCE. 107,4,781-797

Medina MA, Huber WC, Heaney JP (1981b) Modelling stormwater storage/treatment transients: Applications. J. Env. Eng. Div. ASCE. 107,4,799-816

Meirlaen J, Schütze M., Van der Stede D, Butler D., Vanrolleghem, P.A. (2000a) Fast, parallel simulation of the integrated urban wastewater system. WaPUG User Meeting. Birmingham. 9 May 2000

Meirlaen J, Huyghebaert B, Sforzi F, Benedetti L. and Vanrolleghem P. (2000b) Fast, parallel simulation of the integrated urban wastewater system using mechanistic surrogate models. Wat.Sci.Techn. 43,7,301-309.

Meirlaen J, Van Assel J, Vanrolleghem PA (2001) Real time control of the integrated urban wastewater system using simultaneously simulating surrogate models. Second International Conference on Interactions between sewers, treatment plants and receiving waters - INTERURBA II, Lisbon/Portugal. February, *pp.* 175-182.

Meßmer A, Papageorgiou M (1992) Multireservoir sewer-network control via multivariable feedback. J. Water Resources Planning Management. 118,6,585-602

Meßmer A, Papageorgiou M (1996) Verbundsteuerstrategien und lokale Regelungen. In Schilling (1996c). *pp.* 59-73

Méthot JF, Pleau M (1997) The effects of uncertainties on the control performance of sewer networks. Water Sci. Technol. 36,5,309-315

Metropolis N, Rosenbluth AW, Rosenbluth MN, Teller AH (1953) Equation of state calculations by fast computing machines. J. Chem. Phys. 21,1087-1092

Michalewicz Z (1992) Genetic Algorithms + Data Structures = Evolution Programs. Springer, Berlin, Heidelberg, New York

Michas S (1995) The Application of Real Time Control Pronciples in Urban Drainage Systems Using "Hydroworks". MSc dissertation. Department of Civil Engineering. Imperial College of Science, Technology and Medicine, London

Michelbach S, Wöhrle C (1994) Settleable solids from combined sewers: Settling, stormwater treatment and sedimentation rates in rivers. Water Sci. Technol. 291-2,95-102

Milne I, Mallett MJ, Clarke SJ, Flower TG, Holmes D, Chambers RG (1992) Intermittent Pollution - Combined Sewer Overflows, Ecotoxicology and Water Quality Standards. WRc Report No. NR3087/2/4207. R&D Note 123

Milutin D, Bogardi JJ (1996) Application of genetic algorithms to derive the release distribution within a complex reservoir system. In Müller A (ed.) (1986) Hydroinformatics '96. Proceedings of the Second International on Hydroinformatics. Zürich. 9 - 13 September 1996, Balkema, Rotterdam, *pp.* 109-116

Mitchell M (1997) Genetic algorithms: an overview. Bull. Math. Biology. 59,1,197-204

Mudrack K, Kunst S (1994) Biologische Abwasserreinigung. Fischer, Stuttgart Jena New York

Müller S (2001) Einsatz der Gewässergütesimulation im Flussgebietsmanagement. Korrespondenz Abwasser 48,2,214-219

Nelder JA, Mead R (1965) A simplex method for function minimization. Computer J. 7,308-313

Nelen AJM (1990) Control strategies based on water quality aspects. 5th International Conference on Urban Storm Drainage. Osaka, *pp.* 1311-1316

Nelen AJM (1992) Optimized control of urban drainage systems. PhD thesis, Delft Institute of Technology

Nelen F (1993) On the use of optimization techniques for urban drainage operation. 6th International Conference on Urban Storm Drainage. Niagara Falls, *pp.* 1387-1393

Nelen F (1994a) Real time control of urban drainage systems. Seminar on "CSO - a European Perspective". IHE Delft, 24 March 1994

Nelen F (1994b) Optimizing the performance of urban drainage systems by means of better management and real time operation. Hydrotop '94. Colloquium "Better water management". Marseille, 12-15 April

Nelson JK, Buckeyne TE (1985) Computer control at the Metro Denver Sewage District. In Drake RAR (ed.) Instrumentation and Control of Water and Wastewater Treatment and Transport Systems. Oxford, Pergamon, *pp.* 497-503

Neugebauer K (1990) Steuerung von Entwässerungssystemen, Diplomarbeit. Institut für Operations Research, ETH Zürich

Neugebauer K, Schilling W, Weiss J (1991) A network algorithm for the optimum operation of urban drainage systems. Water Sci. Technol. 24,6,209-216

Neumann A (1986) Entwicklung eines lernenden Produktionssystems für die on-line Steuerung eines städtischen Kanalnetzes. Diplomarbeit. Institut für Informatik. Institut für Wasserwirtschaft, Hydrologie und landwirtschaftlichen Wasserbau, Universität Hannover

Neylon KJ (1995) Development of the Hydroworks DM Water Quality Model. Wallingford Software Ltd. Handout at the 2nd International Modelling User Group Meeting Blackpool, November

Nielsen JB, Lindberg S, Harremoës P (1993) Model based on-line control of sewer systems. Water Sci. Technol. 28,11/12,87-98

Nielsen JB, Lindberg S, Harremoës P (1994) Model based on-line control of sewer systems. International User-Group Meeting "Computer aided analysis and operation in sewage

transport and treatment technology" IUGM94. Chalmers Institute of Technology, Göteborg, *pp.* 207-218
Nielsen MK, Carstensen J, Harremoës P (1996) Combined control of sewer and treatment plant during rainstorm. Water Sci. Technol. 34,3/4,181-187
Nielsen PH, Raunkjaer K, Norsker NH, Jensen NA, Hvitved-Jacobsen T (1992) Transformation of wastewater in sewer systems - a review. Water Sci. Technol. 25,6,17-31
Niemczynowicz J (1997) Water profession and Agenda 21. Keynote lecture. 6th IRNES Conference. Imperial College of Science, Technology and Medicine, London, 21-22 September
NMU (1990) Niedersächsisches Ministerium für Umwelt. Runderlaß des MU vom 14.12.89. 205-321510. Niedersächisches Ministerialblatt. Nr. 2. Hannover, *p.* 43
Nollau V (1979) Statistische Analysen. Birkhäuser, Basel, Stuttgart
Norreys R (1991) Water quality river impact model (RIM) for river basin management. PhD thesis. Water Resources Research Group. Department of Civil Engineering,, University of Salford
Norreys R, Cluckie I (1996) Real time assessment of transient spills (RATS). International Symposium Uncertainty, Risk and Transient Pollution Control. Exeter University. 26-28 July 1995. Water Sci. Technol. 33,2,187-198
Novotny V, Jones H, Feng X, Capadaglio A (1991) Time series analysis models of activated sludge plants. Water Sci. Technol. 23,4-6,1107-1116
Novotny V, Capodaglio AG (1992) Strategy of stochastic real-time control of wastewater treatment plants. ISA Transactions. Special Issue on Water and Wastewater Treatment Automation, 31,1
Novotny V, Capodaglio A,,Jones H (1992) Real time control of wastewater treatment operations. Water Sci. Technol. 25,4-5,89-101
Nyberg U, Andersson B, Aspegren H (1996) Real time control for minimizing effluent concentrations during storm water events. Water Sci. Technol. 34,3/4,127-134

O'Connor DJ, Dobbins WE (1958) Mechanism of reaeration in natural streams. Trans. Am. Soc. Civil Eng. 123,641-666
Olsson G (1985) Control strategies for activated sludge process. In Moo-Young M (ed.) Comprehensive Biotechnology The Principles, Applications, and Regulations of Biotechnology in Industry, Agriculture and Medicine. Vol. 4. *pp.* 1107-1119
Olsson G (1992) Process control. In Andrews JF (ed.) Dynamics and Control of the Activated Sludge Process. *pp.* 67-104
Olsson G, Andrews J.F (1978) The dissolved oxygen profile - a valuable tool for control of the activated sludge process. Water Res. 12,985-1004
Olsson G, Chapman D (1985) Modeling the dynamics of clarifier behaviour in activated sludge systems. In Drake RAR (ed.) Instrumentation and Control of Water and Wastewater Treatment and Transport Systems. Pergamon, Oxford, *pp.* 405-412
Olsson G, Rundqwist L, Eriksson L, Hall L (1985) Self-tuning control of the dissolved oxygen concentration in activated sludge systems. In Drake RAR. (ed.) Instrumentation and Control of Water and Wastewater Treatment and Transport Systems. Pergamon, Oxford, *pp.* 473-482
Olsson G, Jeppson U (1994) Establishing cause-effect relationships in activated sludge plants- What can be controlled? Proceedings Workshop Modelling, Monitoring and Control of Wastewater Treatment Plants. Mededelingen Faculteit Landbouwkundige en toegepaste biologische Wetenschappen. Universiteit Gent, 59,4a,2057-2071
Olsson G, Newell B (1999) Wastewater Treatment Systems - Modelling, Diagnosis and Control. IWA Publishing, London

Orhon D (1995) COD fractionation in wastewater characterization - the state of the art. International course on Agro-Industrial waste water treatment. Mexico City, 2-3 October
Orlob GT (ed.) (1983) Mathematical Modeling of Water Quality Stream, Lakes and Reservoirs. Wiley, Chichester
Osborne MP (1995) Contribution to SWMM-USERS email discussion list on 9 August 1995
Osborne MP, Payne JA (1990) Calibration, testing and application of the sewer flow model MOSQITO. 5th International Conference on Urban Storm Drainage. Osaka, *pp.* 377-383
Osidele OO (1992) Analysis of uncertainty and predictability in conceptual simulation models using a controlled random search optimisation procedure. Msc dissertation. Department of Civil Engineering. Imperial College of Science, Technology and Medicine, London
Otterpohl R (1990a) Stormwater treatment in wastewater treatment plants - effects on mechanical and biological efficiency. 5th International Conference on Urban Storm Drainage. Osaka, *pp.* 1111-1115
Otterpohl R (1990b) Dynamische Simulation von Kläranlagen - Grundlagen, Voraussetzungen und Möglichkeiten. Zeitschrift für Stadtentwässerung und Gewässerschutz. 10, 3-10
Otterpohl R (1995) Dynamische Simulation zur Unterstützung der Planung und des Betriebes kommunaler Kläranlagen. Gewässerschutz-Wasser-Abwasser. Vol. 151. Aachen
Otterpohl R, Freund M (1992) Dynamic models for clarifiers of activated sludge plants with dry and wet weather flows. Water Sci. Technol. 26,5-6,1391-1400
Otterpohl R, Freund M, Sanz JP, Durchschlag A (1994a) Joint consideration of sewerage system and wastewater treatment plant. Water Sci. Technol. 30,1,147-153
Otterpohl R, Rolfs T, Londong J (1994b) Optimizing operation of wastewater treatment plants by offline and online computer simulation. Water Sci. Technol. 30,2,165-174
Owens M, Edwards R, Gibbs J (1964) Some reaeration studies in streams. Int. J. Air Water Pollution 8,469-486

Papageorgiou M (1983) Automatic control for combined sewer systems. J. Env. Eng. Div. ASCE. 109,1385-1402
Papageorgiou M, Mayr R (1988) Comparison of direct optimization algorithms for dynamic network flow control. Optimal Control Applic. Meth. 9,2,175-185
Paracampos FJF (1991) Calibration of dynamic models of primary clarifier and storm tank behaviour. MSc dissertation. Department of Civil Engineering. Imperial College of Science, Technology and Medicine, London
Paraskevas P, Kolokithas G, Lekkas T (1993) A complete dynamic model of primary sedimentation. Env. Technol. 14,11,1037-1046
Parkinson J (1996) An appraisal of recreational water quality control procedures. In Schütze M (ed.) Impact of Urban Runoff on Wastewater Treatment Plants and Receiving Waters. Proceedings of the Ninth European "Junior" Scientist Workshop. Kilve/UK, April 1996. Foundation for Water Research, Marlow, *pp.* 97-102
Parkinson, J (2000) Modelling strategies for sustainable domestic wastewater management in a residential catchment. PhD thesis. Department of Civil and Environmental Engineering. Imperial College of Science, Technology and Medicine, London
Parkinson J, Schütze M., Butler D. (2001) Modelling the impact of source control strategies on the sustainability of the urban wastewater system. First National conference on Sustainable Drainage. Coventry, UK. 18 - 19 June
Patry GG, Barnett MW (1992) Innovative computing techniques for development of an integrated computer control system. Water Sci. Technol. 26,5,1365-1374

Patry GG, Takács I (1995) Modelling, simulation and control of large-scale wastewater treatment plants An integrated approach. Mededelingen Faculteit Landbouwkundige en toegepaste biologische Wetenschappen. Universiteit Gent, 60,4b,2335-2343

Paulsen O (1986) Kontinuierliche Simulation von Abflüssen und Schmutzfrachten in der Trennentwässerung. Mitteilungen des Institutes für Wasserwirtschaft, Universität Hannover. Vol. 62

Payne JA., Moys GD, Hutchings CJ, Henderson RJ (1990) Development, calibration and further data requirements of the sewer flow quality model MOSQITO. Water Sci. Technol. 22,10/11,103-109

Pedersen J (1992) Controlling activated sludge process using EFOR. Water Sci. Technol. 26,3-4,783-790

Pedersen J, Sinkjær O, (1992) Test of the activated sludge model's capabilities as a prognostic tool on a pilot scale wastewater treatment plant. Water Sci. Technol. 25,6,185-192

Petersen SO, Hansen OB, Harremoës P (1990) SAMBA-Control. A model for simulation of real time control to reduce combined sewer overflows. 5th International Conference on Urban Storm Drainage. Osaka, *pp.* 1317-1322

Petrie MM, Jack AG (1994) The Development of an Integrated Catchment Management Plan of the city of Perth. WaPUG Blackpool, November

Petruck A (1996) Real-time control of a combined sewer system containing three stormwater storage tanks with overflow with respect to the ecology of the receiving stream. In Schütze M (ed.) Impact of Urban Runoff on Wastewater Treatment Plants and Receiving Waters. Proceedings of the Ninth European "Junior" Scientist Workshop. Kilve/UK, April 1996. Foundation for Water Res. Marlow. *pp.* 121-126

Petruck A, Sperling F (1996) Radar-aided CSO control - criteria for an ecological approach. 7th International Conference on Urban Storm Drainage. Hannover, *pp.* 1091-1096

Petruck A, Cassar A, Dettmar J (1998) Advanced real time control of a combined sewer system. Water Sci. Technol. 37,1,319-326

Pfister A, Sperling F, Verworn H-R (1994) Abflußsteuerung unter Verwendung von radargemessenen Niederschlägen. Zeitschrift für Stadtentwässerung und Gewässerschutz. 29,45-65

Pflanz P (1969) Performance of (activated sludge) secondary sedimentation basins. In Jenkins SH (ed.) Advances in Water Pollution Research. Pergamon Press, London, *pp.* 569-581

Pfister A, Stein A, Schlegel S, Teichgräber B (1998) An integrated approach for improving the wastewater discharge and treatment systems. Water Sci. Technol. 37,1,341-346

Phan L, Herremans L, Delaplace D, Blanc D (1994) Modeling urban stormwater pollution Comparison between British (MOSQITO) and French (FLUPOL) approaches. In Verwey, Minns, Babovic, Maksimović (eds.) Hydroinformatics '94. Balkema, Rotterdam. *pp.* 443-450

Pickles P, de Vries R, Bond R, Williams C, Mann JS (1995) From Lima to London - applying catchment wide automated control in response to new directives. Novatech '95. Lyon. *pp.* 529-537

Pike RW (1986) Optimization for Engineering Systems. Van Nostrand Reinhold, New York

Pintér J (1996) Global Optimization in Action. Continuous and Lipschitz Optimization Algorithms, Implementations and Applications. Kluwer Academic Publishers, Dordrecht

Powell MJD (1964) An efficient method for finding the minimum of a function of several variables without calculating derivatives. Computer J. 7,155-162

Pracejus O (1994) Modellierung des Verhaltens einer Kläranlage und Vergleich der durch lineare Optimierung und durch regelbasierte Optimierung möglichen

Zuflußsteuerungen. Diplomarbeit. Institut für Wasserwirtschaft, Hydrologie und landwirtschaftlichen Wasserbau. Institut für Informatik, Universität Hannover

Prax P, Mičín J, Ošmera P, Šimoník I (1996) Cataloguing characteristic rainfalls by means of genetic algorithms. Proceedings of MENDEL 96. 2nd International Mendel Conference on Genetic Algorithms. Brno, May

Preissmann A (1960) Propogation des Intumescenes dans les Canaue et Rivières - 1èr congress de l'Assoc. Française des Calcul, Grenoble

Press WH, Teukolsky SA, Vetterling WT, Flannery BP (1994) Numerical Recipes in Fortran - The Art of Scientific Computing, second edition. Cambridge University Press, Cambridge

Price R (1994) Urban drainage software in the UK. International User-Group Meeting "Computer aided analysis and operation in sewage transport and treatment technology" IUGM94. Chalmers Institute of Technology, Göteborg, 115-116

Price WL (1979) A controlled random search procedure for global optimisation. Computer J. 20,4,367-370

Price WL (1983) Global optimization by controlled random search. J. Optimization Theory and Applic. 40,3,333-347

Price WL (1987) Global Optimization Algorithms for a CAD Workstation. J. Optimization Theory Applic. 55,1,133-146

Rauch W (1994) A simplified numerical model for continuous simulation of activated sludge plant performance. Paper presented at 7th European Junior Scientist Workshop "Integrated Urban Runoff". Černice Castle/Czech Republic, 2 - 5 June

Rauch W (1995) Integrierte Modelle von Kanalnetz, Kläranlage und Vorfluter als Planungsinstrument der Regenwasserentsorgung. Wasserwirtschaft. 85,4,190-197

Rauch W (1996a) Some general aspects on modelling integrated urban drainage systems. In Schütze M (ed.) Impact of Urban Runoff on Wastewater Treatment Plants and Receiving Waters. Proceedings of the Ninth European "Junior" Scientist Workshop. Kilve/UK. April 1996. Foundation for Water Res. Marlow, *pp.* 79-84

Rauch W (1996b) Modelling, Analysis and control of integrated urban drainage system. Habilitationsschrift. Fakultät für Bauingenieurwesen und Architektur. Universität Innsbruck, September 1996

Rauch,W, Harremoës P (1995) Integrated urban water quality management. Mededelingen Faculteit Landbouwkundige en toegepaste biologische Wetenschappen. Universiteit Gent, 60,4b,2345-2352

Rauch W, Harremoës P (1996a) The importance of the treatment plant performance during rain to acute water pollution. Water Sci. Technol. 34,3/4,1-8

Rauch W, Harremoës P (1996b) Acute pollution of recipients in urban areas. 7th International Conference on Urban Storm Drainage. Hannover, *pp.* 455-460

Rauch W, Harremoës P (1996c) Optimum control of integrated urban drainage systems. 7th International Conference on Urban Storm Drainage. Hannover, *pp.* 809-814

Rauch W, Harremoës P (1996d) Minimizing acute river pollution from urban drainage systems by means of integrated real time control. Proc. of the 1st International Conference on New/Emerging Concepts for Rivers. Chicago, September

Rauch W. Harremoës P (1996e) Integrated design and analysis of drainage systems, including sewers, treatment plant and receiving waters. J. Hydraulic Res. 34,6,815-826

Rauch W, Renner S (1996) Untersuchungen zur Effizienz von Fangbecken in Mischkanalisationen. gwf-Wasser, Abwasser. 137, No. 6

Rauch W, Harremoës P (1998) Correlation of combined sewer overflow reduction due to real-time control and resulting effect to the oxygen concentration in the river. Water Sci. Technol. 37,12,69-76

Rauch W, Aalderink H, Krebs P, Schilling W, Vanrolleghem P (1998a) Requirements for integrated wastewater models - driven by receiving water objectives. Water Sci. Technol. 38,11,97-104

Rauch W, Henze M, Koncsos L, Reichert P, Shanahan P, Somlódy L, Vanrolleghem P, (1998b) River Water Quality Modelling I. State of the art. Water Sci. Technol. 38,11,237-244

Rauch W, Harremoës P (1999a) Genetic algorithms in real time control applied to minimize transient pollution from urban wastewater systems. Water Res. 33,5,1265-1277

Rauch W, Harremoës P (1999b) On the potential of genetic algorithms in urban drainage modelling. Urban Water 1,1,79-90

Rauch W, Bertrand-Krajewski JL., Krebs P, Mark O, Schilling W, Schütze M., Vanrolleghem PA (2001). Mathematical modelling of integrated urban drainage systems. Keynote paper. Second International Conference on Interactions between sewers, treatment plants and receiving waters - INTERURBA II, Lisbon/Portugal. February 2001, 89-106. To be published in Water Sci. Technol.

Reda A (1996) Simulation and Control of Stormwater Impacts on River Water Quality. PhD thesis. Department of Civil Engineering. Imperial College of Science, Technology and Medicine, London

Reda ALL, Beck MB (1997) Ranking strategies for stormwater management under uncertainty Sensitivity analysis. Water Sci. Technol. 36,5,357-371

Reichert P (1994a) Aquasim - a tool for simulation and data analysis of aquatic systems. Water Sci. Technol. 30,2.,21-30

Reichert P (1994b) Concepts underlying a computer program for the identification and simulation of aquatic systems. Schriftenreihe der EAWAG. No. 7, Dübendorf

Reichert P (1995) Design techniques of a computer program for the identification of processes and the simulation of water quality in aquatic systems. Environmental Softw. 10,3,199-210

Reichert P, Schulthess Rv, Wild D (1995) The use of Aquasim for estimating parameters of activated sludge models. Water Sci. Technol. 31,2,135-147

Reichert P, Ruchti J, Simon W (1996) AQUASIM Computer program for the identification and simulation of aquatic systems. Hydroinformatics'96, *pp.* 835-837

Reichert P, Vanrolleghem PA (2001) Identifiability and uncertainty analysis of the river water quality model No. 1 (RWQM1). Water Sci. Technol. 43,7, 329-338.

Reichert P, Borchardt D, Henze M, Rauch W, Shanahan P, Somlyódy L, Vanrolleghem PA (2001) River water quality model no. 1 (RWQM1): II. Biochemical process equations. Water Sci. Technol. 43,5,11-30.

Rheinheimer G, Hegemann W, Raff J, Sekoulov I. (eds.) (1988) Stickstoffkreislauf im Wasser. Oldenbourg, München Wien

Rieß-Dauer R (1998) personal communication. April 1998

Ristenpart E, Wittenberg D (1991) Hydrodynamic water quality simulation - an approximative solution. Water Sci. Technol. 24,6,157-163

Rogers LL, Dowla FU, Johnson VM (1995) Optimal field-scale groundwater remediation using neural networks and the genetic algorithm. Environ. Sci. Technol. 29,1145-1155

Rohlfing R (1993a) Echtzeitsteuerung von Entwässerungssystemen mit Optimierungsverfahren - Durchführbarkeitsanalyse mit hydrodynamischer Simulation der Abflußvorgänge. PhD thesis, Universität Hannover

Rohlfing R (1993b) Feasibility of optimization methods for real time control of urban drainage systems. 6th International Conference on Urban Storm Drainage. Niagara Falls, *pp.* 1381-1386

Rohr RJ (1982) Die Sonnenuhr - Geschichte, Theorie, Funktion. Callwey, München

Roper, RE Jr, Grady CPL Jr (1978) A simple, effective technique for controlling solids retention time on activated sludge plants. J. Water Pollution Control Fed. 50,702-708
Ruan M, Wiggers JBM (1998) A conceptual CSO emission model SEWSIM. Water Sci. Technol. 37,1,259-267
Ryan, D.M. (1974) Penalty and barrier functions. In Gill and Murray (1974). *pp.* 175-190
Ryder A (1972) Dissolved oxygen control of activated sludge aeration. Water Res. 6,441-445

Sachs L (1992) Angewandte Statistik - Anwendung statistischer Methoden, seventh edition. Springer, Berlin Heidelberg New York
Saget A, Chebbo G, Bertrand-Krajewski J-L (1996) The first flush in sewer systems. Water Sci. Technol. 33,9,101-108
Saint Venant AJCB de (1870) Demonstration elementaire de la formule de propagation d'une onde ou d'une intumescence dans un canal prismatique. et remarques sur les propagations du son et de la lumière, sur les ressauts, ainsi que sur la distinction des rivières et des torrents. Comptes Rendus des Séances de l'Academie des Sciences. Vol. 71, *pp.* 186-195
Sartor J (1994) Die Wahrscheinlichkeit des gleichzeitigen Auftretens maßgebender Abflußereignisse in Kanalisationsnetzen und natürlichen Gewässern. Report No. 3. Fachgebiet Wasserbau und Wasserwirtschaft, Universität Kaiserslautern
Saul AJ, Thornton RC (1989) Hydraulic performance and control of pollutants discharged from a combined sewer storage overflow. Water Sci. Technol. 21,8/9,747-756
Savic D (1997) Model Calibration Using Genetic Algorithms. Lecture held at Imperial College of Science, Technology and Medicine. London, 11 February
Savic DA, Walters GA (1997) Genetic algorithms for least-cost design of water distribution networks. J. Water Res. Plann. Man, ASCE, 123,2,67-77
Schaarup-Jensen K, Hvitved-Jacobsen T (1990) Dissolved oxygen stream model for combined sewer overflows. Water Sci. Technol. 22,10/11,137-146
Schaarup-Jensen K, Hvitved-Jacobsen T (1991) Simulation of Dissolved Oxygen depletion in streams receiving combined sewer overflows. In Maksimović Č (ed.) New Technologies in Urban Drainage - UDT 91. Elsevier, London
Schaarup-Jensen K, Hvitved-Jacobsen T (1994) Causal stochastic simulation of dissolved oxygen depletion in rivers. Wat. Sci. Techn. 29,1-2,191-198
Scheer H (1995) Meßverfahren zur Bestimmung der Konzentration an leicht abbaubaren Kohlenstoffverbindungen im Abwasser. awt-abwassertechnik. 5/1995. 31-37
Schilling W, (ed.) (1989) Real-Time Control of Urban Drainage Systems. The state-of-the-art. IAWPRC Task Group on Real-Time Control of Urban Drainage Systems. London
Schilling W (1990) Operationelle Siedlungsentwässerung. Oldenbourg. München, Wien
Schilling W (1994) Smart sewer systems improved performance by real time control. Eur. Water Pollution Control. 4,5, 24-31
Schilling W (1996a) Abflußsteuerung im Kanalnetz zur Entlastung der Kläranlage? In Hahn HH, Trauth R, (eds.) 10. Karlsruher Flockungstage. Wechselwirkung zwischen Einzugsgebiet und Kläranlage. Report No. 78. Institut für Siedlungswasserwirtschaft. Universität Karlsruhe. Oldenbourg, München, *pp.* 129-139
Schilling W (1996b) Potential and limitations of real-time control. 7th International on Urban Storm Drainage. Hannover, *pp.* 803-808
Schilling W (ed.) (1996c) Praktische Aspekte der Abflußsteuerung in Kanalnetzen. Oldenbourg, München Wien
Schilling W, Semke M (1985) Combined sewer control in real time - Bremen, West Germany. In Drake RAR (ed.) 4th Workshop Instrumentation and Control of Water and

Wastewater Treatment and Transport Systems. Pergamon, Oxford, 27 April - 4 May, pp. 183-189
Schilling W, Petersen SO (1987) Real time operation of urban drainage systems - validity and sensitivity of optimization techniques. In Beck MB (ed.) Systems Analysis in Water Quality Management. Pergamon Press, Oxford, pp. 259-269
Schilling W, Hartwig P (1988) Simulation von Reinigungsprozessen in Belebungsanlagen mit Mischwasserzufluß. gwf wasser-abwasser. 29. No. 8
Schilling W, Hartwig P (1990) Combined sewer overflow storage revisited. the significance of treatment plant behaviour under stormwater loading. 5th International Conference on Urban Storm Drainage. Osaka, pp. 1099-1104
Schilling W, Andersson B, Nyberg U, Aspegren H, Rauch W, Harremoës P (1996) Real time control of wastewater systems. J. Hydraulic Res. 34,6,785-797
Schilling W, Bauwens W, Borchardt D, Krebs P, Rauch W, Vanrolleghem P (1997) On the relation between urban wastewater management needs and receiving water objectives. 27th IAHR Congress "Water for a Changing Community". San Francisco, 10-15 August
Schlegel S (1990) Operational experience with different control systems and control loops in sewage treatment plants. In Briggs R (ed.) Instrumentation, Control and Automation of Water and Wastewater Treatment and Transport Systems. Proc. 5th Workshop. Kyoto Yokohama, Pergamon Oxford, pp. 369-376
Schlegel S (1997) Maßnahmen der Betriebskontrolle von Kläranlagen zur Optimierung der Kläranlagenleistung. awt - abwassertechnik. 1,49-51
Schlegel S, Lohmann J (1981) Control of dissolved oxygen in activated sludge plants. Water Sci. Technol. 13,9,225-232
Schmitt TG (1994) Detaillierte Schmutzfrachtberechnung nach ATV-Arbeitsblatt A128. Korrespondenz Abwasser. 41,12,2212-2230
Schmitt TG (1996) Operational flow control of combined sewer systems in long term pollution load simulation. 7th International Conference on Urban Storm Drainage. Hannover, pp. 827-832
Schreider SY, Jakeman AJ, Pittock AB (1996) Modelling rainfall-runoff from large catchment to basin scale The Goulburn Valley, Victoria. Hydrological Processes. 10,863-876
Schütze M (1990) Anwendung von Extrapolationsmethoden zur Rekonstruktion ausgefallener Meßwerte in städtischen Kanalnetzen. Diplomarbeit. Institut für Angewandte Mathematik. Institut für Wasserwirtschaft, Hydrologie und landwirtschaftlichen Wasserbau. Universität Hannover
Schütze M (1991) Anwendung von Methoden der mathematischen Optimierung in der Kanalnetzsteuerung zum Zwecke des verbesserten Gewässerschutzes - Das Programm INTL. Internal report. EAWAG, Dübendorf
Schütze, M. (1994a) Application of quadratic optimization methods for the control of urban drainage systems. 7th European Junior Scientist Workshop "Integrated Urban Runoff". Černice Castle/Czech Republic
Schütze, M. (1994b) SAMBA-Manual. SAMBA Expert System Shell - Program Documentation. Version 1.2e. Internal Report of the project "Knowledge based system 'Measuring Techniques in Civil Engineering'". Curt-Risch-Institut für Dynamik, Schall und Meßtechnik, Universität Hannover
Schütze M (1996a) Development of integrated control strategies for an urban wastewater system. In Schütze M (ed.) Impact of Urban Runoff on Wastewater Treatment Plants and Receiving Waters. Proceedings of the Ninth European "Junior" Scientist Workshop. Kilve/UK, April 1996. Foundation for Water Res. Marlow, pp. 115-120

Schütze, M. (1996b) The Integrated Simulation Tool (IST). Internal report. Department of Civil Engineering. Imperial College of Science, Technology and Medicine. London, September

Schütze, M. (1996c) Application of optimisation methods within the ORTREW project. Internal report. Department of Civil Engineering. Imperial College of Science, Technology and Medicine. London, September

Schütze M (1997) Defining a framework for control strategies using Version 2 of the Integrated Simulation Tool. Internal report. Department of Civil Engineering. Imperial College of Science, Technology and Medicine. London, September 1996

Schütze M (1998) Integrated simulation and optimum control of the urban wastewater system. PhD thesis. Department of Civil Engineering, Imperial College of Science, Technology and Medicine. London

Schütze M, Einfalt T (1999) Off-line development of RTC strategies – A general approach and the Aachen case study. Proc. Eighth International Conference on Urban Storm Drainage, Sydney, Australia, 30 August – 3 September 1999, 410-417.

Schütze M, Doll H, Hildebrandt P (1994) Review on an expert system under development "Measurement techniques in civil engineering" including structural assessment. Proceedings 12th International Modal Analysis Conference. Honolulu, *pp.* 1349-1355

Schütze M, Butler D, Beck MB (1996) Development of a framework for the optimisation of runoff, treatment and receiving waters. 7th International Conference on Urban Storm Drainage. Hannover, *pp.* 1419-1424

Schütze M, Butler D, Beck MB (1999a) Optimisation of control strategies for the urban wastewater system - An integrated approach. Water Sci. Technol. 39,9,209-216

Schütze M, Butler D, Beck MB (1999b) SYNOPSIS - A tool for the development and simulation of real-time control strategies for the urban wastewater system. Eighth International Conference on Urban Storm Drainage, Sydney/Australia. 30 August - 3 September, *pp.* 1847-1854

Schütze M., Butler D., Beck MB (1999c) Integrated simulation of the urban wastewater system and assessment of its performance. Anniversary scientific conference "50 years Faculty of Hydrotechnics". University of Architecture, Civil Engineering and Geodesy. Sofia/Bulgaria, October

Schütze M, Butler D, Beck B (2001) Parameter optimisation of real-time control strategies for urban wastewater systems. Water Sci. Technol. 43,7,139-146

Schütze M, Butler D, Beck B, Verworn HR (in print) Criteria for assessment of the operational potential of the urban wastewater system. In print. Water Sci. Technol.

Schütze M, To TB, Jumar U, Butler D (submitted) Multi-objective control of urban wastewater systems. Submitted to International Federation of Automatic Control 15th IFAC World Congress. Barcelona. 21-26 July 2002.

Schuurmans W (1992) Operation strategies versus control strategies. In Hartong H, Lobbrecht A (eds.) Applications of Operations Research to Real Time Control of Water Resources Systems. Third European Junior Scientist Workshop. Formerum/Terschelling. The Netherlands. 20-25 September 1991, PREDICT. *pp.* 39-50

Schweighofer P, Svardal K (1998) Regelstrategien und Automation. Wiener Mitteilungen. Institut für Wassergüte und Abfallwirtschaft. Universität Wien, 145,379-410

Schwendtner G, Krauth Kh (1992) Einfluß von Stoßbelastungen bei Mischwasserzufluß auf die Ablaufqualität von Belebungsanlagen. BMFT-Verbundprojekt "Schadstoffe im Regenabfluß". Institut für Siedlungswasserwirtschaft. Universität Karlsruhe, Report No. 64, Mai 1992

Shanahan P, Henze M, Koncsos L, Rauch W, Reichert P, Somlódy L, Vanrolleghem P (1998) River Water Quality Modelling II. Problems of the art. Water Sci. Technol. 38,11,245-252.

Shanahan P, Borchardt D, Henze M, Rauch W, Reichert P, Somlyódy L,Vanrolleghem PA (2001) River water quality model no. 1 (RWQM1): I. Modelling approach. Water Sci. Technol. 43,5,1-9.

Sieker F, Harms RW, Durchschlag A (1988) Langzeitsimulation und Mischwasser-Entlastungen - Vergleich der Modelle KOSIM, SMUSI, LWAFLUT. Zeitschrift für Stadtentwässerung und Gewässerschutz. 3,3-9

Sieker F, Lammersen R, Ristenpart E (1993) Bilanzierung von Schmutzstoffeinleitungen aus dem Niederschlags-Abflußgeschehen in einem Gewässer. Research Report 1988 - 1993. Universität Hannover. Institut für Wasserwirtschaft, Hydrologie und landwirtschaftlichen Wasserbau

Simonsen J, Harremoës P (1978) Oxygen and pH fluctuations in rivers. Water Res.12,477-489

Smeets M, Raemdonck N, Bauwens W (1995) A methodology to reduce CSO with additional storage capacity. The Second International Symposium on Urban Stormwater Management. Melbourne, Australia, 11-13 July

Sollfrank U, Gujer W (1991) Characterisation of domestic wastewater for mathematical modelling of the activated sludge process. Water Sci. Technol. 23,4-6,1057-1066

Somlyódy L, Henze M, Koncsos L, Rauch W, Reichert P, Shanahan P, Vanrolleghem P (1998) River water quality modelling III. Future of the art. Water Sci. Technol. 38,11,253-260

Sorooshian S, Arfi F (1982) Response surface parameter sensitivity analysis methods for postcalibration studies. Water Resources Res. 18,5,1531-1538

Spanjers H, Olsson G, Klapwijk A (1993) Determining influent shortterm biochemical oxygen demand by combined respirometry and estimation. Water Sci. Technol. 28,11-12,401-414

Spanjers H, Vanrolleghem P (1995) Respirometry as a tool for rapid characterization of wastewater and activated sludge. Water Sci. Technol. 31,2,105-114

Spanjers H, Vanrolleghem P, Olsson G, Dold P (1996) Respirometry in control of the activated sludge process. Water Sci. Technol. 34,3/4,117-126

Spanjers H, Vanrolleghem P, Olsson G,Dold P (1998) Respirometry in control of the activated sludge process Principles. IAWQ Scientific and Technical Report No. 7, London

Spear RC, Hornberger GM (1980) Eutrophication in Peel Inlet, II. Identification of critical uncertainties via generalised sensitivity analysis. Water Res. 14,43-49

Spildevandskomiteen (1984) Discharge to recipients from overflow structures. Publication No. 21. The Drainage Committee (Spildevandskomiteen). Danish Society of Engineers

Spildevandskomiteen (1985) Forurening af vandløb fra overløbsbygværker. Pollution of watercourses from overflow structures. Publication No. 22. The Drainage Committee (Spildevandskomiteen). Danish Society of Engineers

Spönemann P (1988) Ein adaptives regelbasiertes Sysetm zur Abflußsteuerung. Diplomarbeit. Institut für Informatik. Institut für Wasserwirtschaft, Hydrologie und landwirtschaftlichen Wasserbau. Universität Hannover

Squibbs G, Bottomley M, Norreys R (1997) Using the UPM methodology. WaPUG Spring Meeting

Stanic M, Avakumovic D (1996) Genetic algorithms in optimisation of real sized distribution networks. In Müller A (ed.) Hydroinformatics '96. Proceedings of the Second International on Hydroinformatics. Zürich. 9 - 13 September 1996. Balkema. Rotterdam. *pp.* 511-518

Stenstrom MK, Andrews JK (1979) Real time control activated sludge process. J. Env. Eng. Div. ASCE. 105, 245-260

Stephanopoulos G (1984) Chemical Process Control An Introduction to Theory and Practice. Prentice-Hall, Englewood Cliffs NJ

Stigter JD, Beck MB, Gilbert RJ (1997) Identification of model structure for photosynthesis and respiration of algal populations. Water Sci. Technol. 36,5,35-42

Stöber B (1990) Mathematische Modelle zur Kopplung zwischen einem hydrologischen und einem hydrodynamischen Niederschlags-Abflu8-Simulationsmodell. Diplomarbeit. Institut für Angewandte Mathematik. Institut für Wasserwirtschaft, Hydrologie und landwirtschaftlichen Wasserbau. Universität Hannover

Stokes L, Takács I, Watson B, Watts JB (1993) Dynamic modelling of an activated sludge sewage works - A case study. Water Sci. Technol. 28,11/12,105-115

STOWA (1995) Modelleren van actiefslibsystemen Keuze van model en programmatuur. Report 95-01. Stichting Toegepast Onderzoek Waterbeheer (STOWA), Utrecht

Streeter HW, Phelps EB (1925) A study of the Pollution and Natural Purification of the Ohio River. Vol. III. Public Health Bulletin. No. 146. United States Public Health Service, Washington D.C.

Swann WH (1972) Direct search methods. In Murray W (ed.) Numerical Methods for unconstrained optimization. Academic Press, London New York, *pp.* 191-217

Swann WH (1974) Constrained optimization by direct search. In Gill and Murray (1974), *pp.* 191-217

Takács I, Patry GG, Nolasco D (1991) A dynamic model of the clarification-thickening process. Water Res. 25,10,1263-1271

Takács I, Patry G, Gall B (1995) IC^2S Advanced control for wastewater treatment plants Mededelingen Faculteit Landbouwkundige en toegepaste biologische Wetenschappen. Universiteit Gent. 60,4b,2451-2454

Taylor A (1992) The application of RTC to reduce pollutant spills. In Hartong H, Lobbrecht A (eds.) Applications of Operations Research to Real Time Control of Water Resources Systems. Third European Junior Scientist Workshop. Formerum/Terschelling. The Netherlands. 20-25 September 1991. PREDICT, *pp.* 179-186

Tchobanoglous G (1979) Wastewater Engineering - Treatment, Disposal, Reuse, second edition. Metcalf & Eddy Inc. McGraw Hill, New York

Tchobanoglous G, Burton FL (1991) Wastewater Engineering - Treatment, Disposal, Reuse, third Edition. Metcalf & Eddy Inc. McGraw Hill, New York

Tebbutt THY (1992) Principles of Water Quality Control. Fourth Edition, Pergamon Press, Oxford

Thomann RV (1963) Mathematical model of dissolved oxygen. J. San. Eng. Div. ASCE, 89,1-30

Thomann RV, Mueller JA (1987) Principles of Surface Water Quality Modeling and Control. Harper&Row, New York

Thompson D, Chapman DT, Murphy KL (1989) Step feed control to minimize solids loss during storm flows. Res. J. Water Pollution Control Fed. 61,11/12,1658-1665

Thornberg DE, Eliasson G, Ohlsson A (1994) Use of the EFOR program for optimisation of nitrogen removal at full-scale activated sludge plants. International User-Group Meeting "Computer aided analysis and operation in sewage transport and treatment technology" IUGM94. Chalmers Institute of Technology, Göteborg, *pp.* 63-70

Thornton RC, Saul AJ (1986) Some quality characteristics of combined sewer flows. The Public Health Engineer. 14,3,35-38

Tidswell RG (1996) The Bolton EU SPRINT project - Project synopsis. IAWQ Individual Members Meeting. "Real Time Control", 23-24 January, Warrington

To, TB (1997) Multiobjective evolution strategy with linear and nonlinear constraints. Proc. of 15th IMACS World Congress on Scientifc Computation. *pp.* 357-363

Tomei MC, di Iaconi C, Di Pinto AC, MappaG, Ramadori R. (1996) Development of an expert system for nitrogen removal control. European Water Pollution Control. 6,45-50

Tong RM, Beck MB, Latten A (1980) Fuzzy control of the activated sludge wastewater treatment process. Automatica. 16,695-701

Tracy KD, Keinath TM (1974) Dynamic model for thickening of activated sludge. AIChE Symp. Ser. 136,70,291-308

Tsai YP, Ouyang CF, Wu MY, Chiang WL (1993) Fuzzy control of a dynamic activated sludge process for the forecast and control of effluent suspended solid concentration. Water Sci. Technol. 28,11/12,355-367

Tyson JM, Guarino CF, Best HJ, Tanaka K (1993) Management and institutional aspects. Water Sci. Technol. 27,12,159-172

Udale H, Sedgwick C, Osborne M (1997) The Black Country Trunk Sewer - Hydroworks TM QSIM made simple. WaPUG Spring Meeting

Vaes G (1999) The influence of rainfall and model simplification on combined sewer system design. Dissertation. Katholische Universität Leuven

Vanrolleghem PA (1994) Building blocks for wastewater treatment process control. A review. In Advanced Course on Environmental Biotechnology. Delft. 25 May - 3 June

Vanrolleghem PA (1995) Model based control of wastewater treatment plants. Proceedings ESF Workshop Integrated Environmental Bioprocess Design. Obernai. 6-7 December

Vanrolleghem PA, Verstraete W (1993) On-line monitoring equipment for wastewater treatment processes State of the art. In Proceedings TI-KVIV Studiedag Optimalisatie van Waterzuiveringsinstallaties door Proceskontrole en -sturing. Gent, *pp.* 1-22

Vanrolleghem P, Van Daele M, Dochain D (1995a) Practical Identifiability of a biokinetic model of activated sludge respiration. Wat Res. 29,11,2561-2570

Vanrolleghem PA, Van Daele M, Dochain D (1995b) Optimal experimental design for parameter estimation in activated sludge processes. Proceedings American Control Conference 1995. Seattle. 21-23 June, *pp.* 14-18

Vanrolleghem P, Keesman KJ (1996) Identification of biodegradation models under model and data uncertainty. Water Sci. Technol. 33,2,91-105

Vanrolleghem PA, Fronteau C, Bauwens W (1996a) Evaluation of design and operation of the sewage transport and treatment system by an EQO/EQS based analysis of the receiving water immission characteristics. Proceedings of the WEF Specialty Conference 'Urban wet weather pollution Controlling sewer overflows and water runoff'. Québec Canada, June 16 - 19, *pp.* 1435-1446

Vanrolleghem PA, Jeppson U, Carstensen J, Carlsson B, Olsson G (1996b) Integration of wastewater treatment plant design and operation - a systematic approach using cost functions. Water Sci. Technol. 34,3/4,159-171

Vanrolleghem PA, Dochain D (1997) Bioprocess model identification. In Van Impe J, Vanrolleghem PA, Iserentant D (eds.) Advanced Insturmentation, Data Interpretation and Control of Biotechnological Processes. Kluwer Academic Publishers, Dordrecht, *pp.* 251-318

Vanrolleghem PA, Schilling W, Rauch W, Krebs P, Aalderink H (1998) Setting up measuring campaigns for integrated wastewater modelling. Water Sci. Technol. 39,4,257-268

Vanrolleghem P.A., Borchardt D., Henze M., Rauch W., Reichert P. Shanahan P. and Somlyódy L. (2001) River water quality model no. 1 (RWQM1): III. Biochemical submodel selection. Water Sci. Technol. 43,5, 31-40.

Vazquez J, Bellefleur D, Gilbert D, Grandjean B (1997) Real time control of a combined sewer network using graph theory. Water Sci. Technol. 36,5,301-308

Vazquez-Sanchez E (1996) Dynamic Modelling of The Secondary Clarifier In The Activated Sludge Process Analysis and Development of A New Extended 1-D Model. MSc dissertation. Department of Civil Engineering. Imperial College of Science, Technology and Medicine, London

Veersma AMJ, van der Roest HF, Meinema K. (1995) Stationaire berekeningen met een dynamisch model? H2O. 28,14,432-436

Verbanck MA, Ashley RM (1993) International workshop on origin, occurrence and behaviour of sediments in sewer systems: Outline of technical conclusions. Water Sci. Technol. 27,12,173-176

Verworn HR, Kenter G (1993) Abflußbildungsansätze für die Niederschlags-Abfluß-Modellierung. Zeitschrift für Stadtentwässerung und Gewässerschutz. 24,3-50

Vesilind AP (1968) Discussion of "Evaluation of activated sludge thickening theories". by Dick RI and Ewing BB. J. San. Eng. Div. ASCE. 94,185-191

Viessman W, Lewis G, Knapp. JW (1989) Introduction to Hydrology. 3rd edition. Harper&Row, New York

Vitasovic Z (1989) Continuous settler operation a dynamic model. In Patry G, Chapman D (eds.) Dynamic Modelling and Expert Systems in Wastewater Engineering. Lewis publishers, Chelsea, Michigan, *pp.* 59-81

Vitasovic Z, Andrews JF (1987) A rule-based control system for the activated sludge process. In Beck MB (ed.) Systems Analysis in Water Quality Management. Pergamon Press, Oxford, *pp.* 423-432

Walker LF (1971) Hydraulically controlling solids retention time in the activated sludge process. J. Water Pollution Control Fed. 43,1,30-39

Wallis SG, Young P, Beven K.J (1989) Experimental investigation of the aggregated dead zone model for longitudinal solute transport in stream channels. Proc. Inst. of Civil Eng. Part 2, London, 87,1-12

Walters GA, Savic DA (1996) Recent applications of genetic algorithms to water system design. In Blain WR (ed.) Hydraulic Engineering Software VI., Computational Mechanics Publications, Southampton, *pp.* 143-152

Wan WM (1997) Sensitivity analysis of data simulated using Integrated Smulation Tool-Version 2 and derivation of conversion factors for the use of interfacing. Final Undergraduate Project. Department of Civil Engineering. Imperial College of Science, Technology and Medicine, London

Wang QJ (1991) The genetic algorithm and its application to calibrating conceptual rainfall-runoff models. Water Resources Res. 27,9,2467-2471

Weijers SR (2000) Modelling, Identification and Control of Activated Sludge Plants for Nitrogen Removal. PhD thesis. Technical University of Eindhoven

Weijers SR, Kok JJ, Preisig HA (1995) Control strategies for nutrient removal plants and MPC applied to a pre-denitrification plant. Mededelingen Faculteit Landbouwkundige en toegepaste biologische Wetenschappen. Universiteit Gent, 60,4b,2435-2443

Weijers SR, Engelen GL, Preisig HA, van Schagen K. (1997) Evaluation of model predictive control of nitrogen removal with a carrousel type wastewater treatment plant model using different control goals. 7th IAWQ Workshop on Instrumentation, Control and Automation of Water and Wastewater Treatment and Transport Systems. Brighton, 6 - 9 July, 401-408

Weijers SR, Vanrolleghem PA (1997) A procedure for selecting best identifiable parameters in calibrating activated sludge model No. 1 to full-scale plant data. Water Sci. Technol. 36,5,69-79

Weinreich G, Schilling W (1995) Pollution-based operation strategies for Urban drainage Systems. Presentation given at the Second Annual Network Meeting. MATECH. Exeter, July

Weinreich G, Schilling W, Birkely A, Molland T (1996) Pollution based real time control strategies for combined sewer systems. 7th International Conference on Urban Storm Drainage, Hannover, pp. 821-826

Whitehead PG (1979) Modelling and opertaional control of water quality in river systems. Water Res. 12,377-384

Whitehead PG, Young PC, Hornberger GM (1979) A systems model of streamflow and water quality in the Bedford Ouse river system. 1. Streamflow modelling. Water Res. 13,1155-1169

Whitehead PG, Beck MB, O'Connell E (1981) A systems model of streamflow and water quality in the Bedford Ouse river system. 2. Water quality modelling. Water Res. 15,1157-1171

Wilcox SJ, Hawkes DL, Hawkes FR, Guwy AJ (1995) A neural network, based on bicarbonate modelling, to control anaerobic digestion. Water Res. 29,1465-1470

Willems P (2000) Probabilistic Immission Modelling of Receiving Surface Waters. Dissertation. Katholische Universität Leuven

Williams WD, Tidswell RG (1994) Bolton RTC - A glimpse at the future. WaPUG User Day. November 1994

Wittenberg D (1992) Urbane Gewässer - Das Immissionsprinzip als Planungsansatz für die Stadtentwässerung. PhD thesis, Universität Hannover

Wolf P (1990) Auswirkungen des Mischwasserzuflusses auf Nitrifikation und Denitrifikation in Belebungsanlagen. Korrespondenz Abwasser. 23,3,256-259

Wong KY (1994) Automatic calibration of Environmental Systems Process Models - A Study of a Potential Method - Genetic Algorithms. MSc dissertation. Centre for Environmental Technology. Imperial College, London, September 1994

Wong KY (1996) Methodology for the Optimal Design of Water Distribution systems to Hydraulic, Water Quality and Reliability Targets. Transfer Summary Report. Department of Civil Engineering. Imperial College of Science, Technology and Medicine. London, April 1996

WRc (1994) STOAT Manual. WRc Swindon (no version number given)

Xanthopoulos C, Hahn HH (1993) Anthropogenic pollutants wash-off from street surfaces. 6th International Conference on Urban Storm Drainage. Niagara Falls, pp. 417-422

Yagi S (1990) An application of fuzzy set theory to the computer aided control of pumps in combined sewer systems. 5th International Conference on Urban Storm Drainage. Osaka, pp. 1269-1274

Yakowitz S (1982) Dynamic programming applications in water resources. Water Resources Res. 18,4,673-696

Yin Y (1995) Municipal Wastewater Management - Beyond End-of-pipe Treatment. MSc thesis. Department of Civil Engineering.. Imperial College of Science, Technology and Medicine. London

Young MJ (1988) Programmer's Guide to OS/2. Sybex, San Francisco

Young PC (1992) Parallel processes in hydrology and water quality a unified time-series approach. J. Inst. Water Env. Management. 6,598-612

Zadeh LA (1965) Fuzzy sets. Inf. Control. 8,338-353
Zadeh LA (1979) A theory of approximate reasoning. In Hayes JE,. Mickie D, Mikulich LI (eds.) Machine Intelligence, Vol 9. Wiley, New York
Zhigljavsky AA (1991) Theory of Global Random Search. Kluwer, Dordrecht
Zielke W, Geiger WF, de Haar U, Hoffmann B, Kleeberg H (eds.) (1992) Steuerung in der Wasserwirtschaft. Research report. Deutsche Forschungsgemeinschaft, VCH Weinheim
Zison SW, Mills WB, Diemer D, Chen CW (1978) Rates, Constants, and Kinetic Formulatoins in Surface Water Quality Modeling. U.S. Environmental Protection Agency. ORD. Athens Georgia. ERL, EPA/600/3-78-105
Zunic F, Seus GJ (1993a) On-line-Abflußsteuerung in vermaschten Kanalisationssystemen mit Rückhaltemöglichkeiten - Teil 1 Das Steuerungskonzept. Korrespondenz Abwasser. 40,11,1763-1772
Zunic F, Seus GJ (1993b) On-line-Abflußsteuerung in vermaschten Kanalisationssystemen mit Rückhaltemöglichkeiten - Teil 1 Erprobung des Vermaschungs- und Steuerungskonzeptes und Ergebnisse. Korrespondenz Abwasser. 40,12,1881-1888

Index

A

Activated Sludge Model No. 1 .... 33, 35, 72, 78, 138
Activated Sludge Model No. 2 36
Activated Sludge Model No. 3 36
Advection-Dispersion equation .... 17*f.*
aeration .... 30, 38, 60, 98, 101, 110, 112, 163, 189*f.*
aerobic growth 36
Aggregated Dead Zone 45
algae 46, 54*f.*, 64, 72, 73, 194
AMM-DU criterion 159
AMM-E criterion 159
AMM-M criterion 159
ammonia 21, 31, 53, 59, 64, 69, 97, 104, 106, 136, 153, 186
ammonification 36, 54, 138
ammonium 31, 53*f.*, 60, 63*ff.*, 74*f.*, 75, 84, 100, 106, 129, 132, 136, 143, 153, 159*ff.*, 186, 198, 205, 209*ff.*
Arrhenius 49
assessment criteria 159
attenuation 15

B

base flow .... 44, 62, 71, 110, 149, 193*ff.*
biochemical oxygen demand (BOD) 20, 27, 32, 52, 97, 102, 147, 154*ff.*, 176, 194, 198, 205
    ultimate 52
biofilm 20, 79
BOD-E criterion 159
BOD-M criterion 159

bottom-up 9, 229
branch-and-bound algorithm 90, 124
bulking sludge 38

C

chemical dosage 98, 294
chemical oxygen demand (COD) 4, 21, 25, 26, 75, 84, 97, 103, 154*ff.*, 185*f.*, 204*f.*
clarification 26, 27*ff.*, 37*ff.*, 61, 138, 142*ff.*, 152
CODtot criterion 160
coefficient of variation 265, 266
combined sewer overflow (CSO) 2, 4, 52, 62, 65, 68, 72, 81, 83, 110, 145, 154*ff.*, 193, 204, 223
continuity equation 13, 16, 44
continuous simulation .... 130*f.*
continuously stirred tank reactor (CSTR) .... 18, 28, 45*f.*
control 6, 70, 81*ff.*, 161*ff.*, 209, 233*ff.*
    global 82, 88, 243
    integrated 6*f.*, 25, 70, 82*f.*, 111*f.*, 130, 152, 229*f.*, 243, 247, 251, 262*ff.*, 273*ff.*
    local 82, 243
    static 82
control potential 273
control strategy 6, 85*ff.*, 112*ff.*, 152, 158, 174*ff.*, 230*ff.*, 243*ff.*, 270, 280*f.*
Controlled Random Search 126*ff.*, 163*ff.*, 238*ff.*, 253, 256, 274, 279
controller 81, 94, 96*ff.*, 229*ff.*, 259*ff.*, 281
    multivariable feedback 88
    PID 82, 104, 108
conversion factor 73*ff.*, 150*ff.*, 176

CSODur criterion 161

D

decision matrix 85
degradation of organic matter . 20, 30, 36
delayed oxygen demand 51, 52, 68
denitrification ...30*ff*., 54, 65, 73, 98, 280
deoxygenation .... 51*f.*
Derived Intermittent Standards .... 68*f.*
diffusive wave approximation 15
diurnal pattern 187, 198
DO (dissolved oxygen) 5, 32, 46*ff.*, 68*f.*,
    73*ff.*, 98ff, 110, 132, 146, 154, 158*ff.*,
    205, 210, 214*ff.*, 262, 269, 273
DO-DU criterion .... 158*ff.*
DO-E criterion .... 158*f.*
DO-M criterion .... 158*f.*
domestic appliances 13
dry-weather flow 5, 57*ff.*, 100, 182,
    185*ff.*, 198*ff.*
Dynamic Programming 91

E

end-of-pipe standard 4
Environmental Quality Objective 4
Environmental Quality Standard 4
erosion 58, 65, 280
Eulerian approach 44
eutrophication .... 31, 53*ff.*, 63*ff.*
expert systems 85, 105

F

F/M (food-to-microorganisms ratio) 103,
    108
F2 criterion .... 159*ff.*
F3 criterion .... 160*f.*
faeces 1
feasible domain .... 115, 120*ff.*, 164*ff.*
first flush .... 19, 22, 58*f.*
Fundamental Intermittent Standards 68*f.*
    159

G

genetic algorithm 91, 94, 109, 124*ff.*,
    163, 166*ff.*, 249, 256
gridding 123, 126, 158, 176*ff.*, 235*ff.*,
    249, 254, 259*f.*

grit 24, 59
guidelines 4, 19, 21, 29, 183

H

Horton's infiltration approach 12, 133
hydrogen sulphide 20
hydrological cycle 1
hydrolysis 29, 34*ff.*, 74, 138

I

identifiability 36, 268
infiltration 3, 12, 19, 62, 133
integrated control: see *control–integrated*
integrated Modelling.70*ff.*, 131, 152, 180
integration of information 83
integration of objectives .... 83*f.*

K

Kalinin-Miljukov model 17, 44
kinematic wave 15

L

Lagrangian approach 44
long-term simulation12, 15, 19, 73*f.*, 77,
    130, 152, 179

M

MBPC (model based predictive control)
    108
Metropolis algorithm 124
micro GA .... 168, 254*ff.*
MIMO (multiple input – multiple output)
    108
mixed liquor 29, 112
MLSS (Mixed Liquor Suspended Solids)
    .... 40*f.*, 102*ff.*
momentum equation .... 13*ff.*
morphological changes 66
MPC (model predictive control) ..... 108*f.*
multi-objective optimisation ...268*ff.*, 282
Muskingum-Cunge model 17, 44, 46

N

neural network 109, 282
nitrate 31*f.*, 54, 64, 136, 145, 153, 204

nitrification 30*ff.*, 48, 53*ff.*, 60, 63*f.*, 73, 75, 101, 103, 135, 146, 198, 204
nitrite 31*f.*, 34, 55, 64
Nitrobacter 31
nitrogen cycle .... 53*f.*
Nitrosomonas 31
non-point sources .... 43*f.*

O

objective function 7, 89*ff.*, 105, 110, 114*ff.*, 121*ff.*, 158, 163*ff.*, 175, 177, 233, 240, 250, 270, 278*f.*, 282
off-line development of strategies 85*ff.*, 92, 94, 219
on-line development of strategies ..... 89*ff.*
optimisation
off-line optimisation 7, 113*ff.*, 283
on-line optimisation 88, 109, 113, 270, 283
   global .... 7, 91*f.*, 122*ff.*, 163*ff.*
   linear 86, 89, 120
   local 7, 90, 94, 120, 163, 249, 257, 282
     non-linear .... 120*ff.*
     quadratic 90, 120
oxygen depletion .... 31, 48, 63, 83*f.*
oxygen saturation .... 48*f.*

P

parameter identification 118, 125, 268
pH (*pondus hydrogenii*) 31, 53*f.*, 64, 69, 97*f.*
photosynthesis 48, 55, 75, 148*f.*, 195
pollutant transport .... 17*ff.*, 43*f.*
Powell's optimisation method 120, 126, 163, 170*ff.*, 249
primary clarification .... 24, 26, 27*ff.*

Q

Q-CSO criterion 160
Qoverf criterion 160
Quasi-Newton method 120

R

rain event 7, 12*f.*, 24, 26, 57*ff.*, 73*ff.*, 104, 125, 130, 176, 195, 203, 271*ff.*, 279
reaeration 20, 48*ff.*, 74, 147
real-time control (RTC) 6, 11, 21, 77, 80*ff.*, 109, 111*ff.*, 282
receiving water 1*ff.*, 5, 21, 24, 31, 43*ff.*, 55, 62, 75, 77, 82, 94, 106, 117
redox potential 99
reservoir
   linear .... 15*ff.*
respiration 31, 55, 75, 97
Response Surface Methodology 121*ff.*, 219, 251
resuspension 18, 20, 63, 74*f.*, 134, 187
return activated sludge 38, 98, 102*ff.*, 112, 138, 142, 162*f.*, 190*ff.*, 197
River 1*ff.*, 5, 13, 18, 43*ff.*, 62*ff.*, 109*f.*, 145*ff.*, 192*ff.*, 267
River Water Quality Model No. 1 56, 78, 306
rule 85, 86, 105

S

Saint Venant equations 13*ff.*, 44
saturation concentration .... 48*f.*
screening 24, 324
secondary clarifier 26, 37*ff.*, 61*ff.*, 70*f.*, 75, 104, 112, 135, 138, 142, 162, 189, 204, 217, 280
sensitivity 151, 155, 158, 176, 229, 264*ff.*
sensitivity coefficient .... 265*f.*
set-point 81*ff.*, 98, 101*ff.*, 230*ff.*
Sewer system 1, 11*ff.*, 57*ff.*, 83*ff.*, 133ff., 181*ff.*
Shuffled Complex Evolution 282
simulated annealing 124
SISO (single input – single output) 108
sludge blanket 37*ff.*, 61, 102*f.*, 112, 143*f.*, 198, 205, 217
sludge bulking 38
specific storage volume 184
specific total oxygen uptake rate (STOUR) 98, 104, 245
storage tank 12, 19*f.*, 21*ff.*, 57, 70, 80*f.*
storm tank 57, 62, 99*ff.*, 137, 160, 184, 189, 197, 216, 221, 234, 245*ff.*, 262*f.*

strategy framework 8, 114, 174, 229, 258*ff.*
strategy parameter 8, 114*ff.*, 158, 229*ff.*, 249, 260, 270, 274
Streeter–Phelps model 48
surface runoff 11, 12, 57
suspended solids 19, 25*ff.*, 38, 64, 107, 142
synchronous simulation .... 8, 129*ff.*
SYNOPSIS 8, 79, 129*ff.*, 176*ff.*, 194, 230

T

thickening 37*ff.*, 61, 138, 142
top-down 8, 229

total organic carbon (TOC) 98
treatment plant 1, 13, 21, 24*ff.*, 59*ff.*, 96*ff.*, 135*ff.*, 189

U

uniform emission standard 4
Urban Pollution Management (UPM) procedure 6, 75
urine 1, 31

W

waste activated sludge 38, 98, 102*ff.*, 138, 162, 192, 217*ff.*, 245